T0253602

Nuclear Fuel Cycle

B. S. Tomar · P. R. Vasudeva Rao · S. B. Roy ·
Jose P. Panakkal · Kanwar Raj · A. N. Nandakumar
Editors

Nuclear Fuel Cycle

 Springer

Editors
B. S. Tomar
Homi Bhabha National Institute,
Anushaktinagar
Mumbai, Maharashtra, India

S. B. Roy
Formerly Chemical Engineering Group
Bhabha Atomic Research Centre
Mumbai, Maharashtra, India

Kanwar Raj
Formerly Nuclear Recycle Group
Bhabha Atomic Research Centre
Mumbai, Maharashtra, India

P. R. Vasudeva Rao
Formerly Homi Bhabha National Institute,
Anushaktinagar
Mumbai, Maharashtra, India

Jose P. Panakkal
Formerly Advanced Fuel Fabrication
Facility
Bhabha Atomic Research Centre
Mumbai, Maharashtra, India

A. N. Nandakumar
Formerly Radiological Safety Division
Atomic Energy Regulatory Board
Mumbai, Maharashtra, India

ISBN 978-981-99-0951-3   ISBN 978-981-99-0949-0   (eBook)
https://doi.org/10.1007/978-981-99-0949-0

This Springer imprint is published by the registered company Springer Nature Singapore Pte Ltd.
The registered company address is: 152 Beach Road, #21-01/04 Gateway East, Singapore 189721,
Singapore

Foreword

With the inevitable global thrust towards achieving net zero GHG emissions and the imperative to meet growing energy needs of most of the world as it seeks to improve quality of life of people, a rapid scale up of nuclear energy seems inevitable. Looking at the magnitude of this challenge in run up to clean energy transition, particularly in growing economies as also the unresolved nature of the issue of permanent disposal of spent nuclear fuel that is being used in once through mode, closing the nuclear fuel cycle has become a matter of paramount importance. Leveraging energy from thorium in addition to energy from uranium also necessitates recourse to closed fuel cycle. Thanks to the vision of Dr. Bhabha, consistent and autonomous pursuit of the well thought off three stage nuclear power program and emphasis on self-reliance, India today is among the very few most advanced countries in the world in this sensitive high technology area. Further advances in this area have also opened the possibility of gross reduction in repository requirements for disposal of high-level radioactive waste, reduction in radiotoxicity of such waste to levels comparable with naturally occurring uranium and thorium deposits within a period consistent with institutional life time (say ~ 300 years). As a byproduct of such efforts, several applications of radio-isotopes derived from nuclear waste in the field of health care, industry, agriculture, food technology, etc. have opened up and have been rising steadily. Developments in nuclear fuel cycle are thus destined to cater to global needs in energy, food, healthcare and environmental management.

Activities in nuclear fuel cycle cover a very wide spectrum. The front-end part covers exploration, mining, milling, refining and nuclear fuel fabrication. While the back end comprises of the spent fuel storage, reprocessing, fuel refabrication, byproducts for other applications and radioactive waste management. Both front and back-end activities involve stringent quality as well as nuclear material accounting and control. Although India has a very comprehensive program relating to nuclear fuel cycle it remains an evolving endeavor with many dimensions unfolding progressively. We have large teams of competent and qualified professionals as well as researchers working with synergy in this very specialized area.

In this context, the present book on Nuclear Fuel Cycle will provide the impetus to the professionals working in the field of nuclear science and engineering in updating their knowledge in the field of nuclear fuel cycle. In addition, it will encourage the students in pursuing nuclear science and engineering by providing basic knowledge on various concepts and strategies employed in nuclear fuel cycle. As the book is written by practicing nuclear specialists, it provides an insider's view on various processes in the nuclear fuel cycle. The introductory chapter provides the general information about the nuclear fuel cycle, its basic framework and strategies being followed world-wide. The chapters on exploration, mining, milling and processing of uranium, nuclear fuel fabrication, quality control of nuclear fuels, post irradiation examination, spent fuel reprocessing and radioactive waste management provide the core knowledge on the various stages of the nuclear fuel cycle. The chapter on thermodynamic aspects of fuel behaviour in the reactor provides basic understanding of the interaction between fuel and clad, swelling behaviour of fuel, fuel—fission products compatibility and thereby helps in predicting long term safety aspects of the reactor. The chapters on accounting and control of nuclear materials, and safety in storage and transport of nuclear materials are equally important in terms of the ascertaining the basic standards in safety, security and safeguards in handling nuclear materials.

I am sure the book will generate wide spread interest among the professionals, engaged in nuclear fuel cycle activities world-wide. Further, it will be useful as a reference book for the budding researchers and professionals who are going to make a career in the field of nuclear science and engineering. I compliment Prof. Vasudeva Rao and his team of editors for bringing out this book on Nuclear Fuel Cycle, which probably is the first of its kind in the world encompassing all aspects of the nuclear fuel cycle including safety, security and safeguards.

Dr. Anil Kakodkar
Chancellor
Homi Bhabha National Institute
Mumbai, Maharashtra, India

Chairman
Rajiv Gandhi Science & Technology
Commission
Mumbai, Maharashtra, India

Former Chairman
Atomic Energy Commission
Mumbai, Maharashtra, India

Preface

During the last seven decades, the activities connected with nuclear fuel cycle have grown and spread to several countries. The number of students studying nuclear science and technology and personnel engaged in development of fuel cycle technologies has also increased. A textbook covering all the fundamental aspects of the nuclear fuel cycle for the postgraduate students and researchers in the field of nuclear science and technology has been a long-felt need. A limited number of books are available on nuclear fuel cycle, and these do not have a comprehensive and in-depth coverage. Moreover, they are written more than a decade ago and do not cover the latest developments.

The present book is written keeping in mind the students coming out of universities after their graduation. This book can be prescribed as a textbook for a course on nuclear fuel cycle for postgraduate students. The chapters are designed in such a way that the reader is taken from the fundamentals to the advanced level in a gradual manner. The Chap. 1 explains the nuclear fuel cycle, gives basic information about nuclear fuels and reactors and briefly describes the various aspects of the nuclear fuel cycle and the activities under front end and back end. The Chap. 2 explains the starting point of the fuel cycle covering exploration, mining, milling and processing of uranium, production of metal and advanced fuel materials. Enrichment of uranium using different techniques is also covered in this chapter. The chapter on fabrication of nuclear fuel elements introduces different fuel materials for reactors of various types and describes various types of conventional and advanced fuel elements for commercial (thermal and fast) reactors and research reactors. Quality control techniques, both physical and chemical, used in nuclear manufacturing facilities are explained in the next chapter. The next chapter deals with the phase equilibria, thermophysical and thermochemical properties of nuclear materials with special emphasis on actinide oxides, carbides, nitrides and metal alloys. Chapter 6 explains the various techniques used for postirradiation examination and the data generated useful for the entire nuclear fuel cycle. The chapter on nuclear fuel reprocessing explains the techniques for extraction of valuable plutonium and removal of fission products and minor actinides from the spent (irradiated) fuel.

Radioactive waste is generated in the nuclear fuel cycle by different agencies, and the nature and content depend on various factors. The philosophy and the technology of radioactive waste management are explained in Chap. 8. The philosophy and the techniques used for Nuclear Material Accounting and Control (NUMAC) at various stages of nuclear fuel cycle are explained starting from the fundamentals in Chap. 9. The statistical aspects in nuclear accounting are also covered in this chapter. The transport and storage of nuclear material are governed by international laws stipulated by IAEA, and the safety standards are explained in the next chapter. The various terms used and the classification of radioactive materials and the design criteria for the packages for transport are well explained in Chap. 10. Lastly, Chap. 11 deals with the basic principles of radiation protection, definitions and the effects of exposure to radiation exposure. The safety considerations applicable to mines and mills, fuel fabrication facilities, radiological laboratories, hot cells and waste disposal facilities are also discussed in this chapter.

The authors have also made all efforts to include photographs, schematic drawings and other illustrations for easy understanding of the subject. References, suggestions for additional reading, and wherever possible, exercises are given at the end of the chapters. This book will be useful for any institution (both academic and research) conducting courses on nuclear science and technology and conducting research on related topics. We hope that the international community of academicians and researchers will find this book very useful for updating their knowledge on fuel cycle.

We are grateful to Homi Bhabha National Institute (an institution of Department of Atomic Energy, India, with the status of university) for conceiving this book and bringing together a number of experts both serving DAE and retired from service. We are also thankful to all the authors for recording their valuable knowledge and practical experience in writing this book, and to the reviewers who have gone through the chapters and given their critical inputs to ensure that the book provides a balanced and at the same time adequate details of various aspects of fuel cycle.

Mumbai, India
Mumbai, India
Mumbai, India
Mumbai, India
Mumbai, India
Mumbai, India

B. S. Tomar
P. R. Vasudeva Rao
S. B. Roy
Jose P. Panakkal
Kanwar Raj
A. N. Nandakumar

Acknowledgements

The book on nuclear fuel cycle was conceived by Homi Bhabha National Institute (HBNI) to prepare a textbook for postgraduate students, researchers and practicing professionals. The need for such a book has been felt for quite some time, as such a comprehensive description of the principles and processes involved in the entire nuclear fuel cycle has not been covered in any other textbook on this topic so far. We are grateful to Sri K.N. Vyas, Secretary, Department of Atomic Energy and Chairman, Council of Management, HBNI, for his encouragement for writing such books. We are grateful to Dr. Anil Kakodkar, Chancellor, HBNI and Former Chairman, Atomic Energy Commission of India for readily consenting to write the foreword for the book. We have also received encouragement and support from several senior colleagues from DAE; we are particularly thankful to the Directors of Bhabha Atomic Research Centre (BARC) and Indira Gandhi Centre for Atomic Research (IGCAR).

The authors of individual chapters have made the most valuable contributions towards bringing out the book in the present form. The authors drawn from the different R&D institutions of the Department of Atomic Energy are experts in their fields, and their experience in teaching and research is reflected in the respective chapters. We gratefully acknowledge the valuable contributions of all the authors in bringing out this book.

Several experts, from the Department of Atomic Energy, have provided important inputs towards the preparation of various chapters of this book. We would like to acknowledge in particular, Mr. Rajkumar, Mr. Sandeep Sharma, Dr. A. Rao, Dr. Amit Patel, Dr. Mohammed Sirajuddin, Mr. P. M. Khot, Mr. D. P. Rath, Mr. Chetan Baghra, Mr. Nagendra Kumar, Dr. Chiranjit Nandi, Mr. Somesh Bhattacharya and Mr. Mukesh Choudhary, from BARC and Dr. Sekhar Kumar from IGCAR.

The draft of the book was subjected to internal review by experts from India, namely Dr. D. K. Sinha, Director, Atomic Minerals Directorate, Hyderabad (Chap. 2), Dr. C. Ganguly, Former Chief Executive, Nuclear Fuel Complex, Hyderabad and Dr. S. Majumdar, Former Head, Radiometallurgy Division, BARC, (Chap. 3), Shri H. S. Kamath, Former Director, Nuclear Fuels Group, BARC (Chap. 4), Dr. Mrs. Renu Agarwal, Former Scientific Officer, Fuel Chemistry Division, BARC (Chap. 5), Shri S. Anantharaman, Former Head, Post Irradiation Examination Division, BARC

(Chap. 6), Dr. A. Ramanujam, Former Head, Fuel Reprocessing Division, BARC (Chap. 7), Dr. P. K. Sinha, Former Head, Waste Management Division, BARC (Chap. 8), Dr. K. L. Ramakumar, Former Director, Radiochemistry and Isotope Group, BARC and Head, Nuclear Control and Planning Wing, DAE (Chap. 9) and Dr. M. R. Iyer, Former Head, Division of Radiological Physics, BARC (Chap. 10). We gratefully acknowledge the contribution of all the reviewers which provided critical inputs that helped in improving the content and presentation of the book.

All aspects starting from the planning of the book, finalization of contents, review process, etc., were extensively debated by the core team of editors along with the chapter authors through a number of meetings (even during the pandemic period), for which the support provided by HBNI was invaluable. In this regard, we would like to place on record our deep appreciation for the support and cooperation rendered by Dean and Vice Chancellor, Homi Bhabha National Institute, in bringing out this book. Thanks are also due to Dr. Mrs. P. M. Aiswarya, Research Associate, HBNI, for her editorial assistance by way of improving the quality of figures and tables as well as checking and formatting the references.

B. S. Tomar
P. R. Vasudeva Rao
S. B. Roy
Jose P. Panakkal
Kanwar Raj
A. N. Nandakumar

Contents

Editors and Contributors

About the Editors

Prof. B. S. Tomar is presently Institute Chair Professor at Homi Bhabha National Institute (HBNI), Mumbai. He joined the Radiochemistry Division of Bhabha Atomic Research Centre (BARC), Mumbai, in 1982. He is an expert in nuclear and radiochemistry, radiation detection and measurement, and non-destructive assay of nuclear materials and has published more than 220 research papers in international journals. He has taught the subject of nuclear and radiochemistry at postgraduate level for over 25 years. He superannuated as Director Radiochemistry and Isotope Group, BARC, in December 2017 and later served as Raja Ramanna Fellow, DAE, during 2018–2021. He was Visiting Professor to Technical University, Delft, The Netherlands, during 2007–2008 and Member of the Standing Advisory Group on Safeguards Implementation (SAGSI), IAEA, during 2016–2017.

Prof. P. R. Vasudeva Rao was the Vice Chancellor of Homi Bhabha National Institute (HBNI), a Deemed-to-be University under the Department of Atomic Energy (DAE), India till recently. He earlier served as Scientific Officer with Indira Gandhi Centre for Atomic Research (IGCAR) at Kalpakkam and was its Director, at the time of his superannuation. He obtained his Ph.D. degree in Chemistry from University of Bombay in 1979 for his work on chemistry of actinide elements. Prof. Rao is an expert in Fuel Cycle Chemistry and especially chemistry of fast reactor fuel cycle and is well known for his research on actinide separations. He was responsible for the development of several facilities for R&D on fast reactor fuel cycle at IGCAR. He has over 300 publications in peer-reviewed international journals. He is a Fellow of the Indian National Academy of Engineering as well as the National Academy of Sciences, India.

Dr. S. B. Roy pursued her B.Tech. in Chemical Engineering from Calcutta University in 1981 and obtained doctorate degree in Chemical Engineering from Indian Institute of Technology (IIT) Bombay in 2001. She joined the 25th batch of BARC training school in 1981 and later joined Uranium Extraction Division of BARC in 1982. She was instrumental in development of uranium fuel of different grades for special applications, process for recovery of uranium from secondary sources like monazite, research, facility development and production management of specific quality nuclear fuel materials for various Indian Research Reactors. Post superannuation, she served as Raja Ramanna Fellow, DAE, and later as Visiting Scientist till November 2020, wherein she contributed in the field of Safety Review and Human Resource Development.

Dr. Jose P. Panakkal joined Bhabha Atomic Research Centre from 16th batch of training school and had been working at BARC in various capacities for 40 years. He superannuated as Head, Advanced Fuel Fabrication Facility, Nuclear Fuels Group, BARC, and was awarded Raja Ramanna Fellowship by DAE on superannuation. He also had worked as Guest Scientist at Fraunhofer Institute of NDT, Saarbrucken, Germany (1987–1989). He was recognized as Ph.D. Guide by Mumbai University and Homi Bhabha National Institute (HBNI). He is specialized in the fabrication of nuclear fuels particularly containing plutonium for both thermal and fast reactors and was in charge of an industrial scale MOX (Mixed Oxide) fuel fabrication facility which has made MOX fuels for various types of reactors (BWRs, PHWRs, and fast reactors). He has also developed new techniques for fabrication, quality control, and non-destructive evaluation of nuclear fuels and material characterization as evident from his 280 publications including 130 publications in international journals and conferences.

Shri. Kanwar Raj graduated with honors in Chemical Engineering in 1972 from Indian Institute of Technology Roorkee, India. Since then, he has worked in the various capacities in the field of design, research & development, operations, and safety analysis of radioactive waste management plants/facilities. Shri Raj has extensive expertise in design, commissioning, and operation of high-level radioactive vitrification systems as well as in evaluation of long-term performance of vitrified waste product under geological disposal conditions. As Chief Scientific Investigator, he has led two BARC teams for studies in this area in collaboration with Kfk/INE Karlsruhe and as a part of coordinated research program of IAEA, Vienna. As Head, Waste Management Division, BARC, Mr. Raj supervised operations of Waste Management Facilities at Kalpakkam, Tarapur, and Trombay Centres of the Department of Atomic Energy, India. He has gained hands-on experience in management of radioactive waste generations from the entire nuclear fuel cycle, viz. radiochemical laboratories, fuel fabrication facilities, research/power reactors, reprocessing plants, etc. His expertise also encompasses management of spent radiation sources from various industries, medical centers, and research institutions. Mr. Raj has experience in preparation of National and International Waste Safety Standards for Indian Atomic Energy Regulatory Board and IAEA. He is the author/co-author of about 80 research papers published in various journals.

A. N. Nandakumar a Ph.D. in Physics from Mumbai University, has 34 years of R&D experience in the Bhabha Atomic Research Centre, Mumbai, India, in the fields of safe transport of radioactive material, calculation of radiation shielding, and calculation of radiation dose under normal and emergency conditions involving radioactive material. He was Chief Scientific Investigator from India for the IAEA Coordinated Research Projects on Development of PSA techniques relating to the Safe of Radioactive Material and on Collection of Accident Data for Quantification of Risk in Transport of Radioactive Material. Subsequently, he was appointed Head, Radiological Safety Division, Atomic Energy Regulatory Board, Mumbai. He worked as Transport Safety Specialist and as Consultant in the Radiation, Transport and Waste Safety Division, IAEA, Vienna. He has published over 60 scientific and technical papers in national/international scientific journals and conference proceedings.

Contributors

P. V. Achutan Formerly Fuel Reprocessing Division, Bhabha Atomic Research Centre, Mumbai, India

S. Anthonysamy Formerly Metal Fuel Recycle Group, Indira Gandhi Centre for Atomic Research, Kalpakkam, India

R. K. Bajpai Nuclear Recycle Group, Bhabha Atomic Research Centre, Mumbai, India

Joydipta Banerjee Radiometallurgy Division, Nuclear Fuels Group, Bhabha Atomic Research Centre, Mumbai, India

C. V. S. Brahmananda Rao Materials Chemistry and Metal Fuel Cycle Group, Indira Gandhi Centre for Atomic Research, Kalpakkam, India

D. Das Formerly Chemistry Group, Bhabha Atomic Research Centre, Mumbai, India

A. K. Kalburgi Chemical Technology Group, Bhabha Atomic Research Centre, Mumbai, India

V. Karthik Metallurgy and Materials Group, Indira Gandhi Centre for Atomic Research, Kalpakkam, India

C. P. Kaushik Formerly Nuclear Recycle Group, Bhabha Atomic Research Centre, Mumbai, India

Smitha Manohar Nuclear Recycle Group, Bhabha Atomic Research Centre, Mumbai, India

Prerna Mishra Post Irradiation Examination Division, Nuclear Fuels Group, Bhabha Atomic Research Centre, Mumbai, India

Sudhir Mishra Radiometallurgy Division, Nuclear Fuels Group, Bhabha Atomic Research Centre, Mumbai, India

P. K. Mohapatra Radiochemistry Division, Bhabha Atomic Research Centre, Mumbai, India

A. N. Nandakumar Formerly Radiological Safety Division, Atomic Energy Regulatory Board, Mumbai, India

Jose P. Panakkal Formerly Advanced Fuel Fabrication Facility, Nuclear Fuels Group, Bhabha Atomic Research Centre, Mumbai, India

G. K. Panda Radiological Safety Division, Atomic Energy Regulatory Board, Mumbai, India

Amrit Prakash Radiometallurgy Division, Nuclear Fuels Group, Bhabha Atomic Research Centre, Mumbai, India

Kanwar Raj Formerly Nuclear Recycle Group, Bhabha Atomic Research Centre, Mumbai, India

P. N. Raju Formerly Radiochemistry and Isotope Group, Bhabha Atomic Research Centre, Mumbai, India

P. R. Vasudeva Rao Formerly Homi Bhabha National Institute, Mumbai, India

S. B. Roy Formerly Chemical Engineering Group, Bhabha Atomic Research Centre, Mumbai, India

M. L. Sahu Formerly Uranium Extraction Division, Bhabha Atomic Research Centre, Mumbai, India

Manju Saini Radiological Safety Division, Atomic Energy Regulatory Board, Mumbai, India

D. B. Sathe Fuel Fabrication-Integrated Nuclear Recycle Plant Operation, Nuclear Recycle Board, Bhabha Atomic Research Centre, Tarapur, India

Priti Kotak Shah Post Irradiation Examination Division, Nuclear Fuels Group, Bhabha Atomic Research Centre, Mumbai, India

N. Sivaraman Materials Chemistry and Metal Fuel Cycle Group, Indira Gandhi Centre for Atomic Research, Kalpakkam, India

T. Sreenivas Mineral Processing Division, Bhabha Atomic Research Centre, Hyderabad, India

G. Sugilal Nuclear Recycle Group, Bhabha Atomic Research Centre, Mumbai, India

B. S. Tomar Homi Bhabha National Institute, Training School Complex, Anushak-tinagar, Mumbai, India;

Formerly Radiochemistry and Isotope Group, Bhabha Atomic Research Centre, Mumbai, India

Chapter 1
The Nuclear Fuel Cycle: Introduction

P. K. Mohapatra and P. R. Vasudeva Rao

1.1 What Is a Fuel Cycle?

With the ever-increasing energy demand due to the rising global population and a growing emphasis on technologies those have minimal carbon footprint, nuclear energy has emerged as an important element of any sustainable solution for meeting energy requirements. Production of nuclear energy is mainly through controlled nuclear fission in a reactor where fissile isotopes of actinides such as ^{235}U, ^{239}Pu, and in some cases ^{233}U are used in the form of a suitable compound (or metal alloy) as the fuel, even though fusion is also a long-term candidate as an energy resource.

The amount of energy extracted from nuclear fuel in a nuclear reactor is expressed in gigawatt-days per metric ton of heavy metal (GWd/tHM). "Burn-up" is a measure of fuel depletion and is also measured as the ratio of the fuel atoms fissioned to the number of initial heavy metal atoms, expressed as atom %.

The life cycle of the material used as the nuclear fuel, starting from the mining of the ore to the fabrication of the fuel, its irradiation in reactor, and subsequent processing of the spent nuclear fuel, is termed as the nuclear fuel cycle. Figure 1.1 provides a depiction of various stages of the nuclear fuel cycle.

The nuclear fuel cycle can be of different types, depending on several factors, of which the following are most important:

(a) fissile nuclide being exploited for the generation of energy through fission (^{235}U, ^{239}Pu, or ^{233}U),
(b) the form of the fuel (chemical as well as physical),
(c) the type of reactor in which the fuel is deployed.

P. K. Mohapatra (✉)
Radiochemistry Division, Bhabha Atomic Research Centre, Mumbai 400085, India
e-mail: mpatra@barc.gov.in

P. R. V. Rao
Formerly Homi Bhabha National Institute, Mumbai 400094, India

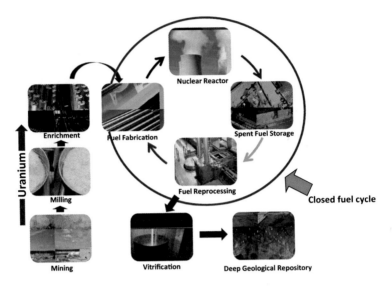

Fig. 1.1 Schematic representation of nuclear fuel cycle

The variations possible in the nuclear fuel cycle based on the factors mentioned above are described below.

(a) The fissile nuclide used, in a way, decides the options for the forms in which it can be used. Natural uranium is used in some reactor systems, while in some others, it is used in an enriched form (uranium with a higher content of U-235 isotope than that present in the natural uranium). Due to neutronic considerations, natural uranium, which consists mainly of U-238 isotope (99.274%) and about 0.72% of U-235 isotope, is used as fuel only in heavy water moderated reactors. The CANDU-type pressurized heavy water reactors (PHWRs) have delivered impressive performance records. However, light water moderated reactors, which use enriched uranium as the fuel, constitute the major fraction of thermal reactor systems, globally. Fast neutron reactors use fuels based on either enriched uranium or plutonium. Recycling of the fuel is especially an important option for reactor systems using fuels based on enriched uranium or plutonium.

(b) Uranium has been used as the fuel in various chemical forms in nuclear reactors. These forms of uranium used in nuclear reactors include the metal, alloy with other elements, and its oxide compound (UO_2), the most preferred form being the oxide (indeed, other forms including aqueous solutions or molten salts are known, but are not discussed in this book). The fuel cycle of the reactor system is influenced by the chemical form of uranium used as the fuel, since uranium does not occur in nature in any of these forms. The uranium obtained from natural resources has to be enriched as per requirement and converted to the desired chemical form, for its use in a reactor. The burn-up that can be reached with the fuel is also influenced by the form of the fuel, in addition to other factors.

The chemical form as well as the burn-up has a great impact on the flow sheet for the recovery of the fuel material from the irradiated fuel, for the purpose of recycling.

(c) Nuclear reactors are of various types, and their most important classification is based on the energy of the neutrons causing fission in the fuel—thermal and fast. World over, with just a few exceptions, nuclear power production is currently based only on reactors where fission is caused by thermal neutrons. The thermal reactors use either natural uranium or slightly enriched uranium as the fissioning material. (Uranium–plutonium mixed oxide has also been used in some cases). On the other hand, fast reactors use fuels of high fissile content (provided by the use of plutonium or enriched uranium) and also reach high levels of burn-up. The fast reactor fuel cycle needs to be closed, in order to be economically sustainable. The fabrication and reprocessing steps for fast reactor fuels need to take into account the presence of high concentration of plutonium. In addition to oxides, other chemical forms such as carbide or metal alloy have been proposed as the fuel for the fast reactor systems, and the fuel cycle has to be designed accordingly to suit the fuel form. For example, metallic fuels are reprocessed through non-aqueous routes.

1.2 Frontend and Backend of Fuel Cycle

1.2.1 Frontend of Fuel Cycle

The nuclear fuel cycle, in a generic sense, starts with the mining of uranium and ends with the disposal of nuclear waste. Some steps may not apply to some fuel cycle types, e.g., for fast reactor fuel cycle, no mining is involved. The steps of mining and milling, conversion, enrichment, fuel fabrication, etc. constitute the "frontend" of the nuclear fuel cycle, while the steps after the discharge of the irradiated fuel from the reactor, such as temporary storage of spent fuel, reprocessing, waste management, etc., are termed as the "backend" of the nuclear fuel cycle.

Uranium-rich minerals are radioactive in nature which is mainly due to the daughter products of uranium derived from radioactive decay processes. The world-wide production of uranium in 2019 amounted to about 54,750 tons [1], and the mined uranium was almost entirely used as the fuel for nuclear power plants. Uranium recovery is done by extraction from ores followed by concentration and purfication. Extraction is done by both excavation and in situleaching (ISL). In general, open-pit mining is used where deposits are close to the surface, and underground mining is used for deep deposits. The mined uranium ores are normally processed by grinding followed by uranium leaching mostly by chemical methods using either an alkali (Na_2CO_3) or an acid (H_2SO_4). The subsequent milling process yields the "yellow cake," which contains uranium as U_3O_8 with >80% uranium content. The milling process concentrates the uranium content by over 800 times from about 0.1% present in the ores. About 200 tons of U_3O_8 are required to keep a 1000 MWe nuclear power reactor generating electricity for one year [2] (This is lower than the requirement for fossil fuel-based power generation by several orders of magnitude).

The U_3O_8 produced from the uranium mill is required to be enriched in the U-235 content from 0.72% to between 3 and 5% (for light water reactors). The CANDU-type pressurized heavy water reactors (PHWRs) do not require uranium to be enriched; fast reactors may need enrichment as high as 90% depending upon the size of the reactor. The isotope enrichment is usually a physical process (though there are reports of some chemical processes as well) and requires the uranium bearing compound to be in a gaseous form. In view of this, uranium is converted to uranium hexafluoride, which is a solid at ambient temperature and sublimes at 56.5 °C. The conversion step is one of the most important steps in the nuclear fuel cycle. The three major processes used in the isotope separation involved in the enrichment of U are: (i) gaseous diffusion, (ii) centrifugation, and (iii) laser separation. While the gaseous diffusion process is not used any more due to large energy consumption, laser isotope separation has not matured enough to be deployed at a large scale. Therefore, the main enrichment process used in commercial plants is centrifugation, where thousands of rapidly spinning vertical tubes exploit about 1% mass difference between the hexafluorides of two uranium isotopes leading to their effective separation.

The enriched UF_6 cannot be used as such as the fuel in the nuclear reactor and, hence, needs to be converted to ceramic pellets of UO_2 (containing LEU) sintered at >1400 °C. Alternatively, mixed U, Pu oxide (MOX) fuel pellets are fabricated. The chemical quality control (CQC) of the nuclear fuel is one of the major tasks entrusted to laboratory chemists who certify the U, Pu, and impurity content that can be tolerated in the reactor fuel. While the major components are measured by electro-chemical methods like potentiometry, bi-amperometry, etc., the trace metallic impu-rities are determined by a host of techniques such as atomic emission and/or atomic absorption spectroscopy-based techniques, after removal of the matrix. Several crit-ical non-metallic trace impurities (hydrogen, chlorine and fluorine, carbon, nitrogen, oxygen, sulfur, total gas analysis, etc.) are also determined by a variety of analytical techniques as a part of CQC to ascertain the desired level of purity of the fuel. The isotopic composition of actinides (U and Pu) used as major constituent of fuels is another important specification of the fuel.

During fuel fabrication, especially for the plutonium bearing fuels and those having highly enriched uranium, the possibility of a criticality incident needs to be taken into consideration carefully, in the design of equipment as well as processes. This issue is non-existent for the PHWR fuels and of lesser importance for LEU-bearing fuel. The dimensions of the fuel pellets and other components of the fuel assembly have to be precisely controlled to ensure satisfactory performance of the fuel in the reactor core.

Chapter 2 of this book deals with the frontend steps in fuel cycle, including uranium mining, milling, refining, enrichment, etc., in a detailed manner, while Chap. 3 deals with various aspects of the nuclear fuel fabrication.

1.2.2 Backend of Fuel Cycle

1.2.2.1 Irradiated Fuel

The composition of the irradiated fuel discharged from a reactor depends on the starting composition, and the extent to which the fuel has been "burnt." The "burn-up" seen by the fuel is a direct measure of the amount of fission energy that has been extracted from the fuel. A burn-up of 1 atom % of the fissile material corresponds to an energy generation of 10,000 MWd from one ton of fuel. It is important to emphasize that unlike in the fossil fuel-based power generation, in nuclear power generation, the fuel is not burnt out completely in one cycle. It is clear that since natural uranium contains only 0.72% of the fissile isotope (U-235), the burn-up of fuel based on natural uranium should be less than 0.72 atom %. However, because of fissions in Pu-239, generated in situ by neutron absorption in U-238, the burn-up can reach up to 1 atom % (10,000 MWd/t). In typical PHWRs, the burn-up level is of the order of 7000 MWd/t. In light water reactors which use enriched uranium, the burn-up levels reach 6–7 atom %, while in fast reactors, it can exceed 10 atom %. In some cases, up to 20 atom % burn-up has been reached.

After a certain level of burn-up, (corresponding to a typical irradiation period of 3–4 years for LWRs), the fuel has to be discharged from the reactor, since the accumulated fission products can affect the neutron economy. For fast reactor fuels, which reach high burn-up levels, other issues such as fuel clad chemical interactions and fuel swelling limit the fuel burn-up. However, a typical LEU-based fuel containing 5% U-235 discharged from the reactor after a burn-up of 55,000 MWd/t still has about 96% of its original U content (of which <1% constitutes U-235), while the fission products and Pu (activation product) constitute ca. 3 and 1%, respectively. Therefore, the fissile content (including Pu) needs to be recovered for its subsequent use as fuel in reactors. The U-235 content in the spent fuel is still higher than that in natural uranium and, hence, can be reused if subjected to the conversion and enrichment steps as mentioned above. In addition, Pu can be used in MOX fuel to increase the fissile content as required. Thus, it is obvious that irrespective of the type of fuel or reactor, the irradiated fuel is still a valuable source of fissile materials. Even if the uranium is depleted with respect to its fissile content, it can be used to prepare the fuel for a fast reactor by the addition of Pu. Thus, the irradiated fuel discharged from a nuclear reactor is not, strictly speaking, "spent fuel," even though the term "spent fuel" is used in literature, and also in this book, to be synonymous with the discharged fuel.

The process of recovering valuable fissile and fertile materials from the irradiated fuel is called reprocessing. This is briefly discussed in the next section but is dealt with in detail in Chap. 7.

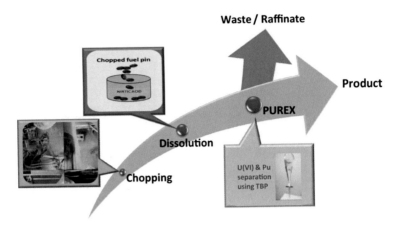

Fig. 1.2 Schematic presentation of spent fuel reprocessing using PUREX process

1.2.2.2 Reprocessing

The process adapted for the separation of U and Pu from the irradiated fuel by different countries is the Plutonium Uranium Redox EXtraction (PUREX) process which is schematically presented in Fig. 1.2. This process uses 30% tri-n-butyl phosphate (TBP) in an aliphatic hydrocarbon diluent (such as kerosene or a mixture of normal paraffinic hydrocarbons) for extracting U and Pu preferentially [3]. The first step involves the chopping of the fuel rods followed by their dissolution in 7–8 M nitric acid. The fuel material dissolves in nitric acid leaving behind the "hull" constituting the insoluble part which traps some amount of the radionuclides. The dissolution step is followed by feed clarification (filtration) and feed acidity adjustment followed by the extraction and partitioning steps. The decontamination factor (DF) from the fission products is very high (typically in the range of 10^4–10^6) and the process can be operated continuously for several cycles as required, without any significant change in the efficiency. Chapter 7 gives a detailed account of spent fuel reprocessing which also includes a discussion on the path of the minor actinides and fission products in the PUREX process. The chapter also describes reprocessing of Th-based spent fuel including the THOREX processes. Fast reactor spent fuel reprocessing is quite challenging due to the high Pu content of the fuel and also higher concentrations of the fission and activation products due to the high burn-up. This is also addressed in Chap. 7 along with a brief description of non-aqueous routes for reprocessing.

1.2.2.3 Waste Management

Minor actinides (MA) such as Np, Am, and Cm are streamed out of the PUREX cycle into the "raffinate" (a waste stream) which when concentrated results in the high-level liquid waste (HLLW). The HLLW is responsible for >95% of radioactivity of all radioactive wastes and, hence, needs safe and acceptable management strategies for the success of the nuclear energy program. The common strategy involves the vitrification (using borosilicate glass) of solid waste oxides in glass blocks and their subsequent burial in deep geological repositories (DGRs). However, such disposal will require a long surveillance period, in view of the very long half-lives of minor actinides and some fission products making the waste management program highly expensive. Also, as the decay of the buried radionuclides takes millions of years, the radiotoxicity of the long-lived radionuclides present in the waste blocks is reduced to a level comparable to that of U from natural sources only after a very long time (Fig. 1.3). This prompted the development of processes for partitioning of minor actinides such as Np, Am, and Cm and fission products such as Cs-137 and Sr-90 which can drastically reduce the average exposure to the operating personnel. The strategy of "Actinide Partitioning" can be followed by the transmutation of the long-lived radionuclides in fast reactors or accelerator driven sub-critical systems (ADS). The "Partitioning & Transmutation" (P&T) strategy is of considerable interest in the long term, as it effectively addresses the concerns about radioactive waste management. However, if "Actinide Partitioning" is linked with the other emerging strategies such as "lanthanide–actinide separation" and "Am–Cm separation," it can have the additional benefit of the recovery of valuables like Am and Cm, which are immensely useful materials.

Apart from specific chemical methods used for various types of radioactive wastes, several physical methods are also used for waste immobilization/compaction prior to their subsurface disposal. The general categorization of radioactive wastes and their classification along with various strategies involved in their safe management are discussed at length in Chap. 8.

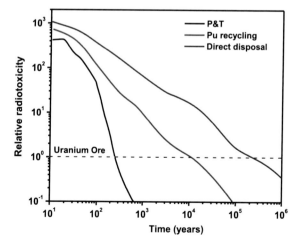

Fig. 1.3 Effect of Actinide Partitioning on the relative radiotoxicity of radioactive waste

1.3   Categories of Nuclear Fuel Cycle Based on Reuse Strategies

Nuclear fuel cycles are categorized into different types based on the different strategies employed on fronts like cost, safety, non-proliferation, and waste management. In general, there are two types of nuclear fuel cycles, viz., the "open" or "once-through" and the "closed" or "recycled." The "twice-through" fuel cycle is a variant that represents limited closure of the fuel cycle. There is also a strategy called "partially closed" fuel cycle which has limited spent fuel recycling option (similar to the "twice-through cycle") and is a concept that is in between the open and the closed fuel cycles.

1.3.1   Open or "Once-Through" Fuel Cycle

In the "open" or "once-through" fuel cycle (Fig. 1.4), the fuel material is used in the nuclear reactor only once; the fuel discharged from the reactor is kept in an "interim storage" facility until most of the short-lived radioactivity has decayed [2, 4]. The fuel can then be disposed into a repository as waste or reprocessed at a later stage for recovery of the fissile material. The open fuel cycle is practiced in the USA, UK, and South Africa where the discharged fuel is subjected to storage for possible vitrification and burial in deep geological repositories (DGRs) at a future date. At present, the irradiated, discharged fuel is stored either at the "at-reactor" (AR) or "away-from-reactor" (AFR) facilities. A block diagram, showing the "once-through" fuel cycle, is given in Fig. 1.4.

1.3.2   Twice-Through Fuel Cycle

In this case, the spent nuclear fuel is reprocessed in order to extract the uranium and plutonium and subsequently recycled once in the reactor. The recovered uranium and plutonium are suitably fabricated as mixed oxide fuel, to replace the enriched uranium oxide fuel. The MA and fission products are sent to the interim storage facility for further processing. The LWR is either fully loaded with MOX fuel or with a combination of MOX fuel bundles (about 30%) and UO_2 fuel bundles. After the second cycle of irradiation, the MOX fuel is not reprocessed and is cooled and stored similar to the "once-through" fuel cycle mentioned above. A schematic representation of the "twice-through" fuel cycle [2, 4] is presented in Fig. 1.5. Such recycling has been implemented in many countries especially with the oxide fuels. The major reason for recycling of the "spent nuclear fuel" (SNF) in the case of the "twice-through" NFC is due to the fact that the SNF still contains ca. 96% of the reusable material (U and Pu) and, hence, is still considered as an energy source due to an

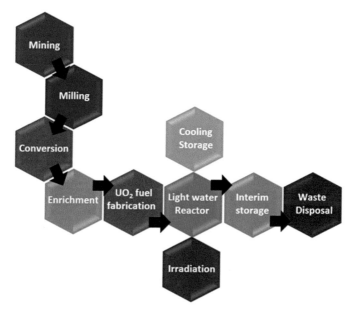

Fig. 1.4 Block diagram of the "once-through" fuel cycle

appreciable fissile material content. Additional cycles in a thermal reactor are not, however, practical due to degradation of the isotopic composition of Pu. As compared to the "once-through" NFC, in which the SNF is stored after use and buried in DGRs after vitrification, the "twice-through" NFC can utilize 17% more natural uranium by the MOX fuel option which can largely discount the heavy enrichment costs. The "twice-through" NFC can further yield a significant reduction in the waste volume (>80%) and radiotoxicity (>90%) of the HLLW.

1.3.2.1 Degradation of Isotopic Composition of Plutonium in Thermal Reactors

It is known that the isotopic composition of plutonium is not significantly changed due to irradiation in fast reactors. Thus, fast reactors can utilize the full energy potential of plutonium through multiple cycles. However, in thermal reactors, the isotopic composition of plutonium is degraded, and the use of plutonium in fuel becomes unviable after a few cycles. As a result, plutonium recovered from irradiated fuel can be used only in one additional cycle of irradiation, in a thermal reactor (twice-through cycle).

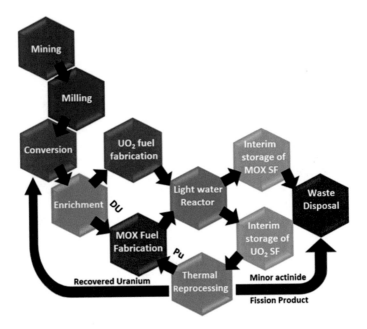

Fig. 1.5 Block diagram of the "twice-through" fuel cycle

1.3.3 Closed or "Recycled" Fuel Cycle

The closed fuel cycle (depicted in Fig. 1.1) is advocated by many countries including France, Japan, India, China, Russia, etc. and has the objective of effective utilization of the fissile material through multiple recycling and reducing the hazardous nature of the radioactive waste to be disposed by recovering the radiotoxic elements. The "closed" or "recycled" fuel cycle requires the development of advanced technologies for a sustainable NFC.

1.4 Open Versus Closed Fuel Cycle Options

The policy choice of the fuel cycle by a country depends on several factors. The common factor is a short-term approach caused by perceptions on economics, proliferation concerns, and assurance of uranium availability. In this approach, recycling of uranium is considered as an economic choice only when the cost of uranium is high. Several countries, therefore, have opted to postpone the implementation of recycling and resorted to store irradiated fuel in interim storages. Another reason for the adoption of open fuel cycle policy by many countries is the concern of nuclear proliferation, as a result of which the recovery of plutonium from the irradiated fuel is discouraged.

On the other hand, it is argued by several countries which advocate the use of a closed fuel cycle that it is a requirement for long-term sustainability of nuclear power. As pointed out in Sect. 1.2.2.1, the fissile content of the irradiated fuel discharged from reactor is quite valuable and does not qualify to be treated as waste material. The effective utilization of the uranium resources in nature is thus possible only through multiple cycles of irradiation. Since uranium resources are non-renewable, one can say that sustainable energy generation through nuclear fission can be realized only through the closed fuel cycle.

However, as discussed above, the number of cycles for which a given fuel material can be irradiated in thermal reactors is limited by the degradation of the isotopic composition of Pu. A high level (near quantitative) utilization of the fuel material through multiple cycles of irradiation is only possible in fast reactors. Fast reactors also need the closed fuel cycle from economic point of view. Thus, it can be argued that fast reactors with closed fuel cycle constitute the long-term sustainable option for nuclear reactors.

The impact of the closed fuel cycle on the long-term management of highly radioactive wastes is another important factor that is in favor of a closed fuel cycle. This aspect is discussed in some detail in the next section.

1.4.1 Impact of Closed Fuel Cycle on Waste Management

To understand the impact of the closed fuel cycle on waste management, one needs to study the composition of the radioactive wastes. The composition of a typical irradiated fuel discharged from the reactor is shown in Table 1.1. From the point of view of the radioactive waste management, one needs to particularly consider the radionuclides with relatively long half-life and significant radiotoxicity. Such nuclides include the isotopes of plutonium and the "minor" actinides (the term minor is used to convey that these elements occur at relatively low concentrations in the irradiated fuel, as compared to the "major" actinides, viz., uranium and plutonium). The radiotoxicity of these radionuclides is such that the wastes containing those are required to be disposed in repositories deep down the earth and not in near-surface repositories. In addition, due to their significant radiotoxicity and long half-lives, the waste form needs to be under surveillance for periods of the order of 10,00,000 years (please recall that the half-life of plutonium-239 isotope, one of the major constituents of the irradiated fuel, is of the order of 24,000 years). These factors not only pose several technological challenges but also considerably increase the cost of the fuel cycle.

The open fuel cycle, where the irradiated fuel as a whole is treated as waste, has a major impact on the radioactive waste management. In the closed fuel cycle, however, uranium and plutonium (in some versions, also the minor actinides) are recovered and recycled, reducing the quantity of the material to be disposed as waste to a very large extent (a factor of four or more). In addition, if the concentration of the radiotoxic long-lived alpha emitters is reduced below certain limits (presently

Burn-up	6700 MWd/T (5 y cooling) (%)	33,000 MWd/T (10 y cooling) (%)
U	98.96	95.5
Pu	0.35	0.85
MAs (Np, Am, Cm)	0.011	0.112
Long-lived FPs	0.044	0.2
Short-lived FPs	0.038	0.17
Stable nuclides	0.60	3.2

Table 1.1 Distribution of principal radioactive elements present in irradiated fuel discharged from a nuclear reactor [5]

accepted limit is 100 nCi/g), the waste can also be disposed near surface. Also, the surveillance period can be reduced to less than thousand years. These factors undoubtedly impact the economics of the fuel cycle.

Further value can be realized from the waste, through separation of radio-cesium (^{134,135,137}Cs), radio-strontium (^{90}Sr), PGMs, etc., which can be used in societal applications.

These variants in the implementation of the closed fuel cycle are discussed in brief in the following section.

1.5 Variants of Closed Fuel Cycle

1.5.1 Closed Fuel Cycle Without MA Recovery

In this type of closed fuel cycle, uranium and plutonium are recovered from the irradiated fuel, and minor actinides (Am, Cm, Np, etc.) and fission products are rejected into the raffinate stream for their interim storage and ultimate vitrification and burial in the DGRs. The plutonium recovered can be recycled in thermal or fast reactors. Depleted uranium can be used in MOX fuels (uranium–plutonium mixed oxide) constituted for thermal (or fast) reactors. This was the principle followed initially in countries with poor uranium reserves where the objective was to recover the depleted uranium and plutonium from the irradiated fuel for their subsequent utilization in the MOX fuel-based program.

1.5.2 Closed Fuel Cycle with MA Recovery

In Closed fuel cycle this type of fuel cycle, the HLLW arising out of the primary separation process (wherein uranium and plutonium are separated from the irradiated fuel through a solvent extraction scheme, for recycling) is subjected to another separation scheme wherein the minor actinides are recovered. The most well-developed schemes for the recovery of minor actinides also recover the lanthanides along with the former because of the similarity in chemical properties of the trivalent actinides and lanthanides. The minor actinides recovered can be subsequently burnt in accelerator-driven sub-critical systems (ADSS) or fast reactors [6]. In this context, lanthanide–actinide separation [7] and subsequently, Am–Cm separation [8] are emerging as important steps for a meaningful transmutation of the long-lived actinides. It is important, however, to point out that in order that waste immobilization and its ultimate disposal in the DGRs are simplified, the recovery of the radiotoxic minor actinides has to be near quantitative and the remnant waste should meet the requirements for categorization as "non-alpha" waste.

1.5.3 Closed Fuel Cycle with MA and Fission Product Recovery

As shown in Table 1.1, the fission products constitute ca. 0.6% of the irradiated fuel out of which the long-lived fission products constitute only approx. 0.04% of all elements. However, the major amount of radioactivity of the HLLW is due to the presence of the highly radioactive fission product nuclides such as ^{137}Cs and ^{90}Sr, which are mainly responsible for the radiation dose burden. Apart from these, there are other radionuclides such as ^{106}Ru with moderate radioactivity and long-lived fission product nuclides such as ^{99}Tc, ^{93}Zr, ^{135}Cs, ^{129}I, and ^{107}Pd. Out of these, ^{137}Cs and ^{90}Sr can find use in societal applications such as in irradiators (^{137}Cs), nuclear medicine (^{90}Y), and as power sources (^{90}Sr), and hence, there have been efforts to separate these radionuclides from the HLLW prior to the radioactive waste vitrification. Also, long-lived fission product nuclides can be separated and transmuted to shorter-lived radionuclides for an effective reduction in the period of surveillance of the vitrified waste blocks. In India, the present strategy is to separate the fission product radionuclides prior to the Actinide Partitioning step (of the P&T strategy) so that the radioactive waste has very low radiotoxicity which can be conveniently managed through vitrification and burial in DGRs.

1.6 Challenges in the Development of Closed Fuel Cycle

The closed fuel cycle, when suitably designed, can enable efficient utilization of the uranium (fissile content) resources (in principle one can visualize utilization of >90% of uranium in fast reactors over several cycles) and thus provide a sustainable option for the development of nuclear energy. However, except for some, most countries have not developed the option of closed fuel cycle. The most important reason for support to the open fuel cycle option arises from the concern of proliferation of nuclear weapons that can be caused by the separation of plutonium in its pure form.

Also, the backend of the fuel cycle involves complex technologies which require chemical engineering operations in hot cells, and very few countries have been able to establish infrastructure to develop the technology. For plutonium utilization through closed fuel cycle, the fuel fabrication facilities also need to be housed in glove box containment, a requirement that does not exist for uranium based fuels. This needs additional infrastructure including human resources trained for remote operations and maintenance. Thus, only those countries with a strong human resource base and mastery of the complex technologies involved in nuclear recycle can opt for the closed fuel cycle.

1.7   International Scenario on Fuel Cycle

Very few countries have programs related to all stages of nuclear fuel cycle. Some countries possess uranium resources but do not have nuclear energy programs; some countries have nuclear reactors but do not have uranium resources and therefore, have to depend on the supply of fuel under international agreements. Only a few countries have nuclear reactors as well as fuel cycle programs. These include France, India, Russia, Japan, UK, and China. The profiles of nuclear fuel cycle programs of various countries have been summarized in Ref. [9].

As mentioned earlier, the most common fuel cycle option internationally is the "open" or "once-through" fuel cycle. The backend of the fuel cycle, however, is a subject of intense research in several countries, indicating the importance of closed fuel cycle as a long-term option. Partitioning of minor actinides from the HLLW is one of the important areas pursued for the development by several countries [10]. Minor actinide separation is proposed through several advanced aqueous partitioning methods using a variety of extraction systems. Several variants of separation schemes are being studied, and these are described in Chap. 8. These include the separation of all the actinides as a group (GANEX), separation of uranium in one cycle followed by the separation of other actinides as a group, etc. Many of these processes also separate the lanthanides along with the actinides, due to the similarity in the properties of the lanthanides and the transplutonium actinides. Variants of the separation processes under development include processes which separate the lanthanides from the minor actinides [7]. Such separations can further reduce the volume of the waste, reducing the demand of space of the repository.

Under the Global Nuclear Energy Partnership (GNEP) program led by the USA, efforts are being made to ensure that countries with developed nuclear technological base provide safe nuclear power to other countries without compromising on proliferation concerns. There is a renewed interest in the development of new separation schemes for the reprocessing of spent nuclear fuels. In this approach, the recycling of spent nuclear fuel is proposed to be carried out without separating out pure plutonium thereby reducing the proliferation concerns.

From the above, one could conclude that the fuel cycle is not only a complex issue that is addressed by country-specific approaches, but will continue to be a subject of intense R&D in countries interested in exploitation of nuclear energy as a sustainable energy resource. The following chapters deal with various stages of nuclear fuel cycle and also various issues in development and implementation of nuclear fuel cycle such as safety issues and nuclear material accounting.

This book aims at giving a comprehensive view of the entire nuclear fuel cycle, with some in-depth description of various aspects of nuclear fuel cycle. Chapters 2 and 3 cover various aspects of the frontend of the fuel cycle. Quality control of nuclear fuels is vital for extracting best performance from the fuel, and various aspects of quality control are dealt with in Chap. 4. A broad overview of the thermochemical and thermophysical properties of nuclear fuels, which influence the choice of the fuel for a particular reactor system and the performance boundaries, is provided in Chap. 5. Post-irradiated examination (PIE) of nuclear fuels is key to the understanding of fuel behavior in the reactor and reasons for fuel failure. Principles and techniques for PIE are described in Chap. 6. Fuel reprocessing schemes and waste management aspects are dealt with in Chaps. 7 and 8. This book also includes chapters on nuclear material accounting (Chap. 9), transportation of nuclear materials (Chap. 10), and a brief description on radiation protection (Chap. 11).

1.8 Summary

Nuclear fuel cycle is an essential element of technology for the development of the nuclear energy program. Accordingly, several countries are actively engaged in research on various aspects of nuclear fuel cycle, in order to evolve efficient as well as economical fuel cycles with minimum impact on the environment and a high degree of safety. Currently, many countries are not in the favor of pursuing the closed fuel cycle, due to economic concerns as well as proliferation issues. However, since nuclear fuel resources are finite, closed fuel cycle is expected to be a necessity at some point in the future, in order to extend the utilization of the resources.

The technologies adopted for nuclear fuel cycle vary based on the fuel and reactor types. However, this book deals with the U–Pu-based fuel cycles in a generic manner, with discussion on some related topics as mentioned above. A broad knowledge of all aspects of the fuel cycle is thus expected to be of high value to all engaged in designing nuclear reactor systems as well as fuel cycle facilities and also to those engaged in operation or maintenance of such facilities.

References

1. www.world-nuclear.org
2. The future of the nuclear fuel cycle. An interdisciplinary MIT study, p. 20 (2011). ISBN 978-0-9828008-4-3
3. D.D. Sood, S.K. Patil, Chemistry of nuclear fuel reprocessing: current status. J. Radioanal. Nucl. Chem. **203**, 547–573 (1996)

4. L. Rodríguez-Penalonga, B.Y.M. Soria, A review of the nuclear fuel cycle strategies and the spent nuclear fuel management technologies. Energies **10**, 1235 (2017)
5. M. Salvatores, G. Palmiotti, Radioactive waste partitioning and transmutation within advanced fuel cycles: achievements and challenges. Prog. Part. Nucl. Phys. **66**, 144–166 (2011)
6. S.S. Kapoor, Accelerator-driven sub-critical reactor system (ADS) for nuclear energy generation. Pramana **59**, 941–950 (2002)
7. K.L. Nash, A review of the basic chemistry and recent developments in trivalent f-elements separations. Solvent Extr. Ion Exch. **11**, 729–768 (1993)
8. P. Matveev, P.K. Mohapatra, S.N. Kalmykov, V. Petrov, Solvent extraction systems for mutual separation of Am(III) and Cm(III) from nitric acid solutions, a review of recent state-of-the-art. Solvent Extr. Ion Exch. **39**, 679–713 (2021)
9. Country nuclear fuel cycle profiles, IAEA technical report series no. 425, 2nd edn., IAEA, Vienna, May (2005)
10. J.N. Mathur, M.S. Murali, K.L. Nash, Actinide partitioning: a review. Solvent Extr. Ion Exch. **19**, 357–390 (2001)

Further Reading

1. https://www.eia.gov/energyexplained/nuclear/the-nuclear-fuel-cycle.php
2. https://www.world-nuclear.org/information-library/nuclear-fuel-cycle/introduction/nuclear-fuel-cycle-overview.aspx
3. https://www.nuclear-power.net/nuclear-power-plant/nuclear-fuel/nuclear-fuel-cycle
4. The future of the nuclear fuel cycle, an interdisciplinary MIT study, MIT (2011). ISBN 978-0-9828008-4-3
5. S. Tachimori, S. Suzuki, Y. Sasaki, A process of spent nuclear fuel treatment with the interim storage of TRU by use of amidic extractants. J. Atomic Energy Soc. Jpn. **43**(12), 1235–1241 (2001) (in Japanese)
6. S. Widder, Benefits and concerns of a closed nuclear fuel cycle. J. Renew. Sustain. Energy **2**, 062801 (2010)
7. Spent fuel reprocessing options. IAEA-TECDOC-1587
8. https://www.radioactivity.eu.com/site/pages/Spent_Fuel_Composition.htm
9. M. Bunn, S. Fetter, J.P. Holdren, B. van der Zwann, The economics of reprocessing vs direct disposal of spent nuclear fuel, JFK School of Government, Harvard University, December 2003
10. C. Poinssot, C. Rostaing, S. Greandjean, B. Boullis, Recycling the actinides, the cornerstone of any sustainable nuclear fuel cycles, in *ATALANTE 2012 International Conference on Nuclear Chemistry for Sustainable Fuel Cycles*, Procedia Chem. **7**, 349–357 (2012)
11. A. Bhattacharyya, P.K. Mohapatra, Separation of trivalent actinides and lanthanides using various 'N', 'S' and mixed 'N, O' donor ligands: a review. Radiochim. Acta **107**, 931–949 (2019)

Chapter 2
Exploration, Mining, Milling and Processing of Uranium

T. Sreenivas, A. K. Kalburgi, M. L. Sahu, and S. B. Roy

2.1 Uranium Ore Exploration and Mining

2.1.1 Geology

2.1.1.1 Occurrence and Minerals

Uranium is a naturally occurring radioactive element very widely distributed throughout the earth's crust [1]. It is present in varied concentration in almost all types of rocks, natural water including sea water, and living organisms. The average concentration of uranium in the earth's crust is about 2.76 ppm in rocks [2], less than 1 ppb in surface water, 0.5–10 ppb in ground water, and 3 ppb in sea water. About 1.3×10^{14} tons of uranium exists in the earth's crust. However, presently, rocks are the only source from which extraction of uranium is viable. The occurrence of uranium in different types of rocks is given in Table 2.1. Uranium does not occur in native elemental form. It is chemically reactive and has great affinity for oxygen. Uranium exists in multiple oxidation states, say, +3, +4, +5, and +6, but the +4 and +6 states are predominant among them [3]. The naturally occurring compounds of uranium are oxides, phosphates, sulfates, vanadates, arsenates,

T. Sreenivas (✉)
Mineral Processing Division, Bhabha Atomic Research Centre, Hyderabad 500016, India
e-mail: tsreenivas@barc.gov.in

A. K. Kalburgi
Chemical Technology Group, Bhabha Atomic Research Centre, Mumbai 400085, India
e-mail: akk@barc.gov.in

M. L. Sahu
Formerly Uranium Extraction Division, Bhabha Atomic Research Centre, Mumbai 400085, India

S. B. Roy
Formerly Chemical Engineering Group, Bhabha Atomic Research Centre, Mumbai 400085, India

Table 2.1 Uranium content in different types of rocks [1, 2]

Rock type	Uranium concentration (ppm)
Ultrabasic	0.02
Gabbro	0.84
Felsic	4.60
Granite	3.50
Albitite	3.0
Basalt	0.6
Rhyolite	8.0
Carbonatite	2.2
Schist	1.5
Phyllite	1.9
Graphitic schist	3.5
Biotite schist	4.7
Gneiss	3.0

carbonates, and silicates [4]. There are about 250 uranium minerals reported till date [5]. Some of the common uranium minerals are given in Table 2.2 [4]. Among them, uraninite, pitchblende, brannerite, davidite, and coffinite are considered as economic minerals. Naturally occurring common secondary uranium minerals are autunite, meta-autunite, torbernite, meta-torbernite, carnotite, uranophane, etc.

The occurrence of uranium in rocks is in different forms such as (i) in discrete form; (ii) embedded into the structure of rock-forming minerals as isomorphic replacement; (iii) adsorbed on surface of crystals and grains; (iv) as liquid inclusions in rock-forming mineral, and (v) in the form of inter-granular liquid [1, 6]. Because of the great variability of the uranium content in these materials, it is not possible to give average abundance. Uranium in natural waters and their precipitates and drainage sediments play a very important role in prospecting of uranium.

2.1.1.2 Classification of Uranium Ore Deposits

An ore deposit is a naturally occurring assemblage of minerals from which one or few of the minerals can be economically extracted. A uranium ore deposit is formed by virtue of mobility of uranium ions from its source rock and its re-deposition or precipitation under suitable redox and chemical conditions [1, 7]. The mobility is facilitated by surface and in situ geological processes. The grade of a uranium ore deposit ranges from few hundred g/t U to more than 20% U. According to classification of International Atomic Energy Agency (IAEA), a high-grade ore contains >2% uranium and a low-grade ore assay <0.1% uranium [8]. Therefore, an ore-forming process is required for enrichment of uranium over its global geochemical background by a factor of 100–10,000.

Table 2.2 Common uranium minerals [4]

Type	Name	Composition
Oxides	Uraninite *	$(U^{+4}_{1-x}, U^{+6}_x)O_{2+x}$
	Pitchblende*	UO_2 to $UO_{2.25}$
Hydrated oxides	Becquerelite	$UO_2.11H_2O$
	Gummite	Altered product of uraninite (may contain silicates, phosphates)
Nb–Ta–Ti complex oxides	Brannerite*	$(U,Ca,Fe,Th,Y)(Ti,Fe)_2O_6$
	Davidite *	$(La,Ce,Ca)(Y,U)(Ti,Fe^{+3})_{20}O_{38}$
Silicates	Coffinite*	$U(SiO_4)_{1-x}(OH)_{4x}$
	Uranophane	$Ca(UO_2)_2(Si_2O_7)0.6H_2O$
	Uranothorite	$UThSiO_4$
	Sklodowskite	$(H_3O_2)Mg(UO_2)_2(SiO_4)_2.2H_2O$
	Uraniferous Zircon	$ZrSiO_4$
Phosphates	Autunite	$Ca(UO_2)_2(PO_4)_2.10–12H_2O$
	Torbernite	$Cu(UO_2)_2(PO_4)_2.10–12H_2O$
	Saleeite	$Mg(UO_2)_2(PO_4)_2.12H_2O$
Vanadates	Carnotite	$K_2(UO_2)_2(VO_4)_2.1–3H_2O$
	Tyuyamunite	$Ca(UO_2)_2(VO_4)_2.5–8H_2O$
Arsenates	Zeunerite	$Cu(UO_2)_2(AsO_4)_2.10–12H_2O$
Carbonates	Schroeckingerite	$NaCa_3(UO_2)_2(CO_3)_3(SO_4)F.10H_2O$
Hydro-carbon	Thucholite*	Uraninite complex with hydrocarbons
	Asphaltite	Many varieties containing uranium-organic complexes

* Primary minerals

Uranium deposits are characterized by their extreme diversity in size, grade, shape, geological environment, mineralogy, etc. [1]. They form at conditions ranging from deep high-grade metamorphic to surficial environments and from Neo-Archean times to the Quaternary Period. The classification of uranium deposits has been improved in the past few years using both geological and genetic classifications. Fifteen main types of deposits and about 36 subtypes are recognized in the latest (IAEA) geological classification of uranium deposits [8]. The different types of uranium ore deposits are given in Table 2.3 [9]. About 40% of the world's uranium deposits are of sandstone type while the other formations like the granite-related, volcanic-related, metamor-phite, proterozoic unconformity, intrusive, and metasomatite constitute another 40% of the total uranium deposits in the world today [10]. However, such a distribu-tion of deposits is not universal and varies country-wise. For instance, in India, the resources at present are predominantly in carbonate-hosted and metamorphite type deposits. The carbonate-hosted deposits constitute 47.4% while the metamorphites form 28.51% and the sandstone-type deposits are about 9.8% only [10].

Table 2.3 Classification of uranium ore deposits [9]

S. No.	Classification	Well-known ore deposits in the category
1	Intrusive-type	Rossing-Namibia, Bingham Canyon-USA Kvanefjeld-Greenland, Catalao-Brazil, Palabora-South Africa
2	Granite-related	La Crouzille District-France, Pribram District-Czech Republic
3	Polymetallic iron oxide breccia complex	Olympic Dam
4	Volcanic-related	Streltsov-Russia, Maureen-Australia, Anderson Mine-USA
5	Metasomatite	Kirovograd District-Ukraine, Elkon District-Russia, Mary Kathleen-Australia, Michelin-Canada
6	Metamorphite	Jaduguda-India, Shinkolobwe-Democratic Rep of the Congo, Rozna-Czech Republic
7	Proterozoic unconformity	Jabiluka-Australia, Millennium-Canada, Cigar Lake and Key Lake-Canada, Lambapur, Chitral-India
8	Collapse breccia type	Northwestern Arizona strip-USA, Wenrich and Titley
9	Sandstone	Beverley-Australia, Arlit District-Niger, Wyoming-USA, Chu-Sarysu type-Kazakhstan, Lodève-Gabon, Westmoreland-Australia
10	Paleo-quartz pebble conglomerate	Witwatersrand Basin-South Africa, Blind River-Elliot Lake area-Canada
11	Surficial	Kamushanovskoye-Kyrgyzstan, Yeelirrie-Australia, Lake Maitland-Australia, Beslet-Bulgaria
12	Lignite coal	Koldzat-Kazakhstan, Freital-Germany, Williston Basin-USA
13	Carbonate	Tummalapalle-India, Sanbaqi-China, Mailuu-Suu-Kyrgyzstan
14	Phosphate	Mangyshlak District-Kazakhstan, Minjingu-Tanzania, Morocco-Florida, Bakouma District-Central African Republic
15	Black shale	Randstad and MMS Vicken-Sweden, Ronneburg district-Germany

2.1.1.3 Exploration for Uranium Ore Deposits

Exploration of uranium anomalies in the early days consisted of the most direct approach such as geological favorability involving places of known uranium minerals and hydrothermal mineralization, especially the existence of pathfinder elements like copper and lead, along with searches in favorable rocks such as granitoids and pegmatites [11]. Major structures like shear zones, boundary faults, and thrust contacts were also explored. In the modern approach, mineral exploration focuses on the processes responsible for evolution of the host rocks, subsequent remobilization, and re-concentration of metallic minerals giving rise to the final form, grade, and mineralogy of the ore body. Ore formation is no longer regarded as a separate event in the earth crust but as a part of crustal evolution. New knowledge on uranium geochemistry and the time-bound character of major uranium deposits of the world have made significant shift in mineral exploration pathway [1].

Most uranium deposits are polyphasic and polygenetic meaning thereby that the ore formation has been going on over millions of years and the deposits have undergone change due to the change in the causative factors in a geological time frame. The origins and near-surface distributions of well-known uranium can be divided into four phases. The first, from ~4.5 to 3.5 Giga-annum (Ga), involved successive concentrations of uranium from their initial uniform trace distribution into magmatic-related fluids from which the first U^{4+} minerals precipitated in the crust. The second period, from ~3.5 to 2.2 Ga, witnessed the formation of large low-grade concentrations of detrital uranium minerals deposited in a highly anoxic fluvial environment. Abiotic alteration of such minerals including radiolysis and auto-oxidation caused by radioactive decay and the formation of helium from alpha particles have resulted in the formation of a limited suite of uranyl oxide-hydroxides. Earth's third phase of uranium mineral evolution, during which most known uranium minerals first precipitated from reactions of soluble uranyl $(UO_2)^{2+}$ complexes, followed the Great Oxidation Event (GOE) at ~2.2 Ga and thus was mediated indirectly by biologic activity. Most simple oxide-type uranium minerals deposited during this phase precipitated from saline and oxidizing hydrothermal solutions (100–300 °C) transporting $(UO_2)^{2+}$-chloride complexes. The onset of hydrothermal transport of $(UO_2)^{2+}$ complexes in the upper crust may reflect the availability of $CaSO_4$-bearing evaporites after the GOE. During this phase, most uranyl minerals would have been able to form in the O_2-bearing near-surface environment for the first time through weathering processes. Later, during 1200–400 million annum (Ma) events like continental splitting, subsequent metamorphism, increased amounts of organic material (non-plant origin) in the oceans, etc., led to the formation of number of types of deposit. During post 400 Ma, uranium deposition is linked with proximity to fertile granite basement as provenance, release of uranium-rich detritus, episodes of acid or alkaline magmatism accounting for of ore-forming fluids and reducing conditions prevailed by carbonaceous material, sulfides, hydrocarbons, etc.

The strategies for uranium exploration are guided by geological criteria and contemporary developments in uranium exploration techniques world over. The stages are schematically shown in Fig. 2.1 [11, 12]. Exploration for uranium

Fig. 2.1 Exploration
strategy for identification of
uranium anomalies and ore
deposits [11, 12]

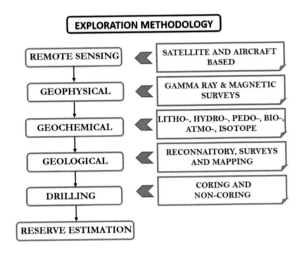

commences with literature survey, followed by study of available satellite images, aerial photos, and geological maps, known radioactivity, geochemical anomalies, etc. This is followed by airborne gamma ray and spectrometric surveys over favorable zones for narrowing down the target areas. Further to this, detailed radiometric survey is taken up on ground (vehicle-borne or on-foot) in the spectrometric surveys over favorable areas for narrowing down the target areas. Subsequently, the target areas are studied using hand-held GM counters and NaI(Tl) scintillometer. The areas having anomalous concentration of uranium and other favorable parameters as identified from the above investigations are evaluated by prospecting (trenching, pitting, geological and structural mapping, shielded probe logging, and sampling) to establish the plan dimension of uranium mineralization. Further, exploration in three stages (reconnoitory, exploratory, and evaluation) is carried out by drilling using conventional as well as high-performing hydrostatic rigs to know the depth of continuity of mineralization. Based on gamma-ray logging of the boreholes and geological considerations, the 3-D configuration of the ore body is mapped and ore reserve is estimated.

The direct methods of detecting uranium anomalies used during earlier phases of exploration relied mainly on the two properties, namely the high density and the radioactivity. However, the signals become weak when the mineralization is at depth and with no surface exposures. The deep buried or concealed deposits are explored using Hyper-spectral remote sensing, followed by integrated geophysical, geochemical, and geological cum radiometric surveys [13–15]. Some of the widely used geophysical techniques in uranium ore exploration and the properties they capture for providing necessary inputs for exploration geologist are given in Table 2.4.

Mapping conductivity/resistivity variations helps unravel complex geological problems and identify areas of hidden potential. Electromagnetic (EM) methods, both ground and airborne, are extensively used to map the conductive ore bodies

Table 2.4 Geophysical techniques for exploration of uranium anomalies and ore deposits [13]

Technique	Information generated
Remote sensing	Maps surface alteration patterns; identifies anomalous vegetation patterns in areas related to abnormal metal content in soil
Gravity method	Measures anomalous density within the earth; positive gravity anomalies are associated with shallow high-density bodies, and gravity lows are associated with shallow low-density bodies. Helps in identification of lithologies structures and at times ore bodies themselves
Magnetic method	Measures variations in magnetic mineralogy among rocks. Useful for deducing subsurface lithology and structure that may indirectly aid identification of mineralized rock, patterns of effluent flow, and extent of permissive terranes and (or) favorable tracts for deposits beneath surficial cover
Gamma-ray method	Identifies the presence of natural radioactive elements like potassium, uranium, and thorium
Electrical methods	Measure earth's electrical impedance or relate to changes in impedance. Identify structures and lithologies
(a) DC measurement	Measures electrical resistivity of earth; applicable to identification of lithologies and structures that may control mineralization
(b) Electromagnetic measurement	Identifies low-resistivity (high-conductivity) massive sulfide deposits. Widely used to map lithologic and structural
(c) Induced polarization measurement	IP responses relate to active surface areas within rocks, disseminated sulfide minerals provide better target for this than massive sulfide deposits

buried in the resistive bed rock which are the potential areas for uranium mineralization. Airborne Electromagnetic (AEM) methods have undergone rapid improvements over the past few decades [16, 17]. Several new airborne Time Domain EM (TDEM) systems have appeared.

The use of natural field (passive) EM surveys has continued to increase, with new or improved systems becoming available for both airborne and ground surveys. The number of large airborne survey systems mounted with combined EM, magnetic, gravimetric, and γ-ray spectrometric capabilities also increased. In data processing, new developments in 3-D inversion and 3-D modeling of EM data on parallel and cloud computing have entered for making exploration both fast and accurate [17].

2.1.2 Mining of Uranium

After a mineral/ore deposit has been identified and established with commercially exploitable grade and resource of metal(s) of interest by exploration geologists and drilling engineers, mining or extracting the ore commences the stages in mining process include mine development, exploitation, and reclamation [18, 19]. Development and exploitation stages are closely related activities. Mine development essentially involves various operations that need to be made for preparing a specific mine conducive for ore extraction. In the exploitation stage, the mine is ready for actual ore production with pre-set mining capacity. Mine closure and reclamation are necessary parts of the mine life cycle.

Modeling of a resource is the first and most important step in selecting a mining method that will optimize exploitation of a mineralized deposit [20]. Geology of the ore body gives various inputs for mining the resource like the dip, shape, and strength of the mineralized zone as well as host rocks. The mining method is determined mainly by the characteristics of the ore deposit and the limits imposed by safety, technology, environmental concerns, and economics. The selection of "economic mining limits" is often supported by a strategic analysis that includes assessing the impact of changes in prices, costs, and recoveries.

The operations in mine development include tunneling, sinking, crosscutting, drifting, and raising [21]. The last stages of mine closure and reclamation acquired importance in recent times with increased emphasis on following the sustainability principles in natural resource utilization. Health and safety of workers under the ALARA (as low as reasonably achievable) and other industrial safety principles have to be holistically adhered to in mine planning process. Regulations for uranium mines demand a higher standard of performance than the rules governing conventional mining and processing operations. All the mine planning steps have to be conducive for well-being of local population and the environment around, besides being commercially attractive.

Mining of metallic ores like uranium is carried out broadly by two methods based on locale. They are (i) surface mining and (ii) underground mining. Surface mining includes mechanical excavation methods such as open pit/open cast (strip mining) and aqueous methods such as placer and solution mining [19, 22]. Underground mining is usually classified into three categories of methods: unsupported, supported, and caving. The most appropriate mining method is selected based on technical, economic, and environmentally accountable considerations.

2.1.2.1 Open Pit Mining

Extraction of a resource by operations exclusively involving personnel working on the surface without provision of manned underground operations is referred to as surface mining. The depth up to which this technology is adopted is generally about 100 m, the actual choice depends on various other factors. In general, higher the grade and larger the deposit, open pit mining is economically employed to greater depths than would be considered for lower grade or smaller deposits.

Surface mining is generally considered to provide better recovery, grade control, flexibility, safety, and working environments than the underground mining. From an environmental point of view, however, surface mines normally cause extensive and more visible impacts on the surrounding landscapes than underground mines, primarily due to the huge amounts of waste and overburden dumped around the mines and the damage to scenery by the surface operations themselves. Noise and vibration also may disturb the surrounding environment, and it is necessary to place a safety zone around the pit to protect people and equipment from fly rocks.

The safety measures cited above are apparent in uranium mines as much as any other industrial-scale mines, and with the exception of additional dust control to reduce radiation exposure, where required, general mining industry planning and machinery are applied at open pit uranium mines. High-grade ores usually necessitate shielding devices be incorporated in loading and hauling vehicles to protect the operators from excessive doses of gamma radiation. Likewise monitoring stations must be established within the pit and in the ore storage areas to determine ambient air quality with analysis for radon and long-lived radioactive dusts associated with uranium ores.

Some of the prominent opencast mines of uranium which form part of the top-ten production centers in the world in recent times are: Ranger in north Australia, Rössing in Namibia, and most of Canada's Northern Saskatchewan mines through to McClean Lake [19]. In India, the Banduhurang mines in Singhbhum East (Fig. 2.2) is the first opencast uranium mine [23]. Banduhurang was commissioned in January 2009. Situated adjacent to Turamdih underground mine, the deposit contains a moderately large reserve of very low-grade ore. It is a conventional opencast mine using excavator-dumper combination maintaining ore benches of 6 m height.

2.1.2.2 Solution Mining

Solution mining or more commonly called in situ leaching or in short ISL involves recovery of minerals from ore zones beneath the surface by in-place leaching and then pumping the pregnant solution to the surface where other downstream operations forming part of separation and purification of dissolved mineral values are carried out [19, 24, 25]. The ISL method of mining, as illustrated in Fig. 2.3, is applicable under certain specific geological and hydrogeological conditions, like sufficient permeability of the host formation to facilitate circulation of mining fluids (usually dominated by sand or sandstone), the ability for multiple recycling of the leaching solution through the ore formation, leachability of the mineral matrix in particular, low abundance of interfering minerals or other constituents, etc.

Both acid and alkaline (carbonate) leachants are used in ISL mining projects internationally. Though the choice depends upon the geology of the host rock, mainly the content of acid-consuming minerals, it also varies from country to country. The relative merits and demerits of both the leaching methods are given in Table 2.5.

Fig. 2.2 The Banduhurang opencast uranium mine in Singhbhum East, Jharkhand, India [23]

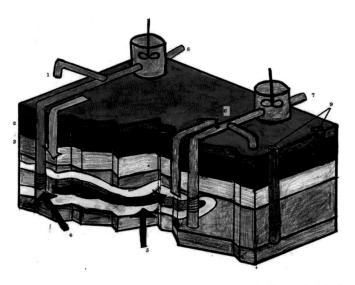

Fig. 2.3 Representation of a roll-front type ISL uranium mine. 1. Chemicals injection pipe, 2. sand, clay, gravel zones, 3. upper clay zone, 4. submersible pump, 5. U-laden leachate, 6. leachate recovery pipes, 7. U-laden liquor for downstream processing, 8. mining liquor from chemical plant, 9. monitoring wells

Sulfuric acid is used as lixiviant in acid leaching plants. The leach solution is fortified with an oxidant like oxygen, or hydrogen peroxide for facilitating conversion of insoluble U^{+4} to soluble U^{+6}. In the case of alkaline leaching, uranium is extracted by adding gaseous oxygen and carbon dioxide to the native ground water.

Table 2.5 Comparison of acid and alkaline leaching in ISL mining of uranium ores

Acid leaching	Alkaline leaching
• Relatively higher recovery of uranium (70–90%) • Favorable leaching kinetics (also depends on the wellfield size, ore permeability, well pattern, etc.) • Possibility of natural attenuation of the remaining leach solution • Radium is not recovered and requires no special restoration • Lower capital and operational project costs	• Uranium leaching from carbonate-bearing ores (i.e., CO_2 content over 1.5–2.0%) • Lower risk of pore plugging by newly formed gypsum and gas bubbles • Lower concentration of dissolved solids in leaching solutions • No corrosion in materials and equipment (pumps)

ISL mining involves the extensive recycling of lixiviant, as only a limited proportion of the uranium is mobilized with each pass of the mining solution (determined by the leachability of the ore, i.e., the kinetic rate of uranium mineral dissolution). Mining solution may be pumped through a particular portion of ore 50–100 times, or sometimes more, over a period ranging from few months to two or more years in length, to achieve the target recovery.

Solution mining is always accompanied with little surface disturbance and no waste rock is generated. Due to the low capital costs (relative to conventional mining), it often proves to be an effective method of mining low-grade deposits. In ISL, the primary risk of contamination for soils, surface waters, and aquifers is from the reagents used for leaching, and from the dissolved metal ions in pregnant solutions. ISL techniques are controllable, safe, and environmentally benign operating under strict operational and regulatory controls. In some cases, (especially in populated areas) it will be necessary to restore the contaminated groundwater and/or establish long-term monitoring programs to ensure that the contamination does not spread into uncontrolled aquifers or other areas. A disposal system for wastewater and other residues also needs to be in place.

2.1.2.3 Underground Mining

When an ore body is located well below the surface, say more than 100 m in depth, underground mining methods are more economical than open pit mining. The first step in underground mining is to access the ore, and the design depends on the topography of the ground surface, the depth of the ore body, rock strength conditions, and possible requirements for road access [18, 22]. First, miners dig vertical shafts to the depth of the ore and then cut a number of tunnels around the deposit. A series of horizontal tunnels offer access directly to the ore and provide ventilation pathways. All underground mines are ventilated, but extra care is taken with ventilation in uranium mines, to minimize the amount of radiation exposure and dust inhalation. The daughter products of uranium isotopes are also radioactive. In the case of ^{238}U, for example, the process of decay will lead to the creation of 13 other radioactive daughter products, ultimately creating a stable isotope of lead (^{206}Pb), all the daughter products like ^{234}Th, $^{230\text{Th}}$, ^{226}Ra, ^{210}Pb, and ^{210}Po are metals excepting ^{222}Rn, which is a radioactive gas. In most underground mines, the ore is blasted and hoisted to the surface for milling.

The access to ore body can be through shafts, adits, and declines (Figs. 2.4 and 2.5). Shafts are vertical openings typically a few meters in diameter and tens or hundreds of meters deep, or over a thousand meter in some cases. Besides accessing the ore, materials and miners are transported in cages or large containers supported on cables operated by a winch beneath a head frame.

Some ore bodies are accessed through adits, which are near horizontal openings, typically from the side of a hill or mountain to access an ore body at a similar elevation. Again, they are typically a few meters in the vertical and horizontal dimensions. A railway line or conveyors systems might be installed in an adit for the removal of ore and waste rock. Some uranium deposits nowadays are accessed through inclined openings called "drifts" or "declines." Stoping then takes place from these levels. Mines using declines for access are impractical for pursuing deep deposits, as the

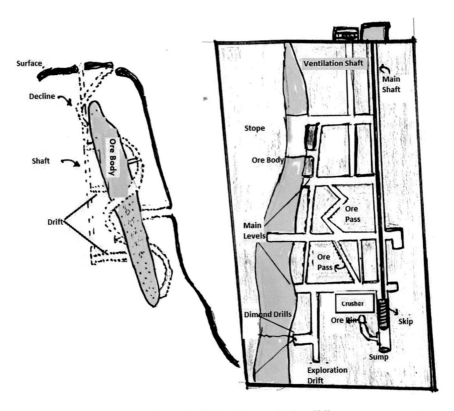

Fig. 2.4 Illustration of approach to ore body in underground mines [26]

Fig. 2.5 Access to underground mines of uranium in Singhbhum district of India **a** shaft and **b** decline (https://www.ucil.co.in, Accessed in March 2021)

declines must be at a gradient that can be traversed by mining equipment, and increased depth equates to increased decline length.

Stoping is the process of extracting the ore or other mineral from an underground mine, leaving behind an open space known as a stope. Stoping is used when the country rock is sufficiently strong not to collapse into the stope, although in most cases artificial support is also provided. There are different types of stoping techniques adopted in underground mines, such as room and pillar, cut and fill, sub-level open stoping, shrinkage stoping, longwall mining, block caving, and sub-level caving. Each of these techniques is specific for different types of ore bodies.

Loading and transportation of broken rock from stopes to surface are completely automated in most of the mines. The deslimed mill tailings (coarser fraction) are used for the mine backfill material. The little amount of waste rock generated in the process of development and stoping are also disposed off in stopes as filling material.

The ventilation requirements of underground uranium mines need to be specially addressed, as irrespective of the grade of the ore mining generates silica dust, diesel fumes, and radiation from both solids as well as gaseous substances like radon [27]. Adequate care needs to be taken for supply of fresh air in each working area at the mine planning stage itself. Ventilation is forced by fans and controlled within a mine by special ventilation shafts and internal mine connections, as well as through general access shafts and tunnels.

2.1.2.4 Mine Waste Treatment

Mine waste consists primarily of waste rock and low-grade ore that must be removed to access the ore [28, 29]. The uranium concentration in the material to be mined varies widely and so does the activity of the waste rock and the ore to be processed. Usually, samples of ore are mixed with waste rocks. For the most part, this material presents essentially no risk of environmental contamination. It may then be disposed off on the surface adjacent to the mine. In some cases, waste rock contains minerals, including sulfides, that may be leached by water passing through waste piles. Oxidation of sulfide minerals is the process which contributes most to the familiar acid mine drainage (AMD) and to the mobilization of metals from mine waste to the environment. The waste rock may also include concentrations of uranium that are elevated above the normal background, but this does not justify processing and recovery. When this is considered likely, steps should be taken to ensure that neither the rock nor leachate leaves the site thereby preventing contamination of either the surface or ground water.

Statutory rules and regulations need to be strictly adhered to with regular monitoring. The uranium ore generally contains all radio-isotopes present in its decay series, and these may assume significance at various stages of operations. The mining personnel are therefore provided with passive personal dosimeters to evaluate individual doses due to exposure to alpha and gamma radiations. In addition, modern personal protective equipment such as respirators, ear-muffs, safety goggles, gumboots, and safety helmet for protection against injuries and harmful exposures are also provided to mine personnel. Many engineering control measures have been adopted to reduce the noise levels of different heavy machinery used underground. Routine monitoring of noise level, air quality, radiation, and radioactivity levels like external gamma radiation, airborne radioactivity, surface contamination, and radiation dosimetry have to be carried out using suitable instruments.

There are several publications from IAEA, Canadian Nuclear Safety Commission (Waste Management, Volume II: Management of Uranium Mine Waste Rock and Mill Tailings Regulatory document REGDOC-2.11.1, Volume II © Canadian Nuclear Safety Commission (CNSC) 2018 Cat. No. CC172-190/2-2-2018E-PDF ISBN 978-0-660-28348-7), US Environmental Protection Agency (EPA), U.S. Nuclear Regulatory Commission (NRC), and Atomic Energy Regulatory Board (AERB—India) which help in defining the environmental protection aspects of uranium mining, processing, and reclamation.

The mine water which is not utilized in mining and milling operations should be treated and disposed off conforming to the norms for disposal of radioactive pollutants.

Mine-exhaust ventilation is contaminated with radon, its daughter products, and to some extent with ore dust, rock-dust, and fumes. The emanation of radon from the uranium ore is the result of radium decay.

The major environmental challenge for the ventilation design is the requirement to dilute radon progeny to below the Working Level, in terms of month per year of worker exposure required by law. The large airflow volumes and the low residence times constitute some of the factors considered by the design engineer with other design-based features, specifically for high-grade ores [30].

2.1.3 Development of Process Flowsheet

Uranium mining and milling projects have several dimensions related to both technical and administrative aspects. Some of the critical aspects which factor into development of process flowsheet are:

(a) **Geological**: the physical nature of the deposit, including its size and shape, the distribution of mineralized areas, the characteristics of the host rocks, and the structural environment;

(b) **Mineralogical**: the chemical and physical characteristics of the uranium minerals as they affect the ease of extraction of the metal from the host rock;

(c) **Infrastructure**: the availability of water, electricity, reagents, spare parts, roads, railways, other means of transportation and communication, other services required to operate the mine and mill, and facilities and services needed by the personnel;

(d) **Technological**: the mining and processing techniques available and applicable to the deposit;

(e) **Economic**: the assurance that a minimum required rate of return on investment capital can be realized, taking into consideration capital and operating costs, taxes, terms of contracts, the current and expected demand and market price of uranium, and the probable economic outlook;

(f) **Human**: the availability of personnel and of technical and administrative skills, and the adoption of an effective operational plan;

(g) **Environmental**: the effects of mining and milling on the environment and the requirements of governmental regulations;

(h) **Legal**: national and local regulations regarding mineral rights, royalty regimes, use of land and water, taxation, and other matters;

(i) **Political**: the international, national, and local political factors which may affect access to the ore body and the conditions under which the project may be implemented and the mine and mill allowed to operate.

There are several specialized TECDOCs and Technical Report Series by IAEA dealing with almost each and every aspect of the above factors [31]. As the technology and regulatory rules are ever-changing in nature, the TECDOCs too are updated with inputs from various countries [32–35]. A close look at the factors listed above emphasizes that the success of any Project in getting clearances not only depends on the economics but also on the technologies adopted for mining, milling, effluent treatment, and post-closure measure taken for safe reclamation of mines and mill area.

2.2 Uranium Ore Processing

2.2.1 *Mineralogy and Comminution*

The sequence of processing steps adopted for the recovery of uranium from the mined material involves basic stages such as liberation of minerals from the host rock, ore pre-treatment, whole-ore or pre-concentrate leaching, separation and purification of desired species from leachate, and precipitation of required metals. Equally important are the protocols used for handling of solids, liquid, and gaseous process streams as well as different types of effluents. The final product obtained at the end in a uranium ore processing sequence is called "yellow cake" which is an intermediate-grade uranium compound [36, 37]. The yellow cake is refined further for preparation of nuclear-grade uranium metal or its compounds. As no two ores have similar mineralogical characteristics in entirety, the ore processing schemes differ from one ore body to the other and at times even within the same ore body if it is a very huge resource. Therefore, process flowsheet development needs detailed characterization of the ore apriori [38]. Figure 2.6 gives generic flowsheet for the recovery of uranium from a typical primary resource. Other factors playing an important role in flowsheet evolution are the size and location of deposit and site-specific environmental requirements.

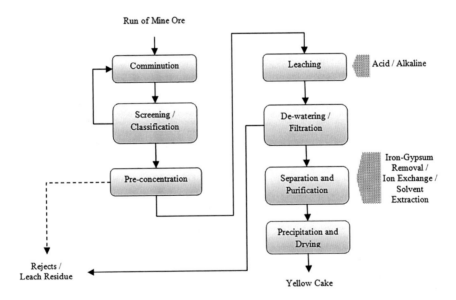

Fig. 2.6 Generic flowsheet for uranium ore processing

2.2.1.1 Mineralogy

Mineralogy of an ore acquires significance in resource processing as it determines the roadmap for extraction of metal/mineral values contained therein. At every stage during the life cycle of an ore deposit, namely exploration to resource evaluation, mine planning, plant design and operation, product quality control, closure, and site rehabilitation, the information generated from process mineralogy is necessary [39–45]. Process mineralogy provides chiefly, ore composition, modal mineralogy, texture, and liberation characteristics.

2.2.1.2 Comminution and Classification

Generally, uranium minerals are finely disseminated in the host rock and are often found intimately associated with the gangue minerals like sulfides, micaceous minerals, carbonates, and phosphates to name a few. Release/liberation of the uranium phases and also the gangue minerals is necessary for any effective recovery process. The emphasis on mineral liberation is more if any physical beneficiation for pre-concentration of uranium values is incorporated in the process flowsheet [18]. Even in chemical processing, liberation either fully or partially is essential for enabling the chemical lixiviants to access the reacting minerals. Another necessity is material handling and/or slurry flow through various technological units. These objectives are achieved in comminution step.

2.2.2 Pre-concentration by Physical Beneficiation

Most of the uranium ore deposits world over are either medium or low-grade variety [18]. Direct chemical processing of an ore naturally invites huge equipment capacity, footprint, chemicals inventory, and inevitably un-favorable economics. Beneficiation of ores helps in pre-concentration of valuable minerals in smaller mass fraction leaving the barren or rejects safer for disposal. Beneficiation of ores uses differences in various physical properties of minerals such as radioactivity, optical property, particle size, specific gravity, magnetic susceptibility, electrical conductivity, and surface-chemical property, for separation of one mineral from the other. However, the success of physical beneficiation of ores depends on nature of mineralization of the desired mineral/s in the ore and the magnitude of difference in the specific physical property under consideration among the assemblage of minerals in the ore [46]. Though recovery of uranium values from the ores is largely carried out by direct application of hydrometallurgical techniques, physical beneficiation processes like sorting, gravity separation, and froth flotation were also used industrially in earlier days and also in some operating plants wherever the implementation yield process benefits [23, 38, 47–50].

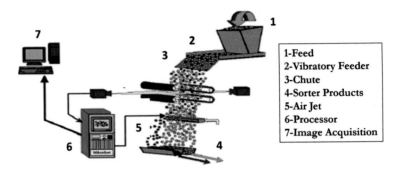

Fig. 2.7 Typical layout of an ore sorter [19, 51, 52]

In the earliest times of uranium mining, sorting was employed based on either visual appearance or simple γ-scanning. In more recent times, sophisticated mechanical sorting (Fig. 2.7) based on physical, mineralogical, or radiometric characteristics are employed [19, 51, 52].

Physical beneficiation techniques like gravity and magnetic separation, played significant role in uranium industry for those resources where the mineral occurs as a by- and co-product. Some of the well-known uranium plants which deployed physical beneficiation in the process circuits are the Olympic Dam, Australia, Palabora mines, South Africa [38], and Copper Plant Tailings of Singhbhum in India [46, 49], to name a few.

2.2.3 Leaching

The transfer of a targeted species present in solid form into an aqueous solution is termed as leaching. This is a critical operation in hydrometallurgy. Uranium is basically a lithophilic element; therefore, all the fundamental studies for understanding the mechanism of its dissolution were carried out on UO_2. Habashi and Thurston, 1967 [53] were the first to propose that the leaching of UO_2 is driven by electrochemical mechanism. Shoesmith and Sunder, 1991 [54] have summarized the five stages in the dissolution of UO_2 as shown in Fig. 2.8. An oxidation potential of about 450mV (vs saturated calomel electrode) is necessary for dissolution of U(IV) to U(VI) in sulfuric acid media in the temperature range of about 35° C. However higher redox potential above 450 mV was found necessary for dissolution of refractory uranium minerals. Similarly cases where effective dissolution of naturally occurring uranium oxides was also found at potentials of 350–400 mV too.

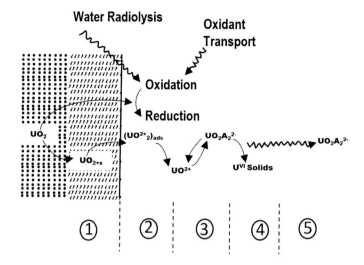

Fig. 2.8 Stages in electrochemical leaching of uranium oxide: Stage 1—oxidation of UO_2 and the formation of an oxidized transient layer on the UO_2 surface; Stage 2—formation of the UO_2^{2+} ion on the oxidized transient layer and its subsequent dissolution. Stage 3—complexation of the UO_2^{2+} with counter anions; Stage 4—precipitation of secondary phases; Stage 5—transportation of uranium species [54] (Stages 1 and 2 are determined by the surface redox conditions, i.e., the balance of kinetics between the anodic reaction (oxidation of UO_2) and the cathodic reaction (reduction of the oxidizing agent). The complexation of UO_2^{2+} with counter-ions in Stage 3 has direct influence on the dissolution (Stage 2) and also affects Stages 4 and 5)

2.2.3.1 Acid Leaching

Though all the three common mineral acids—H_2SO_4, HCl, and HNO_3 are in principle suitable for leaching of uranium minerals, only H_2SO_4 is predominantly used industrially [36, 38]. Other applicable acids, such as HCl and HNO_3, cause more serious environmental pollution than H_2SO_4. The necessary oxidizing conditions for oxidation of U^{+4} to U^{+6} in the leach slurry are provided by both chemical and gaseous oxidants but the Fe^{+3}-MnO_2 combine is the most prevalent one. The role of Fe^{+3} is to act as an electron transfer agent. About 1–2 g/L of Fe^{3+} is usually considered adequate for effective dissolution of uraninite in many industrial units.

2.2.3.2 Alkaline Leaching

The choice of alkaline leaching depends upon the mineralogical composition of the ore. If the acid-consuming minerals are more in quantity say about 5–8% by weight, the alkaline leaching technique is adopted. The reaction mechanism in oxidative alkaline leaching of the uranium minerals is similar to that postulated for acid leaching [36]. The common lixiviants used industrially in this technology are Na_2CO_3, $NaHCO_3$, and $(NH_4)_2CO_3$ and the oxidants are O_2 or air. The advent of pressure swing absorption technology (PSA) has made generation of industrial oxygen very cheap resulting in its use as oxidant in some of the Canadian uranium mills.

2.2.3.3 Comparison of Acid and Alkaline Leaching

Though the mechanism of acid and alkaline leaching follows the same electrochemical model, the similarity among them ends there [36, 55–57]. Alkaline leaching is very selective in comparison with acid leaching. Acid leaching has the advantage that it may dissolve gangue minerals and thereby extract uranium that is not exposed by comminution. Other process conditions followed in the leaching such as the reaction time and temperature also have to be longer and higher, respectively, in alkaline leaching over the levels used in the acid leaching. The major advantage of alkaline leaching is the relatively purer nature or lower concentration of impurities, in the leach liquor which helps in minimizing the number of cleaning stages during the separation and purification step of leach liquor. The yellow cake product precipitated from alkaline leaching circuits is always of higher grade than the final product obtained from acid leaching. As the operating cost of alkaline leaching is relatively high, this technology is traditionally used for uranium ores of higher grade. Table 2.5 gives comparison of acid and alkaline leaching of uranium.

2.2.4 Types of Leaching Based on Operational Mode

2.2.4.1 Atmospheric Leaching

Atmospheric leaching of uranium ores is the commonest technique followed in majority of uranium plants. Continuous stirred type reactors (CSTR) are the equipment used everywhere. Some plants in Canada, Australia, South Africa, and also India were using the "Pachuca" type reactors (pneumatically agitated). Important inputs for design of a leach reactor for ores are the material of construction, operating temperature, solids and solution characteristics, slurry density, impeller design, and rotational speed. Rubber-lined steel reactors with pitched blade turbine-type impellers are used in many uranium mills using sulfuric acid for leaching when the reaction temperature is lower than 70 °C. The pulp is heated using steam. Tanks are often fitted with a cover to reduce the emission of fumes (and radon for high-grade ores), which are extracted through a small vent pipe attached to the cover. Alternatively, a half-cover roof is provided to protect the steelwork. The tanks are usually baffled and fitted with an outlet upcomer to minimize stratification and short circuiting of the slurry. Adoption of large, high displacement, axial flow propellers has significantly reduced the power requirements for many operations. All necessary instrumentation for process control, viz. pH, Eh, and lixiviant concentration (residual SO_4^{2-}) and solids concentration are integrated into the circuit. Carbonate leaching circuits also use three phase interaction; gas (O_2/Air)–Ore–Carbonate salts. Leach recovery therefore depends on efficient and appropriately designed mixing system. Though Pachuca reactors were used in earlier days in some plants like in Moab Mill, Cotter Corporation, Beaverlodge mill at Eldorado, and others, more recently alkaline leaching plants at Langer Heinrich, Namibia, and the Tummalapalle plant in India use mechanical agitation only.

2.2.4.2 Autoclave Leaching

Deposits containing uranium minerals in chemically refractory form, viz. brannerite, thucholite, davidite, and betafite, use autoclave leaching technology. Autoclaves facilitate reactions to be carried out at elevated temperature and pressure. Besides increasing the leach efficiencies and kinetics, leaching in autoclave reactors also gives stabilized leach residues, which can be safely disposed. Autoclave leaching at moderately high temperature and pressure may also be required if the uranium minerals are disseminated in the gangue in extremely finer sizes. Such conditions necessitate fine grinding for breaking the gangue (like sulfides) and help in release of uranium values. Examples for such occurrence are the uranium values present in pyrite of the Tummalapalle uranium deposit and the Gogi deposit in southern India [58].

Table 2.6 gives list of some important uranium mills where autoclave processing [50] was used. At present the Tummalapalle Uranium mill in India adopts oxidative pressure alkaline leaching in its process circuit. Uranium mills use either Pressure OXidizing autoclaves (POX) or High Pressure Acid leach autoclave technology (HiPAL) depending upon the mineralogical characteristics of the ore, type of leaching followed—acid or alkaline, and on the type of oxidant as well as the level of oxidizing environment necessary during the leaching reaction. Similarly, different types of autoclaves—vertical or cigar type/horizontal or in some cases, pipe autoclave reactors were investigated, for their suitability in uranium leaching. As the autoclaves operate at elevated temperatures and pressures, the corrosion and erosion processes too are very severe, necessitating appropriate materials of construction. Autoclaves are always operated in conjunction with heat exchangers for effective heat recovery and energy conservation. The design of spargers for gas induction and dispersion into the reactors is a critical aspect specifically for uranium ores where oxidation is a pre-requisite for leaching. Care needs to be taken for pumping system which has to operate against pressure at the entry point and even at the exiting part as it controls the levels of slurry within the reactor. In order to minimize the capital expenditure, industrial units use refractory bricks over which the desired alloy sheets of required thickness are cladded.

2.2.4.3 In Situ Leaching (ISL)

ISL extraction (Fig. 2.3) is conducted by injecting a suitable leach solution into the ore zone existing below the water table. The injected leachant oxidizes, complexes,

Table 2.6 List of uranium mills with autoclave processing

Name of the plant	Year of commencement	Type of leaching
Eldorado Nuclear Ltd. Beaverlodge Mill, Uranium City	1953	Alkaline
Saskatchewan Atlas Corp., Minerals Division, Moab Utah	1956	Alkaline
United Nuclear-Homestake, Partners, Grants New Mexico	1958	Acid
Cotter Corporation Cannon City Colorado	1970 1980	Alkaline Acid
Lodeeva Mill, France	1980	Alkaline
Rio Algom Corp., Lisbon Mill, Moab Utah	1980	Alkaline
Key Lake Mining Corp., Key Lake, Saskatchewan	1972	Acid
COGEMA Resources Inc., McClean Lake, Saskatchewan	1983	Acid
Uranium One Inc., Dominion Reefs Mill, S. Africa	2008	Acid

and mobilizes the uranium values through production wells which are pumped to the surface for further processing. The lixiviant can be either dilute sulfuric acid or alkaline solution. Though geological disturbance and disposition of tailings above surface are minimal, ISL is tagged with lower recovery and contamination of ground water, as containment of the lixiviant within the intended subsurface volume may be difficult and costly. Pre-shutdown remediation and restoration of the leached zone by washing out chemicals and metal-bearing solutions can prove protracted and expensive. ISL is suitable for sandstone-type deposits which satisfy the criteria discussed in solution mining section. ISL generally offers uranium at lower cost. ISL is very popular in Kazakhstan. This technology is also adopted for some low-grade uranium mines of Australia and US.

2.2.4.4 Pug Cure Leaching

This technique consists of impregnating a coarsely ground (1–3 mm) dry ore with small volume of highly concentrated lixiviant and curing the same at moderately high temperature (<100 °C) for some time depending on the mineralogy of the ore followed by leaching. The pugging process is generally carried out at 85–90% solids content and leaching at 50% solids. The Somair and Cominak uranium mills in Niger adopt this technology. This technology is suitable for ores with refractory uranium minerals and those requiring very high acid concentration for dissolution.

2.2.4.5   Heap Leaching

This is one of the simplest among the leaching techniques followed for treatment of either lean tenor or/and below cut-off grade ores which possess good permeability (Fig. 2.9). At times, heap leaching is also adopted for overburden material containing some values of desired metal. In heap leaching, the ore is mined and piled over an impermeable lining of the floor which is further covered with a layer of washed gravel at next level to ensure a homogenous regular circulation of the leaching solution while protecting the lining. Leach liquors sprayed at the upper surface of the heap percolates downward through the bed of the ore. The floor of the pads must be impermeable, to prevent contamination of ground and water table. Contact of the lixiviant during percolation process with surrounding rocks results in leaching of the dissolvable constituents. Heap leaching is ascribed with advantages likes low capital and operating costs as it is carried out with coarsely crushed feed at atmospheric pressure and there is no necessity of solid–liquid separation for separating metal-laden liquor from passive gangue. However, heap leaching is plagued with low recovery, more footprint, and issues of compaction due to reaction products like jarosite and longer residence time. Heap leaching of uranium was carried out on ROM projects in France (Bessines, Langone), Niger (Somair), Canada (Agnew Lake), and other locations [51].

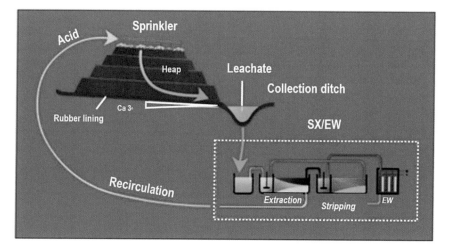

Fig. 2.9   General layout of a heap leaching pad for ores and minerals

2.2.5   Solid–Liquid Separation and Clarification

The broad classes of solid–liquid (S/L) separation equipment finding use in uranium mills are the thickeners, vacuum and pressure filtration units, and centrifuges (Fig. 2.10). The choice of inclusion of any of the above unit operations in the process circuit depends primarily on the duty envisaged and the process economics. For example, in a uranium ore processing flowsheet, it is the solids which are of interest in the dewatering stage preceding the comminution stage; for post-leaching slurry stream, the solid–liquid separation aims at maximum recovery of leach liquor and effective washing of the leach residue. Similarly, if the interest is in removing the suspended solids from the leach liquor before the separation and purification by ion exchange (IX) or solvent extraction (SX), and after the precipitation of yellow cake, it is the solids recovery again which is the primary objective. Therefore, a variety of equipment are deployed for achieving the desired duty conditions. The equipment choice also depends on engineering concerns like the footprint, required infrastructure and associated costs, lead times, and maintenance costs and labor. The solid–liquid separation operations represent up to 40% of the mill capital costs, and more importantly once solubilized, the uranium values carry with them 90% of the total production costs also. Separation of liquid phase from the solids is essential for conservation of water and reagents and improves the effluent quality too.

Counter-current decantation is used for solid-liquid separation from leach slurry of high grade uranium ores. Though the cost of operation is cheaper they consume large amounts of water and surface foot-print. Among the various filtration units, in uranium milling, horizontal vacuum belt filter (HVBF) is most widely used in solid-liquid separation of lean tenor leach slurry as well as in yellow cake product recovery stages. HVBF is superior in performance with respect to rate of filtration, filtrate clarity, availability in continuous mode, lower infrastructural requirements, and easy scale-up procedure.

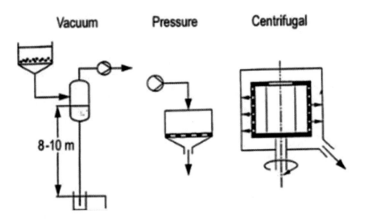

Fig. 2.10 Solid–liquid separation units in hydrometallurgy

2.2.6 Separation and Purification

The process conditions during the leaching of uranium ores are optimized for maximum solubilization of uranium values during which some of the gangue minerals like chlorite, biotite, pyrite, vanadates, calcite, dolomite, and magnesite too are also leached [59, 60]. The ions commonly found in the leach liquors of uranium ores include $V_2O_7^{4-}$, Cu^{2+}, Al^{3+}, K^+, Fe^{3+} or Fe^{2+}, Mn, Na^+, Ca^{2+}, Mg^{2+}, Th^{4+}, SiO_4^{2-}, Rare earths (RE^{3+}), MoO_4^{2-}, SO_4^{2-}, Cl^-, PO_4^{3-}, HCO_3^-, and CO_3^{2-} [19, 36]. The concentration of these impurities in the leach liquor depends on the nature of gangue minerals, externally added reagents, and also on the leaching conditions adopted [59, 60]. Composition of a typical leach liquor obtained during sulfuric acid leaching of low-grade uranium ore and a carbonate-hosted low-grade uranium ore with alkaline leachants is given in Table 2.7 [61]. Removal of the impurities from the leach liquor to the maximum extent is essential for meeting the stringent purity specifications desired in yellow cake.

Table 2.7 Partial chemical composition of leach liquors in typical acid and alkaline leach circuits [61]

Sulfuric acid process		Sodium carbonate–bicarbonate process	
Analyte	Concentration	Analyte	Concentration
Turbidity	2–3 ppm	U_3O_8	1.6 g/L
pH	1.7–1.8	Na_2CO_3	22.8 g/L
EMF	457 mV	$NaHCO_3$	62.6 g/L
U_3O_8	0.3–0.37 g/L	TDS	175 g/L
Fe^{2+}	0.507 g/L	Si	0.015 g/L
Fe(T)	2.46 g/L	Ca	0.04 g/L
SO_4^{2-}	19.07 g/L	Mg	0.012 g/L
Cl	1.26 g/L	SO_4^{-2}	76 g/L
Cu	0.007 g/L	PO_4^{-3}	0.022 g/L
Mo	0.00055 g/L	Cl^{-1}	4.3 g/L
Ni	0.016 g/L	Na	103 g/L
P_2O_5	0.345 g/L	Fe	0.006 g/L
Mn	1.1–1.38 g/L	Mn	<0.005 g/L
SiO_2	0.89 g/L	Mo	<0.05 g/L
CaO	1.05 g/L	V	<0.025 g/L
MgO	1.33 g/L	Cu	<0.001 g/L
Al_2O_3	1.37 g/L	Ni	0.001 g/L
		Pb	0.02 g/L
		Zn	<0.005 g/L
		Cd	<0.01%

The two techniques used for separating and purifying the U-laden leach liquors are ion exchange (IX) and solvent extraction (SX) [36, 38, 62]. The choice of either of the techniques or combination of both depends on the leach liquor composition and tenor of uranium. IX is favored over SX when large volumes of leach liquors with low uranium concentrations, say <1 g of U_3O_8/L, are to be treated. This is because solvent losses are primarily related to the volume of solution handled [18, 63–65]. In cases where the leach liquor is of high acidity or low pH, SX is preferred technology, as it can treat a greater range of acidic feed solutions without major changes in its chemical behavior. Some plants use IX/SX for upgrading the uranium content in the liquor [55]. It is an accepted fact that higher uranium content in the concentrated pregnant leach liquor gives good quality precipitate or yellow cake.

2.2.6.1 Ion Exchange

The mechanism of leach liquor purification by IX process, a specific chemical process in which dissolved ions are exchanged for other ions with a similar charge, consists of three basic stages—loading, backwashing, and elution. In the loading stage, the leach liquor is charged into the column filled with resin (solid) beads and during the plug-flow type passage of liquor in the column, the exchange of ions takes place between the liquor and the solid adsorbent or the resin. In the second stage of backwashing, the resin bed is dilated to wash off any entrained solid particles originated from the leach liquor during loading. The wash process also loosens the bed for ensuring proper flow for the subsequent stages. In the elution step, the loaded ions are desorbed using an eluent. Both gel or macroporous type resins are used in uranium industry [66]. Resins are generally characterized based on their morphology, sorption kinetics and loading capacity, selectivity, moisture retention capacity, desorption efficiency and kinetics, resistance to fouling by organic and inorganic contaminants, porosity, osmotic shock, and radiation resistance [67, 68].

The leach liquors generated in sulfuric acid leaching (pH 1.5–2.0) and alkaline leaching (pH 8–10) with carbonates contain dissolved U complexes predominantly in anionic form say $[UO_2(SO_4)_3]^{4-}$, $[UO_2(SO_4)_2]^{2-}$, $[UO_2(CO_3)_3]^{4-}$, and $[UO_2(CO_3)_2]^{2-}$, respectively [66, 67, 69, 70]. The two different types of anion exchange resins used in U processing plants are the strong-base and weak-base type. Strong base anionic exchange (SBA) resins contain quaternary ammonium functional group or sulfonic acid, with an associated anionic labile ion like Cl^-. SBA resins operate over a wider pH range of 0–13. The affinity trend of various anions to a SBA resin in acid and alkaline solutions generally follows the sequence: (i) sulfuric acid leach process: $V_2O_7^{4-} > MoO_4^{4-} > [UO_2(SO_4)_3]^{4-} > [UO_2(SO_4)_2]^{2-} > Fe(OH)(SO_4)_2^{2-} > SO_4^{2-} > Fe(SO_4)_2^- > NO_3^- > HSO_4^- > Cl^-$ and (ii) sodium carbonate leach process pH 9–10: $V_2O_7^{4-} > [UO_2(CO_3)_3]^{4-} > MoO_4^{4-} > [UO_2(CO_3)_2]^{2-} > SO_4^{2-} > CO_3^{2-} > NO_3^- > Cl^- > HCO_3^-$. However, the affinity series may show variation based on relative concentrations and the chemical environment of the liquor [71].

Uranium has been commercially recovered by IX since the 1950s, using fixed-bed (FBIX) in many plant [18, 38, 72]. This was followed by introduction of "moving bed" columns where the loaded resin was transferred to a separate column for elution [18]. The FBIX or fluidized bed ion exchange system (Higgins Loop, Himsley column) was prevalent in US, South Africa, and Canada [18, 73]. Several key modifications have been introduced in the later designs to suit present-day requirements of uranium industry. IX is still the main workhorse for many low-grade ores. It has been adopted by three of the most recently built tank leaching operations, Langer Heinrich, Kayelekera, and Husab and continues to be used in all operating ISL projects in the USA, Kazakhstan, and Australia [74].

2.2.6.2 Solvent Extraction

Solvent extraction (SX) exploits the differences in relative solubility of chemical compounds in two different immiscible liquids, mostly an aqueous solution and an organic solvent, for the separation and purification task [75]. SX technology is used in uranium industry for handling leach liquors of relatively higher uranium content (>1 g/L) such as those occurring in Saskatchewan province, Canada, plants in Australia, and also in Niger (Table 2.8). Instances of SX use in conjunction with IX as in Eleux process are also known (Table 2.8) [18, 36].

As in the case of IX, the SX process involves the following stages, (i) extraction: the leach liquor containing dissolved uranium is equilibrated with a solvent which selectively complexes with the moiety of interest and transfers it into the organic phase or solvent phase, (ii) scrubbing: the loaded organic phase is equilibrated with plain water or pH conditioned water for removal of entrained solids and/or weakly held ions, and (iii) stripping: the loaded solvent is equilibrated with an aqueous phase containing various salts (sodium sulfate/sodium carbonate/ammonium carbonate/ammonia-ammonium sulfate/pH modifier) or even plain water, during which the loaded moiety is re-transferred from solvent to aqueous phase. A schematic diagram of the stages in SX process is given in Fig. 2.11. The barren stream obtained in the extraction stage is called raffinate.

Table 2.8 Uranium plants with IX and SX technology [10]

Mine (operational since)	Country	Purification process
McArthur River (1983), McClean Lake (1999), Rabbit Lake (1975)	Canada	SX
Ranger (1981), Olympic Dam (1988), Honeymoon (2011)	Australia	SX
Rössing (1976), Husab (2016)	Namibia	IX, SX IX, SX
Taukent Mining Chemical Plant (1982)	Kazakhstan	IX, SX
Arlit (Somaïr) (1971, 2009), Akouta (Cominak) (1978)	Niger	SX, SX
White Mesa Mill (1980)	USA	SX

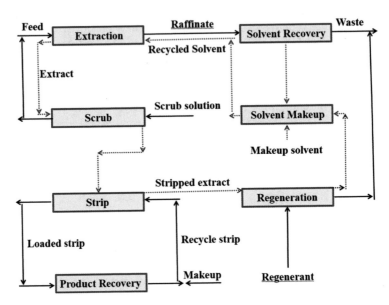

Fig. 2.11 A standard solvent extraction circuit for uranium separation and purification from leach liquor

The choice of solvent (diluent + extractant) is critical for the success of the separation process. It should be non-volatile, non-flammable, and non-toxic as well as economical for industrial use. The extractant in SX is of three different types—anionic, cationic, and neutral [18, 76, 77]. Diluent is generally the major part of the solvent phase in which the extractant is dissolved. SX systems also have modifiers added to the solvent to improve solvent properties like increasing the solubility of an extractant, changing interfacial parameters, or reducing adsorption losses. Because the diluent composition affects the solvency and therefore the chemical and physical properties of the extraction, scrubbing, and stripping, many diluents are now supplied in a blended form so as to contain n-paraffin with a small amount of aromatics. As each leaching solution is peculiar to the particular ore and process conditions, choice of diluent must be based on tests with selected modifiers on a particular leaching solution. The selection of a diluent should not be arbitrary, or be based on costs alone, or chosen because another seemingly similar operation is using a particular diluent.

A great variety of extractants have been used for the extraction of uranium from aqueous solutions [16, 77–80]. Majority of them are nitrogen-based, phosphorous-based, and sulfur-based molecules. However, in the sulfuric acid system, only the nitrogen and phosphorous-based extractants are industrially used. The DAPEX process with D2EHPA (Di-2-ethylhexylphosphoric acid) as the extractant is the first developed and commercialized SX process to recover uranium from sulfuric acid leach solutions. The mechanism of extraction and stripping using the alkyl phosphoric acid extractants is given in Eqs. 2.1–2.3.

Extraction

$$UO_2^{2+} + 2(HDEHP)_2 \rightleftharpoons UO_2(HDEHP.DEHP)_2 + 2H^+ \qquad (2.1)$$

Alkaline Stripping

$$UO_2(HDEHP.DEHP)_2 + 4Na_2CO_3 \rightleftharpoons UO_2(CO_3)_3^{4-} + 4NaDEHP + H_2O$$
$$+ CO_2 + 4Na^+ \qquad (2.2)$$

Acid Stripping

$$UO_2(HDEHP.DEHP)_2 + 2H^+ \rightleftharpoons 2(HDEHP)_2 + UO_2^{2+} \qquad (2.3)$$

However, the DAPEX process was superseded by the AMEX process with amines as the extractant because amine is more selective for uranium extraction than D2EHPA for sulfuric acid leach solutions. The extractability of UO_2^{2+} occurs in the order: tertiary > secondary > primary amine. Extraction and stripping steps in the AMEX process are given in Eqs. 2.4, 2.5, and 2.6:

Extraction

$$UO_2^{2+} + SO_4^{2-} + 2(R_3NH)_2SO_4 \rightleftharpoons (R_3NH)_4UO_2(SO_4)_3 \qquad (2.4)$$

(R = alkyl amine with alkyl group of 8–10 carbon atoms).

Stripping

Acidic stripping

$$((R_3NH)_4UO_2(SO_4)_3 + 4HX \rightleftharpoons 4R_3NHX + UO_2^{2+} + 3HSO_4^- + H^+ \qquad (2.5)$$

(HX = HCL or HNO_3).

Neutral stripping

$$(R_3NH)_4UO_2(SO_4)_3 + 7Na_2CO_3 \rightleftharpoons 4R_3N + UO_2(CO_3)_3^{4-} + 4HCO_3$$
$$+ 3SO_4^{2-} + 14Na^+ \qquad (2.6)$$

Most impurities present in the uranium leachate are not extractable with tertiary amine as they do not form or even if formed would be very weak anionic complexes with SO_4^{2-}. Tricaprylyl amine, the tertiary alkyl amine sold as Alamine® 336 (Cognis, now BASF) or Armeen® 380 (Akzo Nobel), is widely used, usually in conjunction with an alcohol phase modifier (typically iso-decanol) to prevent third-phase formation and inhibit the emulsion formation which can occur as a consequence of the high molecular mass of the organic phase complexes. The tertiary amine is a weak-base reagent and hence is capable of treating feed solutions with a wide range of acidities: Acidities can vary from pH 2 for pregnant leach solutions (PLS) to >100 g/L H_2SO_4 when treating IX eluate in an Eluex process.

Although tertiary amines have been widely accepted in industrial practices, problems still exist in some applications some of them being (a) crud formation from soluble silica; (b) degradation of tertiary amine by oxidation and/or higher redox potential (550 mV); (c) removal and recovery V and Mo; (d) separation of Th, and (e) Process improvements for chloride-bearing liquors. Addressing these issues necessitates (i) new active interfacial modifiers to eliminate or alleviate crud formation caused by soluble silica; (ii) SX processes to recover V and Mo before uranium extraction; (iii) possible use of primary amines to recover and then store thorium; (iv) new extractants or synergistic systems to extract uranium in the presence of chloride; and (v) new SX systems to recover uranium from carbonate solutions.

2.2.6.3 Eluex Process

The Eluex process is a combined technology which makes the processes of IX and SX link together. This consists of extraction of uranium at first by the process of IX from the leach liquor. The saturated resin is eluted by sulfuric acid for collecting loaded U in the eluate. The uranium values in the acidic eluate are extracted by the process of SX . The combined IX/SX process resulted in significant savings in direct operating costs.

The Rössing Uranium Mine in Namibia, commissioned in 1976, and the Husab Plant in Namibia, commissioned in 2016, use sulfuric acid leach and a combined IX/SX process [74]. A schematic representation of the Rössing flowsheet is presented in Fig. 2.12.

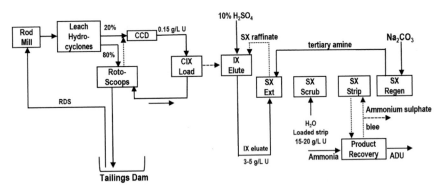

Fig. 2.12 Combined IX-SX flowsheet for uranium separation and purification at Rossing plant [74]

2.2.7 Selective Separation of Uranium: Precipitation

The dissolved uranium in the leachate or in the pregnant leach liquor generated after IX or/and SX is recovered as "yellow cake" (solid) during the precipitation process. Precipitation of yellow cake can materialize either by addition of solid chemical reagents [36, 81] or by using gaseous reductant like hydrogen [82]. Besides the tank precipitation method traditionally followed in the uranium industry, fluid bed precipitation technology is also adopted of late [38, 83]. The preference for a given precipitant and process is driven by uranium assay in the liquor and its pH, the tendency for co-precipitation of other dissolved ions, need of recycling the post-precipitated liquor or barren liquor, chemical purity desired, and morphology of the precipitate [84, 85], possible effect on environment due to the effluents (barren liquor) containing the added precipitants, and process economics.

2.2.7.1 Precipitation Chemistry and Methods

The salts of $U_2O_7{}^{2-}$ are represented by the general formula, $M_xU_2O_7$, where M denotes the alkali or alkaline metals like Na^+, K^+, Mg^{2+}, Ca^{2+}, and "x" has a value of 1 or 2 depending on the valency of the metal cation, say $x = 2$ for the alkali metals and $x = 1$ for the alkaline earth metals.

Precipitation Using Magnesia, Ammonia, Sodium Hydroxide, and Hydrogen Peroxide

Complexes of uranium in leachates can be precipitated by adding milk of magnesia after the leachates are adjusted to a final pH between 7.0 and 7.5. The reaction proceeds as shown in Eq. 2.7:

$$UO_2SO_4 + MgO + H_2O \rightarrow UO_3xH_2O + MgSO_4 \tag{2.7}$$

where x takes a value of 1 or 2. The precipitated oxides of uranium may convert to diuranate in the presence of excess base or to basic uranyl sulfate when in contact with sulfate solutions. The basic uranyl sulfate thus formed converts to diuranate upon reaction with $Mg(OH)_2$ according to reactions given in Eqs. 2.8–2.10:

$$2[UO_3.xH_2O] + 2MgOH_2 \rightarrow MgU_2O_7 + (2x + 1)H_2O \tag{2.8}$$

$$2[UO_3.xH_2O] + MgSO_4 + 4H_2O \rightarrow (UO_2)_2SO_4(OH)_2.4H_2O \\ + 2Mg(OH)_2 + 2(x - 1)H_2O \tag{2.9}$$

$$(UO_2)_2SO_4(OH)_2.4H_2O + 2Mg(OH)_2 \rightarrow MgU_2O_7 + MgSO_4 + 7H_2O \tag{2.10}$$

In the case of precipitation with ammonia, the U-bearing pregnant liquor is heated to 70 °C and precipitated as crude ammonium diuranate $(NH_4)_2U_2O_7$ at pH $\approx$ 7.5 (Eq. 2.11) by drop-wise addition of ammonium hydroxide or by addition of gaseous air-ammonia mixture.

$$2[UO_2(SO_4)_3]^{4-} + 6NH_3 \rightarrow (NH_4)_2U_2O_7 + 4SO_4^{2-} \qquad (2.11)$$

Precipitation by ammonia or its compounds is followed in major uranium-producing mines like Key Lake/McArthur River, Rabbit Lake, Canada; Nuclear Fuels Corporation (NUFCOR), South Africa; Ranger and Olympic Dam, Australia; and Rössing, Namibia and Priargunski/Krasnokamensk, Russia [38, 86].

Precipitation of dissolved uranium from alkaline leach liquors is predominantly carried out using NaOH or caustic essentially to preserve the expensive Na values already present for recycle in the circuit [18, 32]. The reactions in this case are given in Eqs. 2.12 and 2.13, respectively. Caustic addition first converts the bicarbonate present in the leach liquor to carbonate and then brings about a reaction with sodium uranyl tricarbonate to precipitate the dissolved uranium as sodium diuranate or as polyuranates:

$$NaHCO_3 + NaOH \rightarrow Na_2CO_3 + H_2O \qquad (2.12)$$

$$2Na_4UO_2(CO_3)_3 + 6NaOH \rightarrow Na_2U_2O_7 + 6Na_2CO_3 + 3H_2O \qquad (2.13)$$

Precipitation with hydrogen peroxide (H_2O_2) gained considerable attention owing to the environmentally benign nature of the reaction process and also the high quality, both physically and chemically of the uranium oxide product it yields. Uranyl ions are known to precipitate as the uranyl peroxide—UO_4, in acidic solutions (pH = 3–4) with the addition of excess H_2O_2 according to Eq. 2.14.

$$UO_2^{2+} + H_2O_2 + nH_2O \rightarrow UO_2(O_2).nH_2O + 2H^+ \quad n = 2 \text{ or } 4 \qquad (2.14)$$

As seen in Eq. 2.14, the acidity increases during the progress of precipitation with corresponding decrease in the quantity of the precipitated uranium peroxide (the reaction reaches the equilibrium). Monitoring and control of the pH is of vital importance due to its high influence on the precipitation kinetics [83, 87]. The pH is generally adjusted to remain around 3 by adding ammonium/sodium hydroxide.

Precipitation with Gaseous Reductant

Uranium may be precipitated from alkaline solution by reduction with gaseous hydrogen according to the following Eq. 2.15:

$$Na_4UO_2(CO_3)_3 + H_2 \rightarrow UO_2 + Na_2CO_3 + 2NaHCO_3 \qquad (2.15)$$

This process was first studied in Canada and employed at a small plant in Yugoslavia. The reduction process required high temperature and pressure for achieving quantitative recovery and grade leading to high capital expenditure and operating expenses. Further, the UO_2 produced was very fine leading to slower settling and more losses in product separation stage.

2.2.7.2   Practical Considerations in Yellow Cake Precipitation

The methods of precipitation of uranium from various liquors depend upon the concentration of uranium and the solution chemistry of liquor. If the leach liquors are sufficiently purer or clear from impurity ions and the assay with respect to uranium is high say, above 3–4 g/L, direct precipitation is followed either for acidic or alkaline liquors using suitable precipitant.

2.2.7.3   Quality of Yellow Cake

The yellow cake produced in the precipitation stage has to satisfy stringent specifications not only with respect to concentration of neutron poisons but also other metals which have detrimental effect either during refining stage or even in nuclear fuel properties (Table 2.9) [18]. For example, cadmium and boron are undesirable because they are strong neutron absorbers. Halogens are undesirable as they cause corrosion problems in the refining process. Other major impurities include thorium, iron, vanadium, zirconium, molybdenum, rare earths, phosphate, and sulfate.

Table 2.9 Typical specifications of yellow cake product. Minimum U_3O_8: 65% by weight [18]

Na	7.5	Halogens (Cl, Br, I)	0.2
H_2O	5.0	V_2O_5	0.23
SO_4	3.5	Rare earths	0.2
K	3.0	F	0.15
Th	2.0	Mo	0.15
Fe	1.0	As	0.10
Ca	1.0	B	0.10
Si	1.0	Extractable organic	0.10
CO_2	0.50	*Insoluble U	0.10
Mg	0.50	Ti	0.05
Zr	0.50	^{226}Ra	7.4 Bq/Kg
PO_4	0.35	Max. particle size	6.35 mm

Note U concentrates must be of natural non-irradiated origin assaying 0.711% by weight of ^{235}U
* Uranium insoluble in HNO_3

2.2.8 Product Separation and Drying

The uranium precipitate (yellow cake) slurry formed by various routes is generally thickened and then filtered in well-covered horizontal belt filter for collection of yellow cake which is subsequently dried in any of the following equipment say, multi-hearth furnace, rotary tube dryers and calciners, hollow flite dryers, and spray dryer [19, 23].

Uranium oxide concentrate (UOC) is packaged in sealed 200-L steel drums meeting IP-1 industrial package requirements as set out by the IAEA [88].

Equipment and processes producing airborne dust, radon gas, and radon daughters are segregated and enclosed in confinement as far as possible and wherever necessary mechanical exhaust is used in occupied areas. All packaging and handling of uranium ore and compounds shall be mechanized to avoid direct handling and exposure. Wherever necessary, personal protective equipment (PPE) should be used in such operations. Respirable dust concentration should be kept far below the threshold limit values (TLV).

2.2.9 Mill Waste Treatment

Solid: The coarser fraction of the tailings from the mill after leaching should be used in the underground mine for stowing. The fines generated from the mill tailings after solid–liquid separation and neutralization are sent to tailings pond in the form of a slurry. The main radioactive materials remaining are those from the uranium decay series, mainly ^{230}Th and ^{226}Ra. The ore processing has a significant effect on the amount and type of tailings that are produced. A significant proportion of the solids being disposed in a uranium tailings impoundment consist of gypsum and iron hydroxide sludges generated during neutralization of the leach residues. One concern regarding the impoundments of tailings is the potential for release of radon gas, and impoundments are monitored to ensure that radon does not pose a hazard. Radon can be controlled by limiting the amount of tailings exposed during operations by maintaining only small parts of an impoundment cell open at any one time, or by use of a water cover. The tailings dam is invariably provided with impermeable lining at the bottom for retention of radioactivity within the pond. In most cases, tailings are now impounded in purpose-built lined cells, placed in a mined-out pit, or sent to an engineered facility. For final closure, the tailings mass is required to be below the surrounding ground.

A new development in tailings disposal is the thickened tailings disposal system. In this method, the tailings which usually contain 20–30% solids are thickened in high rate/high-density deep thickeners producing a highly viscous slurry (in the form of paste) which can be pumped and deposited in dry stacking area. The paste, because of its high yield stress value, spreads all around at a gentle slope and forms a heap. As deposition continues, the heap grows in area and height. At the periphery, small dykes are built to contain tailings within the disposal area. Once the desired height is attained, the deposition point is shifted to a nearby suitable location to form an adjacent heap. This kind of deposition technique helps to utilize the stacking area volume to the fullest possible extent. The decanted liquid from the tailings pond is taken to the effluent treatment plant (ETP) for treatment, and the sludge generated in ETP is also disposed off in the tailings pond.

Liquid: The decant solution from the tailings pond is treated chemically in the ETP. The quality of the treated effluent from the ETP should be monitored to ensure that it meets the prescribed discharge standard and then only released to the environment [89].

2.2.10 Reclamation and Closure

Although reclamation and closure have always been considered during mine development, current practice has advanced to the point where the reclamation and closure plan is an important element for any mine's ultimate success. These plans consider all disturbances associated with the mine and processing plant areas. Closure activities may involve some post-closure water treatment where a treatment facility is required, and long-term sampling is undertaken. Modern mine practice is to carry out continuous rehabilitation during the life of an operation. Appropriate reclamation and closure are guaranteed by a bond to ensure that sufficient resources are available should the operating company fail prior to final reclamation and closure.

2.3 Refining of Yellow Cake and Uranium Metal Production

Diuranates of sodium (SDU), magnesium (MDU) or ammonium (ADU), and uranium peroxide, precipitated from uranium leach liquor, are known as yellow cake. Yellow cake is the basic input material for uranium oxide and metal production. The yellow cake of sodium or magnesium diuranates produced from uranium ore in the uranium recovery (milling) process contains several impurities and needs purification before it is taken for further processing to produce nuclear-grade oxide or uranium metal. Ammonium diuranate and uranium peroxide are relatively pure and occasionally can be taken for calcination to produce nuclear-grade oxide, without refining. Various steps in the refining of yellow cake are shown in Fig. 2.13.

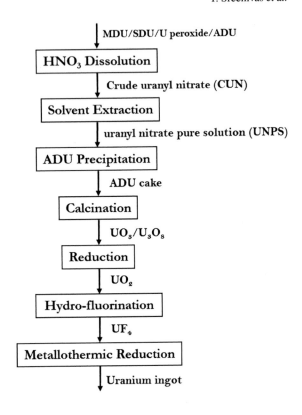

Fig. 2.13 Typical process flowsheet for the uranium ingot production

MDU/SDU/U peroxide/ADU

HNO₃ Dissolution

Crude uranyl nitrate (CUN)

Solvent Extraction

uranyl nitrate pure solution (UNPS)

ADU Precipitation

ADU cake

Calcination

UO₃/U₃O₈

Reduction

UO₂

Hydro-fluorination

UF₄

Metallothermic Reduction

Uranium ingot

2.3.1 Dissolution of Diuranate Cake

A concentrated solution of crude uranyl nitrate (CUN) is obtained by dissolving diuranate of sodium or magnesium in commercial-grade nitric acid.

$$Na_2U_2O_7 + 6HNO_3 \rightarrow 2UO_2(NO_3)_2 + 2NaNO_3 + 3H_2O \qquad (2.16)$$

$$MgU_2O_7 + 6HNO_3 \rightarrow 2UO_2(NO_3)_2 + 2Mg(NO_3)_2 + 3H_2O \qquad (2.17)$$

The slurry obtained after dissolution is called crude uranyl nitrate slurry (CUNS). Required quantity of water or acid is added to make feed (called UNF) for solvent extraction. Uranium concentration and free acidity in UNF are generally maintained at 200–250 g per liter and 2–2.5 N, respectively.

2.3.2 Solvent Extraction

Uranyl nitrate feed (UNF) is purified by solvent extraction process and uranyl nitrate pure solution (UNPS) is produced. ~32% tributyl phosphate (TBP) diluted with kerosene or dodecane (generally called lean solvent (LS)) is used as solvent in solvent extraction and DMW is used as stripping agent. TBP is periodically processed to remove degraded products.

The extraction system consists of various units (as shown in Fig. 2.14) and is described below:

- **Extraction Unit**: In this unit, UNF is contacted counter currently with dilute TBP (Lean Solvent). Uranium in aqueous phase is transferred to organic phase. The organic phase containing uranium (called fat solvent or extract) flows to next scrubbing unit. Aqueous flows out as a raffinate.
- **Scrubbing Unit**: In this unit, fat solvent from the extraction unit is counter currently washed with small amount of pure dilute nitric acid to remove the traces of the impurities to obtain extract pure (EP) which flows to stripping section. Scrubbed raffinate from this unit contains traces of impurities but substantial amount of uranium. This is recycled with feed solution.
- **Stripping Unit**: Pure extract from the scrubbing unit is contacted with DM water to transfer uranium from organic phases to aqueous phase, called UNPS. The solvent free of uranium (Lean Solvent) is recycled back to extraction unit.

$$[UO_2(NO_3)_2]_{aq} + 2TBP_{org} \rightarrow [UO_2(NO_3)_2.2TBP]_{org} \qquad (2.18)$$

$$[UO_2(NO_3)_2.2TBP]_{org} + DMW \rightarrow [UO_2(NO_3)_2]_{aq} + 2TBP_{org} \qquad (2.19)$$

Fig. 2.14 Slurry extractor used in the Solvent extraction system

No.	Element	Maximum limit (ppm)
1	Boron	0.12
2	Iron	100
3	Gd	0.04
4	Dy	0.1

Table 2.10 Major impurity levels in UNPS

Major impurity levels in UNPS are maintained below the limit as mentioned in Table 2.10 to make it nuclear grade.

- The UNPS is generally recycled back to UNF, if purity of the UNPS is not matching with the technical specification mentioned in Table 2.11.
- Degradation of TBP is identified based on the concentration of uranium retained in lean TBP. TBP is periodically treated with carbonate solution to remove degraded products from TBP.

2.3.3 Ammonium Diuranate (ADU) Precipitation

The uranyl nitrate pure solution (UNPS) is reacted with gaseous ammonia diluted with air to obtain ADU as precipitate.

$$2UO_2(NO_3)_2 + 6NH_3 + 3H_2O \rightarrow (NH_4)_2U_2O_7 + 4NH_4NO_3 \qquad (2.20)$$

Subsequent to precipitation, ADU slurry is filtered using vacuum filter and then cake is washed with DM Water to remove entrapped ammonium nitrate. The cake is dried under vacuum, and Dry ADU powder is stored in engineered container.

2.3.4 Calcination of ADU

Ammonium diuranate is calcined at ~450–600 °C in furnace to produce a mixture of UO_3 and U_3O_8. The product obtained has a composition in between UO_3 and U_3O_8 and is found to produce acceptable-grade uranium dioxide during hydrogen reduction.

$$(NH_4)_2U_2O_7 \xrightarrow{350\,°C} 2UO_3 + 2NH_3 + H_2O \qquad (2.21)$$

$$9(NH_4)_2U_2O_7 \xrightarrow{700\,°C} 6U_3O_8 + 14NH_3 + 15N_2O + 2N_2 \qquad (2.22)$$

2.3.5   *UO₃ Reduction*

In this process, UO_3 obtained from calcination is reduced to UO_2 in the cracked ammonia gas ($N_2 + H_2$ mixture) environment.

A mixture of nitrogen and hydrogen is produced by passing ammonia at a low pressure through a Fe/Ni catalyst bed (retort) heated at 850 °C temperature. Uranium oxide powder (UO_3) obtained after calcination is reduced in a current of nitrogen–hydrogen mixture, obtained by the cracking of ammonia. The reduction operation is carried out in an electrically heated rotary tubular furnace at a temperature of 600 °C. A photograph of a typical reduction reactor is shown in Fig. 2.15.

$$2NH_3 \overset{850\,°C}{\rightarrow} N_2 + 3H_2 \quad \Delta H = +5.45\,\text{kcal/mol NH}_3 \tag{2.23}$$

UO₃ Reduction

$$UO_3 + H_2 \overset{600\,°C}{\rightarrow} UO_2 + H_2O \quad \Delta H = -25.3\,\text{kcal/mol U} \tag{2.24}$$

$$U_3O_8 + 2H_2 \overset{600\,°C}{\rightarrow} 3UO_2 + 2H_2O \quad \Delta H = -13.1\,\text{kcal/mol U} \tag{2.25}$$

Fig. 2.15 Reduction reactor

2.3.6 Hydrofluorination of UO₂

Uranium dioxide is converted to uranium tetrafluoride, using anhydrous HF gas, in an electrically heated rotary tubular furnace at 450 °C temperature.

$$UO_2 + 4HF \rightarrow UF_4 + 2H_2O \quad \Delta H = -43.2\,kcal/mol\ U \qquad (2.26)$$

UF$_4$ (Green Salt) is produced by reaction of UO$_2$ and anhydrous hydrogen fluoride (AHF) gas in inconel rotary tubular reactor at 320–450 °C. AHF is heated prior to its feeding to the reactor. Excess HF along with produced water is condensed and collected in monel carboy below the reactor. The off gas from the reactor is scrubbed in primary wet scrubber with alkaline solution. UF$_4$ produced is heated in a tubular expulsion reactor to remove entrapped HF and moisture.

UF$_4$ samples are sent to quality control laboratory for analysis of UO$_2$F$_2$, Ammonium oxalate insolubles (AOI), moisture, and free acidity for each batch. The limits are 2.5, 1.0, and 0.3%, respectively, for UO$_2$F$_2$, AOI, moisture, and free acidity.

2.3.7 Uranium Metal Production by Metallothermy

2.3.7.1 Magnesiothermic Reduction (MTR) of UF₄

Uranium metal ingot is produced by the reduction of UF$_4$ at high temperature with magnesium metal chips or calcium granules.

$$UF_4 + 2Mg \rightarrow U + 2MgF_2 \quad \Delta H = -85.0\,kcal/mol\ U \qquad (2.27)$$

The reaction is carried out in an MgF$_2$ inside lined boiler-quality vessel, generally called Magnesiothermic Reduction (MTR) Reactor. The reactor is lined with magnesium fluoride to protect it from damage due to heat of reaction and prevent contamination of uranium metal with the material of the reaction vessel. Figures 2.16 and 2.17 show the photographs of MTR reactor and Furnace, respectively.

A mold with an electrical vibrator is used for making the lining. Graphite crucible is placed at the bottom of the lined reactor. A mixture of required quantity of UF$_4$ and Mg is charged in the lined reactor, capped with MgF$_2$. The reactor is heated to a maximum temperature of 700 °C, and magnesiothermic reduction reaction is carried out. After completion of the reaction and cooling, the slag blocks, lose powders, and ingot are separated. The specifications of uranium ingot are tabulated in Table 2.11.

Fig. 2.16 MTR reactor

Fig. 2.17 MTR furnace

Table 2.11 Uranium Ingot specification

Sr. No.	Element	Maximum limit (ppm)
1	Boron	0.12
2	Iron	200.0
3	Gd	0.04
4	Dy	0.1
5	Overall uranium content	>99.85%

2.3.7.2 Calciothermic Reduction of Uranium Oxide

Metallothermic reduction of the uranium dioxide yields uranium metal and metal oxide as shown in the following reactions:

$$UO_3 + 3Ca \rightarrow U + 3CaO \quad \Delta H = -164.0 \, kcal/gram \, atom \, U \qquad (2.28)$$

$$UO_2 + 2Ca \rightarrow U + 2CaO \quad \Delta H = -44.0 \, kcal/gram \, atom \, U \qquad (2.29)$$

$$UO_2 + 2Mg \rightarrow U + 2MgO \quad \Delta H = -35.0 \, kcal/gram \, atom \, U \qquad (2.30)$$

After the reduction reaction, the metal droplets solidify in separate pockets surrounded by a matrix of metal oxide slag and thus ending up as powder. The metal oxide slag is separated from uranium by chemically leaching out the oxide slag. Calcium is preferred over magnesium for the reduction process due to its higher heat of reaction and better leachability of the oxide. The reduction reaction is conducted in an oxide-lined metallic reduction reactor. The oxide lining is done to avoid the contact of molten uranium from the metallic reactor. This is primarily to protect the metallic reactor as well as to avoid the contamination of the molten uranium.

2.3.8 Waste Management in Uranium Metal Production

2.3.8.1 Raffinate Treatment and Disposal

The raffinate leaving from extraction unit contains impurity, activity, and trace amount of uranium. This is precipitated with magnesia and treated further to recover active cake containing trace amount of uranium. It is to be treated either as a recycled source or active waste for further disposal after accounting, depending on uranium concentration.

$$2UO_2(NO_3)_2 + Mg(OH)_2 \rightarrow MgU_2O_7 + 2Mg(NO_3)_2 + H_2O \qquad (2.31)$$

2.3.8.2   Treatment and Disposal of ADU Filtrate

The alkaline filtrate from ammonium diuranate precipitation is collected in a tank and allowed to settle, so that any cake carried by the filtrate can be collected in the tank and processed later for uranium recovery. The supernatant is transferred into a separate tank and is treated with trisodium phosphate and ferric chloride for decontamination. The decontaminated solution is transferred for proper disposal after accounting.

2.3.8.3   HF Scrubber Water Treatment

Non-condensed HF gas with cracked gas is scrubbed in hydro-fluorination section for removing the HF from exhaust gases. After scrubbing, the gas is allowed to let out to atmosphere through solution bubbling. The condensed HF along with condensed moisture is collected in a container from where condensed HF is drained. Condensed HF is treated with ferrous sulfate and lime to remove soluble uranium as from UO_2F_2 to UF_4 and allowed overnight to settle suspended UF_4.

$$UO_2F_2 + 2FeSO_4 + H_2SO_4 + 2HF \rightarrow UF_4 + Fe_2(SO_4)_3 + 2H_2O \quad (2.32)$$

The slurry containing uranium tetrafluoride is sent to the waste management section for further treatment and to recover the uranium value by complexing and precipitation process as discussed above.

2.3.9   Advanced Methods

Other methods of uranium powder production are described below:

a. **Hydriding–Dehydriding**: The hydriding–dehydriding process is a reversible chemical reaction that first creates uranium hydride by heating (200–250 °C) the bulk uranium in hydrogen atmosphere. As the hydrogen is adsorbed by the uranium, changes in the lattice parameter between the larger UH_3 crystal and uranium crystal introduce stresses in the bulk U metal that breaks it apart into fine powder. The hydride powder is further heated (275–400 °C) under vacuum to remove hydrogen from it and yield uranium powder. The powder is however very fine and often pyrophoric.

b. **Grinding and Cryogenic milling**: As uranium is ductile, grinding is found to be a very inefficient process for powder production. Moreover, it introduces impurities in the metal. Similar problem is encountered with cryogenic milling as well.

c. **Atomization**: Atomization involves passing the molten metal through a nozzle and disintegrating it with a high velocity jet of inert gas. The process involves more number of process steps than that of powder production by oxide reduction. However, in this process, the effluent generation is less compared to oxide reduction.

d. **Sodium vapor reduction of UCl₄ in the presence of KCl**: This reaction leads to uranium metal powder in one step. However, UCl_4 is quite hygroscopic and difficult to handle. Sodium handing poses challenges too. The process has only been demonstrated on experimental basis and is nowhere being used in production scale.

2.4 Development of Uranium Supported Advanced Fuel Materials

Thorium-based oxide and alloys are potential fuel for breeder as well as high temperature reactors. However, thorium being fertile needs to be combined with fissile material to be used as a nuclear fuel. Th-U oxide and alloys are of particular interest as uranium is required to be combined with thorium for the latter to be used as fuel. Also, ^{232}U is produced from irradiation of thorium which is a hard gamma emitter and makes it inherently proliferation resistant. However, this poses technological challenges also in the reprocessing of the fuel. Th-U fuel cycle offers many advantages, e.g., possible use in thermal as well as fast reactors, inherent proliferation resistance, high temperature of operation, etc. More details about thorium oxide fuel have been discussed in Chap. 3. The present chapter discusses about the metallic fuel.

The major source of thorium oxide is monazite, essentially a phosphate of the cerium earths. This mineral occurs associated with silica and other rare earths in the monazite sands. Monazite deposits are mainly located in India, Brazil, Australia, Sri Lanka, Africa, and Canada.

2.4.1 Processing and Production of Thorium

Several methods have been developed for production of thorium from monazite. Monazite is recovered as by-product during beach sand processing. Monazite is orthophosphate of rear earths containing about 9% of thoria (ThO_2). The world reserve of thorium is about 1.2 million tons.

The process of thorium production followed by M/S Indian Rear Earths Limited, India, as well as many other plants of the world, follows hydrometallurgical methods, using sulfuric acid or sodium hydroxide digestion. At the starting of the process, beach sands are subjected to physical beneficiation process, for increasing the concentration of monazite to about 85%. Various selective separation processes, namely gravity, magnetic and electrostatic methods, are utilized in beneficiation. Indian monazite, along with thorium, contains about 0.35% uranium, the balance being rear earths.

During sulfuric acid digestion, rare earths, thorium, uranium, and phosphates get solubilized leaving contaminating materials such as silica, rutile, and zircon as residue. Solubilized uranium, thorium, and rare earths can be recovered by hydroxide precipitation or by double sulfate precipitation or oxalate precipitation.

In alkali digestion process, monazite is treated with sodium hydroxide to solubilize phosphate and separate out rare earths, thorium, and uranium as hydroxide. After decantation and washing off phosphate solution, residual hydroxide cake is treated with HCl at controlled temperature and p^H, when solubilized rear earths is separated out for further recovery of RE leaving a solid residue rich in thorium concentration. For nuclear applications, thorium and uranium both need to be purified from neutron-absorbing impurities. Th-U separation and purification are done by selective extraction using TBP in suitable concentration from nitrate medium.

Thorium oxide is produced by thorium oxalate precipitation from pure thorium nitrate solution and successive calcination of thorium oxalate. Uranium follows the ammonium diuranate precipitation route. Thorium metal can further be obtained from pure thoria by calciothermic reduction at 1100 °C. The detailed flowsheet of the thorium powder preparation is shown in Fig. 2.18. The thorium powder thus produced can be consolidated by pelletization and melting the pellets.

2.4.2 Process Metallurgy

Thorium powder is consolidated by compacting it in pelletizing press. The pellets thus produced could be melted in either vacuum arc melting furnace (Fig. 2.19) or EB melting furnace. Although induction melting is better for alloying due to better homogenization due to magnetic stirring, melting of thorium and its alloys is not preferred in vacuum induction melting furnace. Induction melting causes the pickup of impurities from the crucible material by the thorium. Thorium metal is readily fabricated by casting, powder metallurgy, extrusion, rolling, and other methods. Cold worked material starts to recrystallize and anneal at slightly above 500 °C, although temperatures up to 700 °C may be required to ensure complete annealing for slightly worked material. Thorium ingots can be easily rolled in the temperature range of 740–850 °C with reductions up to 25–30% per pass being feasible. The details of Th-U alloy as fuel are described in Chap. 3.

2.4.3 Process Engineering

Thorium and uranium, due to their reactive nature in powder form as well as due to the associated radioactivity, should be handled under inert atmosphere in glove boxes/fume hoods. The glove boxes/fume hoods should be maintained under negative pressure.

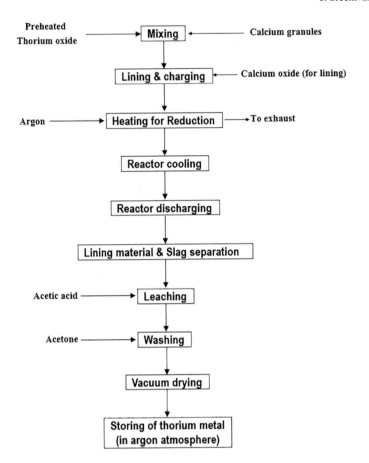

Fig. 2.18 Flow sheet for thorium metal powder preparation

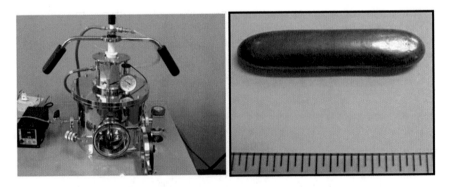

Fig. 2.19 Vacuum/inert arc melting setup (left) and Thorium metal finger (right)

2.4.4 Waste Management

The preparation of the thorium metal and alloys leads to several waste streams, viz. leach liquor, lining material, and alloy scrap. These waste streams contain uranium/thorium. The waste streams have to be suitably treated for recovery of uranium as well as to make waste streams worthy for disposal as per the applicable guidelines.

2.5 Uranium Enrichment

2.5.1 Isotope Separation

2.5.1.1 Isotopes of Uranium

All isotopes of uranium are radioactive. However, the half-lives of some of them are of the order of age of earth and hence these isotopes are present in earth's crust, viz. ^{234}U, ^{235}U, and ^{238}U. For all practical purposes, we treat these uranium isotopes as stable.

2.5.1.2 Principles of Isotope Separation

The isotopic composition of uranium that makes it useful in nuclear industry is often much higher than its natural concentration. For uranium isotope ^{235}U, the desired isotopic concentrations are between 2 to 4% (as fuel for LWR). In fast reactor fuels, the enrichment of uranium in ^{235}U is much higher. Thus, a need arises for separation (either complete or partial) of these isotopes in order to make them useful for their intended applications. Separation of uranium isotope (^{235}U) generally being only partial (from 0.71% to 2–4%) is termed as uranium enrichment. For an isotope separation process, the element or the compound that contains the isotopes must necessarily be in gaseous state (for physical processes) or in liquid state (for chemical processes), obviously because isotope separation cannot take place in solid state. The measure of qualitative separation is defined in terms of separation factor "α". Separation factor "α" of an isotopic mixture having only two isotopes (say isotopes ^{1}A and ^{2}A having molecular/atomic weights M_1 and M_2), a so-called binary isotopic mixture, is defined as in Eq. 2.33 and is schematically represented in Fig. 2.20.

$$\alpha = \frac{[xp/(1 - xp)]}{[xw/(1 - xw)]} \qquad (2.33)$$

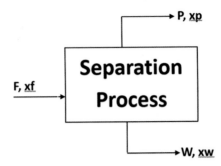

Fig. 2.20 Definition of separation factor, where *xp*, *xw*, and *xf* are mole/weight fractions and P, W, and F are molar/mass flow rates of Product, Waste and Feed, respectively. Product is the stream enriched in desired isotope, while Waste stream is depleted in desired isotope

(i)   Isotopes have different nuclear properties, but their physical and chemical properties are nearly the same. However, there is an infinitesimally small but finite difference between their physical and chemical properties. This small difference in the properties is made use of, in separating the isotopes.

(ii)   The order of difference in physical/chemical properties between the desired isotope and the other undesired isotopes dictates the difficulty of separation. Higher the difference, lesser the difficulty in separation and vice versa.

(iii)   There are methods of separation that depend on the difference in physical properties of the isotopes such as mass, inertia, and diffusivity. The separation factor of such processes depends on either ΔM or $\Delta M/M$ (where: $\Delta M = M_2 - M_1$ and $M = [M_1 + M_2]/2$), whereas for methods of separation based on difference in chemical properties (i.e., chemical potential-based properties such as vapor pressure, equilibrium constant, and solubility), the separation factor is a function of $\Delta M/M^2$.

(iv)   From the above, it is clear that chemical processes of separation are more suitable for isotopes of lighter elements, whereas physical processes are more effective for isotopes of heavier elements. Thermodynamically, chemical processes are more reversible than physical processes. Thus, chemical processes are more energy efficient than physical processes. The separation of isotopes of hydrogen (such as ^1H, ^2H, and ^3H) is more economical than the separation of isotopes of uranium (^{235}U and ^{238}U).

Another way of categorizing the separation methods is as follows:

A.   Methods based on difference in Free energy (dynamics of constituent molecules): The difference in internal vibration and rotation of molecules containing isotope species gives rise to difference in free energy. The properties based on free energy are vapor pressure, reactivity or equilibrium constants, solubility, etc.

(a)   The separation factors for such methods depend on $\Delta M/M^2$.

(b)   Separation factors are relatively larger for isotopes of lighter elements.

(c)   These methods are thermodynamically reversible.

      (d) Examples are distillation, chemical exchange, and liquid extraction and are suitable for separating isotopes of lighter elements such as hydrogen, lithium, and boron.

B. Methods based on difference in kinetics of constituent molecules:
      (a) The separation factors for such methods depend on $\Delta M/M$.
      (b) These methods are thermodynamically irreversible and energy intensive.
      (c) These methods (e.g., Gas diffusion and thermal diffusion) are suitable for heavier elements.

C. Methods based on external force field such as centrifugal and electromagnetic force field:

The separation factor for such methods depends on ΔM. Examples are gas centrifuge, plasma centrifuge, and nozzle separation which are suitable for isotopes of heavier elements.

(v) Selective methods:

These methods are independent of interaction between constituent molecules and are not governed by the thermodynamics. In these methods, the atoms/molecules of desired isotope are selectively excited and separated. Examples are electromagnetic separation and laser isotope separation. Though the separation factors for such methods are very large, the throughputs are generally very low. Therefore, such methods are commercially not very attractive.

Different methods of separation of isotopes and their characteristics are provided in Table 2.12.

2.5.2 Separation Processes for Uranium Enrichment

While selecting a process for isotope separation, obviously the first choice is a reversible process such as distillation or chemical exchange. However, since the separation factors of these processes are a function of $\Delta M/M^2$, and therefore, for uranium enrichment, (M being high), these processes would yield extremely low separation factors. Hence, separation processes based on physical properties are more suitable for uranium enrichment. As already mentioned earlier, physical process for isotope separation needs the isotopic element or compound to be in gaseous state. Uranium being a metal, one of its compounds, i.e., uranium hexafluoride (UF_6), is chosen as feed gas for an enrichment process. Although it is solid at room temperature, it has sufficiently high vapor pressure and thus can be easily converted into vapor. Also, fluorine is monoisotopic and thus UF_6 is essentially a binary isotopic mixture. This makes enrichment easier, since separation of binary isotopes is easier than that of multiple isotopes.

Although gas diffusion was the first commercially deployed enrichment process in United States and France, it was later abandoned in favor of gas centrifuge due

Table 2.12 Different methods of separation of isotopes

Category	Processes	Characteristics
Reversible	(i) Distillation (ii) Chemical exchange	• Based on chemical properties • High separation factor • High energy efficiency • Simple construction and less expensive
Irreversible	(i) Gas centrifuge $ (ii) Gas diffusion (iii) Thermal diffusion (iv) Aerodynamic separation	• Based on physical properties • Low separation factor • Low energy efficiency • Difficult construction and more expensive
Selective *	(i) Electromagnetic isotope separation (ii) Laser isotope separation	• Very high separation factor • Energy intensive process • High cost • Low throughput (capacity)

* Category not governed by thermodynamics
$ Gas centrifuge is the least irreversible process among the category

to its better commercial viability. Nevertheless, the following processes have been exploited for uranium enrichment either at laboratory scale, pilot plant level, or even to some extent at commercial level.

a. Thermal diffusion
b. Aerodynamic isotope separation
c. Chemical exchange
d. Laser isotope separation
e. Electromagnetic isotope separation
f. Plasma centrifuge

2.5.2.1 Gas Diffusion

The principle of this method is based on the phenomenon of molecular diffusion. In a closed cell in thermal equilibrium with its surroundings, all molecules of a gas mixture have the same mean kinetic energy ($\frac{1}{2} mv^2$). Hence, lighter molecules (with lower "m") travel faster and strike the cell walls more frequently than do heavier ones. A hole in the wall, if it is small enough to prevent the outflow of the gas as a whole (viscous flow), will allow the passage of a larger proportion of lighter molecules, in relation to their concentration, than of the heavier ones. In gas diffusion, for the separation of uranium isotopes, the gas (UF_6) is made to pass through a very fine barrier (of pore size about 200 Å) such that in the pores a molecular flow

Fig. 2.21 Schematic representation of gas diffusion

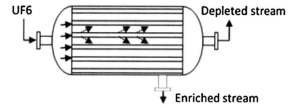

regime is attained. In molecular flow, no collisions take place between molecules, and the molecules move independent of each other, unlike in viscous flow. Under such condition, the flow depends on molecular weight of the gas. The schematic representation of gas diffusion is shown in Fig. 2.21.

The separation factor of this process is low, and it requires a very large number of stages for a given separation as compared to other processes. Typically, to enrich natural uranium into product having 3% assay and waste having 0.2% assay of ^{235}U, it would require about 1300 stages of gaseous diffusion. Also, capital investment and specific power consumption are very high for this process.

2.5.2.2 Thermal Diffusion

A thermal diffusion column for isotope separation consists of two tall vertical concentric pipes with enriched and depleted gas outlets at the top and bottom. The inner pipe is maintained at a high temperature (typically few hundreds of degrees Celsius), while the outer pipe is kept cooled at ambient temperature. The binary gas mixture is fed at the center of the column, whereas the enriched and depleted streams are withdrawn from the top and bottom of the column, respectively. The process is generally not used commercially, as it offers much lower separation factor than gas diffusion and is quite energy intensive.

2.5.2.3 Gas Centrifuge

The centrifugal force acting on a particle is proportional to its mass, and it is given as $F = m\omega^2 r$. When a gas (say UF_6) containing isotopic molecules of ^{235}U and ^{238}U is made to rotate in a chamber at a very high angular speed, then the isotopic molecules will experience different centrifugal forces depending on their masses or molecular weights.

In this process, gaseous UF_6 is fed into a cylindrical rotor that spins at very high speed inside an evacuated casing. The heavier isotope tends to move closer to the rotor wall than the lighter one, resulting in partial radial separation. An axial countercurrent created over and above the radial one enhances the separation further of the rotor. The schematic representation of gas centrifuge is shown in Fig. 2.22.

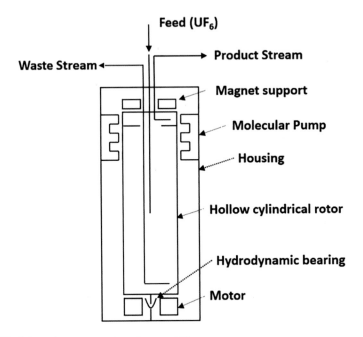

Fig. 2.22 Schematic representation of gas centrifuge

In a gas centrifuge, the elementary separation factor is exponentially proportional among other parameters, to the difference in molecular weights of the two isotopes to be separated. This elementary separation factor is further enhanced many times, by creating an additional axial counter-current separation. The separation factor is also exponentially proportional to the square of the linear velocity of the rotor and its length, thus offering the process with high potential in terms of separation factor. Thus, the separation factors of gas centrifuge are very high as compared to the gaseous diffusion. Hence, very small number of stages is required for a given separation as against a large number of stages required in gas diffusion. Capital investment for the process is moderate, and specific power consumption is much lower than that of gas diffusion process.

2.5.2.4 Aerodynamic Separation Process

In this process, a partial separation of isotopes is obtained in a flowing gas stream that is subjected to a very high centrifugal acceleration. The molecules depending on their mass number will experience different centrifugal forces, resulting in separation.

2.5.2.5 Laser Isotope Separation Process

The laser isotope separation process is based on selective excitation of uranium atoms or molecules through laser beams. The laser isotope separation process is a selective process unlike the other processes, which are equilibrium processes, in which the interactive forces between the molecules are not to be overcome. Thus, in this process no such forces are required to be overcome; instead, the molecules are selectively excited and separated out. Therefore, these processes are not governed by thermodynamics. The principle of laser isotope separation process is to excite a desirable atom or molecule to a high-energy state using an exactly tuned frequency of laser. The excited atom/molecule in higher energy state can be photo-ionized as in Atomic Vapour Laser Isotope Separation (AVLIS) or can be photo-dissociated as in Molecular Laser Isotope Separation (MLIS).

In these laser isotope separation processes, theoretically the separation factor is infinite. But practically, the separation factors obtained are much lower mainly due to the following reasons:

(1) Due to Doppler effect, the atoms/molecules see a band of frequencies rather than a unique frequency.
(2) There is a phenomenon called charge transfer, wherein the desired isotopic species (excited) transfer their charge to the undesired isotopic species.
    In AVLIS, the uranium metal vaporizes at 2600 K and the system requires use of a material of construction that is suitable for the high temperature operation. The throughput in the process is extremely low, and therefore, the process is commercially not viable.

2.5.2.6 Electromagnetic Isotope Separation (EMIS)

The basic principle behind this process is same as that of an electromagnetic mass spectrometer, that a charged particle will follow a circular trajectory while passing through a uniform magnetic field. Two ions having same electrical charge, but different masses ($^{235}U^+$ and $^{238}U^+$) will have different trajectories, with heavier $^{238}U^+$ having larger radius of trajectory. In this process, the UCl_4 molecules are bombarded with electrons to produce U^+ ions, which are electrically accelerated and made to pass through a strong magnetic field. The charged particles while passing through the magnetic field follow a curved trajectory. The difference in radius of curvature allows the separation of isotopes. Figure 2.23 shows a schematic representation of electromagnetic isotope separation.

The process has theoretically very high separation factor, but allows very low throughput and consumes large amount of electrical energy. A major problem with the EMIS process is that less than half of the UCl_4 feed gets actually converted to the desired U^+ ions, and less than half of these desired ions actually reach the collector. Thus, the effective output of the process is very low.

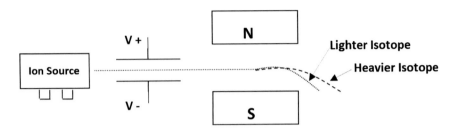

Fig. 2.23 Schematic representation of electromagnetic isotope separation process

2.5.2.7 Chemical Exchange

This process is more suitable for separation of isotopes of lighter elements such as hydrogen, lithium, and boron. In the process, two immiscible compounds, say "AB" and "AC" of a given element say "A" (with two isotopes "A1" and "A2"), are made to intimately contact with each other in a plate column (in chemical exchange or exchange distillation) or in a packed tower (in ion exchange). In chemical exchange or exchange distillation, the two compounds are both liquids (immiscible) and in case of ion exchange, one compound is solid while the other is in liquid form. It may be noted that in all these processes, no chemical reactions take place and there is only an exchange of isotopes across the two compounds being physically contacted with each other. The tendency of the exchange of the two isotopes is due to the difference in their free energies. Generally, the heavier isotope tends to concentrate in the heavier compound. In general, the chemical exchange process is given as;

$$A_1B + A_2C \leftrightarrow A_2B + A_1C \tag{2.34}$$

The separation factors for these processes are equal to the equilibrium constants of the exchange reactions involved. The separation factor decreases with the increase in isotopic molecular weight. For isotopes of lighter elements, α is very high in comparison with that of heavier elements. Typically, in case of uranium, the separation factors are between 1.002 and 1.003.

2.5.2.8 Plasma Centrifuge

In this process, the principle of ion cyclotron resonance is used to selectively energize the ^{235}U isotope in the plasma that consists of ^{235}U and ^{238}U ions. The plasma is subjected to a uniform magnetic field along the axis of a cylindrical vacuum chamber as the plasma flows from source to collector. The high strength magnetic field (produced by super-conducting magnets) produces helical motion of ions with lighter ^{235}U spiraling faster and having higher ion cyclotron frequency than the heavier ^{238}U ions. As the ions move toward the collector plate, they pass through an electric field produced by an excitation coil oscillating at the same frequency as the ion cyclotron frequency of the ^{235}U ions. This causes the helical orbit of the ^{235}U ions to increase in radius while having minimal effect on heavier ^{238}U ions. Thus, a difference in the helical radii of lighter and heavier ions allows the separation and consequent collection of the desired isotope. The plasma separation process for uranium isotope separation is still in R&D stage, while in France it is being used for stable isotope separation.

2.5.3 Requirements of an Ideal Enrichment Process

An ideal process is the one that gives large separation, both from quality and quantity point of view, at minimum energy input. The following are the four crucial process parameters that must be considered together to decide on the most economic process:

1. **Large separation factor (α)**: It is always desirable to have the separation factor as high as possible. This would reduce the number of stages required for a given separation, and thereby the capital cost of the plant.
2. **High throughput**: A process should be able to handle large quantities of feed material for separation. Although the separation factor for the electromagnetic separation (Calutron, USA) was 10^4 times that of gaseous diffusion, it was high throughput of gas diffusion process that made it more preferable over EMIS.
3. **Low inventory**: The holdup of the process equipment should be low. This offers simplicity and low equilibrium times and allows quick switchover from one feed to another.
4. **High energy efficiency**: Energy efficiency of the separation process should be high, which means that the process should consume less energy for a given separation. In another way, the specific energy consumption of the process that defines energy consumed for a given separation should be as low as possible.

A comparison of various enrichment processes is given in Table 2.13.

Table 2.13 Comparison of various enrichment processes

Process	Separation factor	High throughput	Low inventory	Energy efficiency
Electromagnetic separation	Excellent	Very poor	Poor	Poor
Gaseous diffusion	Poor	Excellent	Poor	Poor
Gas centrifuge	Very good	Excellent	Very good	Very good
Thermal diffusion	Very poor	Good	Poor	Very poor
Nozzle separation	Good	Good	Good	Good
Laser separation	Excellent	Very poor	Excellent	Very good

2.5.4 Process Flowsheet

As the enrichment process needs uranium in UF_6 form, the MDU/SDU as obtained from the mines has to undergo a series of processes for UF_6 production. Post-enrichment, the enriched UF_6 gas is required to be de-converted to obtain UO_2

Fig. 2.24 Process flowsheet for UF_6 production: **a** overall process for enrichment, **b** refining and conversion, **c** uranium enrichment, **d** de-conversion (*for de-conversion, dry routes are also available*)

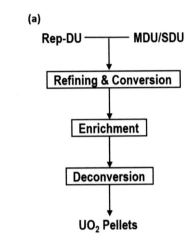

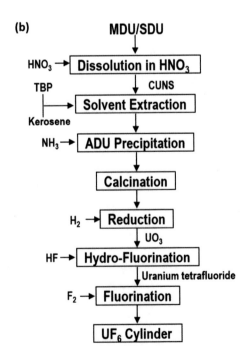

Fig. 2.24   (continued)

(c)

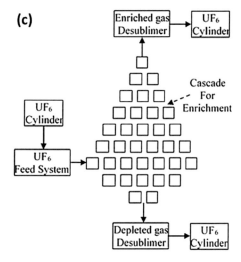

Uranium Enrichment

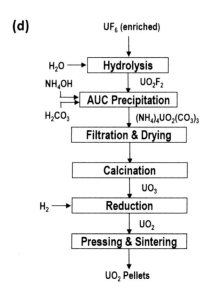

powder that finally goes into the making of fuel pellets. The main processes involved are schematically shown in flowsheet form in Fig. 2.24.

2.5.5 Management of Depleted Stock

More than 90% of the feedstock (natural; 0.72% assay) to an enrichment facility comes out as depleted stock (<0.35% assay). A large amount of depleted stock has practically no utility. Due to high level of toxicity (chemical as well as radiological) of HEX gas, long-term storage of the depleted stock has been one of the major concerns of global nature. As a solution to this, the depleted gas is converted into a non-corrosive solid powder, i.e., either UF_4 or U_3O_8. Long-term storage of both UF_4 as well as U_3O_8 is much easier and requires less surveillance in comparison with that of UF_6. Although from long-term storage point of view, conversion (of UF_6) to U_3O_8 is preferred, in some specific cases reduction to UF_4 offers a flexibility in terms of ease in its conversion to either uranium metal or back to UF_6.

The waste management in enrichment plants are similar to that of yellow cake refining processes which have been discussed in Sect. 2.3.8. Additionally, in enrichment plants, nuclear criticality is taken care of by suitably designing the vessels/cylinders for the process and storage and by following specific storage layout based on sub-criticality analysis.

Exercises

1. What is an ideal cascade? Compare characteristics of an ideal cascade with that of a square cascade.
2. What are the merits of using uranium hexafluoride as feed material in most uranium enrichment processes?
3. Explain the working principles of AVLIS and MLIS processes for isotope separation.
4. Write and explain the mathematical expression of Separation Factor for a Gas Centrifuge process for isotope separation.
5. Describe the characteristics of an Ideal cascade.
6. What is CUT? What is its significance in ideal cascade?
7. What is the minimum work of separation required for separating a mixture of two isotopes, say A and B, into their pure forms by a thermodynamically reversible process?
8. What is Mixing Loss? Describe some of the salient features of an ideal cascade.
9. A separation unit takes feed of natural uranium (0.72%) at the rate of 600 kgU/yr and operates under half cut ($\theta = 0.5$) condition. The product stream obtained from the unit was found to be 0.85% in U-235 concentration. Estimate the separation factor of the unit.
10. It is proposed to make a cascade using several identical separation units (as specified in Question no. 9), to give enriched uranium product of 3.0% concentration. Find out the number of stages required in the enricher section of the cascade, if the external feed concentration is same as 0.72%. Assume ideal cascade configuration.

11. In a uranium isotope separation process, the weight percent of U-235 in product is 0.90%, while that of depleted stream is 0.60%. Work out the Total Separation Factor of the process.

12. Globally, if the cost of enrichment is $100 per SWU, then calculate the price to be paid for enriching natural uranium to get 1 ton of enriched uranium of 3.0% concentration. Assume natural feed concentration as 0.72% and waste concentration as 0.25%. All concentrations are in weight percentages.

13. Estimate the number of stages required in Enricher section, if a cascade for the above job (as given in Question 9) is made of identical separating units each having a total separation factor of 1.44. Consider the cascade to be an ideal cascade.

14. Down blending is a mixing process in which enriched uranium is mixed with natural uranium in order to get a specific desired concentration of uranium-235. In one such operation, it is required to make 50 tons of 2.6% enriched fuel by blending 3.0% enriched uranium with natural uranium. Estimate the quantities of enriched uranium (3.0%) and natural uranium (0.72%) for the specified requirement. All specified concentrations are in weight percentages.

15. Calculate the Separative Power of the above process (given in Q-8 above) for the following parameters:

   (i) $xf = 0.72\%$ (weight percent of U-235 in Feed)
   (ii) $F = 1000$ kgU/yr (Feed flow rate in terms of uranium flow)
   (iii) $CUT = 0.435$ (Ratio of Product to Feed flow rate)

References

1. P. Bruneton, M. Cuney, *Uranium for Nuclear Power: Resources, Mining and Transformation to Fuel* (Woodhead Publishing, Cambridge, 2016), pp. 11–52
2. J.S. Herring, Uranium and thorium resources, in *Encyclopaedia of Sustainability Science and Technology*, ed. by R.A. Meyers (Springer, New York, 2012)
3. D. Hobart, *Uranium in Periodic Table of Elements* (LANL Publications, United States, 2013)
4. *Manual on Laboratory Testing for Uranium Ore Processing*, IAEA Technical Report Series 313 (IAEA, Vienna, 2013). ISBN 92-0-145190-3
5. R.M. Hazen, R.C. Ewing, D.A. Sverjensky, Evolution of uranium and thorium minerals. Am. Miner. **94**, 1293–1311 (2009)
6. V.I. Gerasimovskii, On the modes of occurrence of uranium in rocks. Sov. J. At. Energy **3**, 1407–2141 (1957)
7. B. Lehmann, Uranium ore deposits. Rev. Econ. Geol. AMS On-line **2**, 16–26 (2008)
8. *Geological Classification of Uranium Deposits and Description of Selected Examples* (International Atomic Energy Agency, Vienna, 2018), p. 430. ISSN 1011-4289
9. *Uranium Resources as Co- and By-products of Polymetallic, Base, Rare Earth and Precious Metal Ore Deposits*, IAEA TECDOC series no. 1849 (IAEA, Vienna, 2018). ISSN 1011-4289
10. *Uranium: Resources, Production and Demand*, A Joint Report by the Nuclear Energy Agency (Boulogne-Billancourt, France) and the International Atomic Energy Agency (Vienna), NEA No. 7413 (OECD, 2018)
11. A.C. Saraswat, Uranium exploration in India: perspectives and strategy. Explor. Res. Atomic Miner. **1**, 1–11 (1988)

12. K. Kyser, *Uranium for Nuclear Power: Resources, Mining and Transformation to Fuel* (Woodhead Publishing, Cambridge, 2016), pp. 53–76
13. D.B. Hoover, D.P. Klein, D.C. Campbell, Geophysical methods in exploration and mineral environmental investigations (1995)
14. G.V. Keller, Rock and mineral properties, in *Electromagnetic Methods in Applied Geophysics, Volume 1, Theory: Investigations in Geophysics*, ed. by M.N. Nabighian, vol. 3 (Society of Exploration Geophysics, 1987)
15. E. Auken, L. Pellerin, N.B. Christsen, K. Sorensen, A survey of current trends in near surface electrical and electromagnetic methods. Geophysics **71**, 249–260 (2006)
16. *Advances in Airborne and Ground Geophysical Methods for Uranium Exploration*, IAEA Nuclear Energy Series (IAEA, Vienna, 2013), 72 pp. ISSN 1995-7807; No. NF-T-1.5. STI/PUB/1558 ISBN 978-92-0-129010-6
17. A.K. Chaturvedi, Heliborne, *Time Domain Electromagnetic Surveys for Uranium Exploration, Indian National Seminar & Exhibition on Non-destructive Evaluation NDE*, Hyderabad, India (2015)
18. *Uranium Extraction Technology*, IAEA technical report series no. 359 (IAEA, Vienna, 1993)
19. P.H. Woods, Uranium mining (open cut and underground) and milling, in *Uranium for Nuclear Power: Resources, Mining and Transformation to Fuel*, ed. by I. Hore-Lacy (Woodhead Publishing, United Kingdom, 2016), pp. 125–156
20. R. Marjoribanks, *Geological Methods in Mineral Exploration and Mining* (Springer, Berlin/Heidelberg, 2010), p. 238
21. J. de la Vergne, *Hard Rock Miner's Handbook*, 5th edn. (Stantec Consulting, Alberta, 2003), p. 310
22. *Methods of Exploitation of Different Types of Uranium Deposits.* IAEA-TECDOC-1174 (IAEA, Vienna, 2000), pp. 16–23
23. A.K. Suri, T. Sreenivas, Developments in processing of conventional uranium ore resources of India, in *6th International Mineral Processing Congress (IMPC)*, New Delhi, India, October (2012), p. 271
24. *Manual of Acid In Situ Leach Uranium Mining Technology*, IAEA-TECDOC-1239 (IAEA, Vienna, 2001), p. 283
25. *In Situ Leach Uranium Mining: An Overview of Operations*, IAEA nuclear energy series (IAEA, Vienna, 2016), p. 60. ISSN 1995-7807; No. NF-T-1.4
26. Anon, *Environmental Code of Practice for Metal Mines* (Mining Section, Mining and Processing Division, Public and Resources Sectors, Directorate Environmental Stewardship Branch Environment Canada, 2009), p. 29
27. M.J. McPherson, *Subsurface Ventilation and Environmental Engineering* (Chapman and Hall, London, 1993), p. 905
28. *Technologies for the Treatment of Effluents from Uranium Mines, Mills and Tailings*. IAEA-TECDOC-1296 (IAEA, Vienna, 2002), pp. 1–127
29. K.T. Thomas, Management of wastes from uranium mines and mills. IAEA Bull. **23**(2), 33–38 (1981)
30. C. Gherghel, E. Souza, Ventilation requirements for uranium mines, in *Proceedings 12th US-North American Mine Ventilation Symposium* (2008)
31. *Guidebook on the Development of Projects for Uranium Mining and Ore Processing*, IAEA-TECDOC-595 (IAEA, Vienna, 1991), p. 159
32. D. Yan, D. Connelly, Implications of mineralogy in uranium ore processing, in *ALTA 2008 Uranium Conference*, Perth, Australia, May 2008
33. H. Schnell, Uranium processing practices, innovations and trends, in *Mineral Processing and Extractive Metallurgy: 100 Years of Innovation*, ed. by C.G. Anderson, R.C. Dunne, J.L. Uhrie (Society for Mining, Metallurgy & Exploration (SME), 2014), pp. 457–465
34. H. Schnell, Uranium Chapter 12.14, in *SME Handbook on Mineral Processing and Extractive Metallurgy*, ed. by K. Kawatra, C. Young (Society of Mining, Metallurgy and Exploration, Colorado, 2019), pp. 2151–2184

35. P. Woods, R. Edge, M. Fairclough, Z. Fan, A. Hanly, I.M.D. Angoula, H.M. Fernandes, P.P. Haridasan, M. Phaneuf, H. Tulsidas, O. Voitsekhovych, T. Yankovich, Initiatives supporting good practice in uranium mining worldwide, in *Uranium Past and Future Challenges*, ed. by B.J. Merkel, A. Arab (Springer, Cham, 2014), pp. 31–40
36. R.C. Merritt, *The Extractive Metallurgy of Uranium* (Colorado School of Mines Research Institute, 1971)
37. J. Clegg, D. Foley, *Uranium Ore Processing* (Addison Wesley, Reading, MA, 1958)
38. C.R. Edwards, A.J. Oliver, Uranium processing: a review of current methods and technology. JOM **52**, 12–20 (2000)
39. M. Okrusch, H. Frimmel, *An Introduction to Minerals, Rocks, and Mineral Deposits* (Springer, Berlin, Heidelberg, 2020), p. 981. ISBN 978-3-662-57314-3
40. D.J.T. Carson, The vital role of process mineralogy in the minerals industry. JOM **41**, 40–42 (1989)
41. W. Baum, Ore characterization, process mineralogy and lab automation a roadmap for future mining'. Miner. Eng. **60**, 69–73 (2014)
42. P. Gottlieb, The revolutionary impact of automated mineralogy on mining and mineral processing, in *Proceedings of the 24th International Mineral Processing Congress* (Science Press, Beijing, 2008)
43. D.J. Bradshaw, The role of 'process mineralogy' in improving the process performance of complex sulphide ores, Keynote paper 932, in *XXVII International Mineral Processing Congress*, Santiago, Chile (2014)
44. N.P.H. Padmanabhan, T. Sreenivas, Utility of process mineralogy in ore processing. Indian Mineralogist. **44**(1), 236–257 (2010)
45. E.M. Wightman, C.L. Evans, Representing and interpreting the liberation spectrum in a processing context. Miner. Eng. **61**, 121–125 (2013)
46. N.P.H. Padmanabhan, T. Sreenivas, Application of physical separation techniques in uranium resources processing. Metals Mater. Process. **20**(2), 85–106 (2008)
47. I.J. Corrans, J. Levin, Wet high intensity magnetic separation for concentration of Witwatersrand gold-uranium ores and residues. J. S. African Min. Metall. **79**, 210–228 (1979)
48. S.V. Muthuswami, S. Vijayan, D.R. Woods, S. Banerjee, Flotation of uranium from uranium ores in Canada: Part I-Flotation results with Elliot Lake uranium ores using chelating agents as collectors. Can. J. Chem. Eng. **61**(5), 728–744 (1983)
49. R. Natarajan, T. Sreenivas, N. Krishna Rao, Pre-concentration of low grade uranium ores by gravity and magnetic methods: a case study with copper tailings from Singhbhum, Bihar, India. Explor. Res. Atomic Miner. **5**, 93–103 (1992)
50. N.P.H. Padmanabhan, T. Sreenivas, Process parametric study for the recovery of very-fine size uranium values on super-conducting high gradient magnetic separator. Adv. Powder Technol. **22**, 131–137 (2011)
51. H. Schnell, *Building a Uranium Heap Leach Project*, URAM 2014 Conference—Vienna, June 2014
52. D. Marsh, Development and expansion of the Langer Heinrich operation in Namibia, in *Proceedings of the International Symposium on Uranium Raw Material for the Nuclear Fuel Cycle: Exploration, Mining, Production, Supply and Demand, Economics and Environmental Issues (URAM-2009)*, Vienna, Austria June 22–26 (2009)
53. F. Habashi, G.A. Thurston G.A, Kinetics and mechanism of dissolution of uranium oxide. Energia Nucleare **14**(4), 238–244 (1967)
54. D.W. Shoesmith, S. Sunder, *An Electrochemistry-Based Model for the Dissolution of UO_2*, Atomic Energy of Canada Limited Report (Atomic Energy of Canada Limited (AECL), 1991), p. 10488
55. D. Lunt, P. Boshoff, M. Boylett, Z. El-Ansary, Uranium extraction: the key process drivers, The. J. South Afr. Inst. Min. Metall. **107**, 419–426 (2007)
56. D. Lunt, A. Holden, Uranium extraction: the key issues for processing, in *ALTA Uranium 2006* (ALTA Metallurgical Services, Melbourne, 2006)

57. E. Hunter, *On the Leaching Behaviour of Uranium—Bearing Resources in Carbonate Bicarbonate Solution by Gaseous Oxidants* (Faculty and the Board of Trustees of the Colorado School of Mines, USA, 2013)
58. T. Sreenivas, J.K. Chakravartty, Alkaline processing of uranium ores of Indian origin. Trans. Indian Inst. Met. **69**, 3–14 (2016)
59. D. Chetty, Acid-gangue interactions in heap leach operations: a review of the role of mineralogy for predicting ore behavior. Minerals **8**(47), 1–11 (2018)
60. B.J. Youlton, J.A. Kinnaird, Gangue–reagent interactions during acid leaching of uranium. Miner. Eng. **52**, 62–73 (2013)
61. Unpublished Internal Investigation Reports, Mineral Processing Division, Bhabha Atomic Research Centre, Department of Atomic Energy, Government of India (2014 and 2016)
62. J. Lehto, The nuclear industry: ion exchange, in *Encyclopaedia of Separation Science* (Academic Press, 2000), pp. 3509–3517
63. D. Van Tonder, M. Kotze, Uranium recovery from acid leach liquors, in *ALTA 2007 Uranium Conference* (ALTA Metallurgical Services, Melbourne, Australia, 2007)
64. K.C. Sole, P.M. Cole, A.M. Feather, M.H. Kotze, Solvent extraction and ion exchange applications in Africa's resurging uranium industry: a review. J. Solv. Extr. Ion Exch. **29**(5–6), 868–899 (2011)
65. A.J. Brown, B.C. Haydon, Comparison of liquid and resin ion exchange processes for the purification and concentration of uraniferous solutions. CM Bull. **72**(805), 141–149 (1979)
66. E. Zaganiaris, *Ion Exchange Resins in Uranium Hydrometallurgy* (Books on Demand GmbH, Paris, 2009), 191 pp. ISBN 13: 9782810601882
67. M. Mikhaylenko, J. Van Deventer, Notes of practical application of ion exchange resins in uranium extractive metallurgy, in *ALTA 2009 Uranium Conference* (ALTA Metallurgical Services, Melbourne, Australia, 2009)
68. V. Yahorava, J. Scheepers, M.H. Kotze, D. Auerswald, Evaluation of various durability tests to assess resins for in-pulp applications, in *The Sixth Southern African Conference on Base Metals* (The Southern African Institute of Mining and Metallurgy, 2009)
69. Y. Xie, C. Chena, X. Ren, X. Wang, H. Wang, Emerging natural and tailored materials for uranium-contaminated water treatment and environmental remediation. Prog. Mater. Sci. **103**, 180–234 (2019)
70. E.C. Avelar, C.G. Alvareng, G.P.S. Resende, C.A. Morais, M.B. Mansur, Modelling of solvent extraction equilibrium of U (VI) sulfate with alamine 336. Braz. J. Chem. Eng. **34**(1), 355–362 (2017)
71. F.X. McGarvey, J. Ungar, The influence of resin functional group on the ion-exchange recovery of uranium. J. South. African Inst. Min. Metall. **81**, 93–100 (1981)
72. K.K. Beri, Jaduguda uranium mill—rich experience, in *Proceedings of International Seminar on Uranium Technology*, vol. 1 (Bhabha Atomic Research Centre, Department of Atomic Energy, Mumbai, India, 1989), pp. 431–462
73. D. van Tonder, D. Lunt, D. Donegan, *Selecting the Optimum IX System for Uranium Recovery*, ALTA Metallurgical Services, May 27–28, 2010, Perth, Australia (2010)
74. K. Soldenhoff, Uranium IX—Past, Present and Future, Keynote address in ALTA 2019, May 2019, Perth, Australia (2019)
75. M. Cox, J. Rydberg, Introduction to solvent extraction, in *Solvent Extraction Principles and Practice*, ed. by J. Rydberg, M. Cax, C. Musikas, G.R. Choppin (CRC Press, 2004), p. 480. ISBN 9780824750633
76. G.M. Ritcey, *Solvent Extraction-Principles and Applications to Process Metallurgy, Chapter 3.25, Uranium*, vol. 2 (G.M. Ritcey & Associates Incorporated, Ottawa, 2006)
77. J. Rajesh Kumar, J.S. Kim, J.Y. Lee, H.S. Yoon, A brief review on solvent extraction of uranium from acidic solutions. J. Sep. Purif. Rev. **40**(2), 77–125 (2011)
78. Z. Zhu, C.Y. Cheng, A review of uranium solvent extraction: its present status and future trends, in *ALTA 2011 Uranium*, May 26–27, Perth, WA, Australia (ALTA Mettalurgical Services, 2011)
79. I. Ivanova, K.S. Fraser, K.G. Thomas, M. Mackenzie, Uranium hydrometallurgy circuits-an overview, in *ALTA Uranium Conference 2009*, May 25–27, Perth Australia (2009)

80. J.E. Quinn, D. Wilkins, K.H. Soldenhoff, Solvent extraction of uranium from saline liquors using DEHPA/Alamine 336 mixed reagent. Hydrometallurgy **134–135**, 74–79 (2013)
81. A.W. Ashbrook, *Basic Uranium Extraction Chemistry, Extractive Metallurgy of Uranium Short Course* (University of Toronto, Toronto, 1978)
82. F.A. Forward, J. Halpern J, Developments in the carbonate processing of uranium ores. J. Metals **6**(12), 1408–1414 (1954)
83. S. Planteur, M. Bertrand, E. Plasari, B. Courtaud, J.P. Gaillard, Thermodynamic and crystal growth kinetic study of uranium peroxide. Cryst. Eng. Comm. **15**, 2305–2313 (2013)
84. R.J. Woolfrey, The Preparation and Calcinations of Ammonium Urinates (Australian Energy Commission, 1968). AAEC/TM 476
85. S. Manna, S.B. Roy, J.B. Joshi, Study of crystallization and morphology of ammonium diuranate and uranium oxide. J. Nucl. Mater. **424**, 94–100 (2012)
86. L.K. Kim, *U Mining and Milling, LLNL-TR-747582* (Lawrence Livermore National Laboratory, 2018), p. 17
87. R. Gupta, V.M. Pandey, S.R. Pranesh, A.B. Chakravarty, Study of an improved technique for precipitation of uranium from eluted solution. Hydrometallurgy **71**(3–4), 429–434 (2004)
88. *Safe Transport of Radioactive Material*, IAEA safety standards series no. SSR-6 (Rev. 1) (IAEA, Vienna, 2018), p. 165. ISBN 978-92-0-107917-6
89. N.K. Sethy, V.N. Jha, A.K. Sutar, P. Rath, P.M. Ravi, R.M. Tripathi, Dissolved radionuclide in the industrial effluent of uranium facilities, Jaduguda, India. Int. J. Low Radiat. **9**(3), 189–198 (2014)

Further Reading

1. W.D. Wilkinson, *Uranium Metallurgy*, vol. 1, Uranium Process Metallurgy (Interscience Publishers)
2. C.K. Gupta, H. Singh, *Uranium Resource Processing-Secondary Resources* (Springer, 2003)
3. K. Cohen, *The Theory of Isotope Separation as Applied to the Large-Scale Production of U-235*, Natl. Nuclear Energy Ser. Div. III, vol. 1B (McGraw-Hill Book Co., New York, 1951)
4. H.R.C. Pratt, *Counter-Current Separation Processes* (American Elsevier Publishing Company Inc., New York, 1967)
5. G. Avery, E. Davies, *Uranium Enrichment by Gas Centrifuge* (Mills & Boon Ltd., London, 1973)
6. S. Villani, *Isotope Separation* (American Nuclear Society, 1976)
7. Ratz, *Aerodynamic Separation of Gases and Isotopes, Lecture Series 1978* (Von Karmen Institute for Fluid Dynamics, Belgium, 1978)
8. M. Benedict, T. Pigford, H. Levi, *Nuclear Chemical Engineering*, 2nd edn. (McGraw-Hill Book Co., New York, 1981)
9. F.L. Cuthbert, *Thorium Production Technology* (Addison-Wesley Publishing Company, 1958)
10. Raj Kumar, S. Das, S.B. Roy, A. Renu, J. Banerjee, S.K. Satpati. Effect of Mo addition on the microstructural evolution of ϒ-U stability in Th-U alloys. J. Nuclear Materials **539**, 152317 (2020)

Chapter 3
Fabrication of Nuclear Fuel Elements

Sudhir Mishra, Joydipta Banerjee, and Jose P. Panakkal

3.1 Introduction

Fabrication of nuclear fuel elements is the step that transforms nuclear grade low-enriched uranium (LEU: <20% ^{235}U) in the form of UF_6 or natural uranium, plutonium, and thorium as nitrate or oxide into a highly engineered and encapsulated product that are tailored to specific needs of nuclear power or non-power reactors. Nuclear fuel assembly or fuel element designs are specific to a given reactor type and fuel management strategy. The components of "nuclear fuel elements" are as follows:

(i) Nuclear Fuel: which is a combination of a fissile material {$^{235}U_{92}$, $^{239}Pu_{94}$ or $^{233}U_{92}$) and a fertile material ($^{238}U_{92}$ or $^{232}Th_{90}$) in the chemical form of oxide, carbide, or nitride as small cylindrical "pellets" or tiny "microspheres" or as metal alloys, silicide and composites in the shape of pin or plate.

(ii) Cladding Material: which encapsulates the radioactive fuel and acts as containment and barrier for radioactive and health hazardous materials in the fuel, present initially and highly radioactive and health hazardous fission products and actinides formed during reactor operation, and prevents the fuel to come in direct contact with the coolant. Fuel cladding is considered to be an integral part of fuel element. Integrity of cladding material and zero fuel failure

S. Mishra · J. Banerjee
Radiometallurgy Division, Nuclear Fuels Group, Bhabha Atomic Research Centre,
Mumbai 400085, India
e-mail: sudhir@barc.gov.in

J. Banerjee
e-mail: joydipta@barc.gov.in

J. P. Panakkal (✉)
Formerly Advanced Fuel Fabrication Facility, Nuclear Fuels Group, Bhabha Atomic Research
Centre, Mumbai 400085, India
e-mail: jpanakkal@yahoo.com

during reactor operation is the common goal of fuel design and fabrication. The fabrication of clad materials for fuel elements is not covered in this book.

(iii) Fuel Element/Assembly: for clustering or packing the encapsulated fuel pins, fuel rods, or fuel plates with spacers and other structural components in specific geometric configurations for mechanical stability and efficient transfer of fission heat energy from fuel to coolant.

The cladding and other structural components of fuel element should have very low neutron absorption cross-section, high corrosion resistance against coolant, high resistance to irradiation damage, and excellent chemical compatibility with fuel and coolant.

The present chapter deals with fabrication of conventional and advanced fuel elements for:

(i) nuclear power reactors, including light-water-cooled and moderated reactors (LWRs), CANDU-type pressurized heavy-water-cooled and moderated reactors (PHWRs), sodium-cooled fast breeder reactors (SFRs) and high-temperature gas-cooled reactors (HTGRs),

(ii) non-power reactors with focus on reduced enrichment research and test reactors (RERTRs), highlighting the conventional and advanced manufacturing processes of fuel elements based on natural uranium isotopes (^{238}U + ^{235}U), mixed uranium plutonium (^{238}U + ^{239}Pu) and mixed thorium uranium (^{232}Th + ^{233}U) isotopes in the form of oxide, carbide nitride, metal alloys, silicide fuels, and composites.

3.2 Nuclear Reactors and Their Fuels

The design of nuclear fuels is very important for reliable, safe, and economic operation of nuclear reactors. The design and specifications are carefully made after considering various factors like type of reactors, fuel, material properties, power, reactor physics, safety, and other aspects. Nuclear fuel elements are the fundamental units of a nuclear reactor, and the core of a nuclear reactor consists of several groups of fuel elements (subassemblies or bundles). Optimum design, zero defect manufacturing technology, six sigma quality control of nuclear fuel and fuel elements, controlled operating parameters of the reactor and in-core fuel management should be synchronized in order to have zero fuel failure in reactor and in turn reliable, safe and economic operation of reactors. Radiological safety and nuclear criticality safety are also of paramount importance in nuclear fuel fabrication facilities in order to keep radiation risks to plant personnel and the public and to the environment to a minimum and avoid inadvertent criticality accident. This aspect is covered in detail in Chap. 11.

The fissile isotopes commonly used in nuclear reactors are ^{235}U, ^{239}Pu, and ^{233}U. Naturally occurring uranium containing only 0.7% ^{235}U is used in pressurized heavy-water reactors (PHWRs/CANDU). Uranium enriched in ^{235}U is used in light-water reactors (LWRs) and pressurized water reactors (PWRs). Sodium-cooled fast reactors (SFRs) use ^{239}Pu or ^{235}U-based fuel. Naturally occurring fertile isotopes ^{238}U and ^{232}Th are converted to fissile ^{239}Pu and ^{233}U, respectively, during the operation of reactors.

The common types of fuel used are in the form of high-density cylindrical pellets, e.g., natural UO_2, (U–Pu) MOX (Mixed Oxide) fuel encased in thin clad tubes (zirconium alloy, stainless steel). The fuel commonly used for research reactors which operate at lower temperature and power levels is metallic uranium and Al matrix dispersion fuel with aluminum alloy clad.

3.3 Nuclear Non-power Reactors and Power Reactors

Nuclear reactors are classified into two categories, viz., non-power or research reactors and power reactors [1, 2]. The reactors used mainly for testing fuel, cladding, training etc., besides producing isotopes for medical or industrial applications are research reactors. The power reactors which are operated for producing power are classified into thermal and fast reactors. The thermal neutron reactors use low-energy neutrons (thermal neutrons) for fission reaction. The thermalization of neutrons is achieved by using moderators. The fast neutron reactors do not use any moderator, and the fission is by fast neutrons.

Nuclear fuel for any type of reactor should meet a number of basic design criteria. The fuel should deliver the designed power expected from the reactor during the irradiation cycles (time interval of the operation of the reactor between successive loadings of the fuel). The fuel must be designed for adequate removal of heat maintaining required reactivity during its residence in the reactor core. The fuel should have high burn up capability, high density, high melting point, high thermal conductivity, and moderate coefficient of linear expansion. The fuel should have good irradiation stability. There should not be any phase change till the melting point. It should have good compatibility with clad and coolant. The fuel should have easy fabricability. The cladding of the fuel element which is the first containment barrier should be chosen such that the fission products are confined within the fuel element during normal operating conditions. The cladding material should have low neutron absorption cross-section, resistance to high temperature oxidation, and enough high temperature strength and ductility. The commonly used cladding materials are listed in Table 3.1. The other barriers are the primary circuit and the containment building. The corrosion properties of the clad and neutron economy are also considered in finalizing the material and dimensions of the clad. Cladding failure may occur at extreme accidental conditions. The fuel material is chosen to have optimum utilization of the fissile material ensuring economic operation of the reactors. In short, many factors are considered in designing the fuel depending on the type of reactors.

Table 3.1 Various types of fuel and cladding used in different reactors [3]

Type	Conventional fuel	Advanced fuel	Cladding	Burn up (GWd/tHM)
Non-power thermal reactor	1. Al-13–15%HEU or Pu 2. Al-50%U_3O_8(HEU) 3. U metal	1. Al–U_3 Si_2(LEU) 2. Al-U_6Fe(LEU) 3. U–Mo alloy dispersed in aluminum	Al	1–2
PWR BWR	3–4% enriched UO_2	$UO_2 \leq 4\%$ PuO_2	Zircaloy-4 Zr–Nb Zircaloy-2	40–50 25–30
PHWR	Natural UO_2	1. SEU or DEU (0.9–1.2% ^{235}U) 2. UO_2-0.4% PuO_2 3. ThO_2- $\leq 4\%$ PuO_2 4. ThO_2- $\leq 2.5\%$ $^{235}UO_2$	Zircaloy-4	8–10
LMFBR	(UPu)O_2 (20–25% PuO_2)	He or Na-bonded (U, Pu)C or (U, Pu)N Na-bonded U–Pu–Zr (15–20% Pu, 6–10% Zr)	D-9	70–100

3.3.1 Research Reactors

The common design of research reactors is generally swimming pool or tank type. Plate-type dispersion fuel or cylindrical fuel elements are immersed in a pool of water. The research reactors operate at low temperature and are compact in size. The ratio of the fuel surface area to volume is large for such reactors, and hence, neutron losses are high. So, the fuel has high heavy metal density and high enrichment to compensate neutron losses. The fuel chosen for such reactors is therefore generally metallic or dispersion type. The enrichment of the fuel is up to 20% ^{235}U, and it may be as high as 93% ^{235}U in some of the older reactors. The fuel operates at low surface temperature and moderate neutron flux. The moderators used for reducing the energy of neutrons in such reactors are generally heavy water or graphite. These reactors usually operate with thermal neutrons, but research reactors with fast neutrons are also being operated in small number in different countries. Some of them are namely BOR-60 of Russia and FBTR of India. These types of reactors are used for irradiation studies of different types of fuel and cladding materials at higher fluence. Fuels in liquid form have also been used in some of the research reactors called aqueous homogenous reactors (AHRs). Here, a fuel material like uranyl nitrate is mixed with moderator, and the fission products produced during the operation of reactors are removed from the aqueous solution intermittently. Isotopes for various applications are obtained by separation from the liquid fuel. Different types of fuels and cladding materials used in the reactors are as listed in Table 3.1 [3].

For most non-power reactors using highly enriched ^{235}U fuel (HEU), the main challenge is to ensure proliferation resistance by replacing the use of nuclear weapon grade (>90% ^{235}U) high-enriched uranium (HEU: $\geq 20\%$ ^{235}U) metallic fuel with low-enriched uranium-based (LEU: <20% ^{235}U) dispersion or monolithic fuels with high uranium density that will match the high neutron flux and other parameters of like HEU-based fuels. Accordingly, a global Reduced Enrichment Research and Test Reactor (RERTR) program is underway. Fabrication of dispersion fuels containing metallic and ceramic fuel bearing particulates and uranium metal fuel for non-power reactors are covered under dispersion fuels and metallic fuels later in this chapter.

3.3.2 Thermal Neutron Power Reactors

Thermal reactors are either pressure vessel or tube type in design and use thermal neutrons for fission. The neutrons produced during fission have high energy associated with them, and their higher probability of interaction with fuel is achieved by reducing the energy. This is carried out by using moderators like light water, heavy water, or graphite. Different types of thermal reactors, namely boiling water reactors (BWRs), pressurized water reactors (PWRs), pressurized heavy-water reactors (PHWRs), advanced gas-cooled reactors (AGRs), and light-water graphite-moderated reactors (RBMKs) are being operated in different countries. The heavy-water and graphite-moderated reactors use natural uranium whereas light-water-moderated reactors use slightly enriched uranium as fuel. The heat extraction in water-moderated thermal reactors is accomplished by water. In light-water-cooled graphite-moderated reactors (RBMK), the cooling is achieved by water where as in AGR, CO_2 gas is used as the coolant. PHWR uses natural uranium oxide pellets encapsulated in zircaloy cladding as fuel. The fissile content in this fuel is only 0.7% ^{235}U. The use of heavy water both as coolant and moderator in PHWRs facilitates neutron economy that makes it possible for the scantly available U^{235} atoms of natural uranium to undergo fission. Mixed U–Pu oxide (MOX) fuel assemblies are also being used to increase the burn up in some BWRs and PHWRs by replacing some of the UO_2 assemblies with MOX assemblies.

Pressurized water reactors (PWRs) use low-enriched UO_2 as fuel. The enrichment is generally in the range of 3–4% of ^{235}U in UO_2. Both the moderator and coolant are maintained as light water because of increased availability of fissile uranium atoms in such fuel. PuO_2 in the range of 3–4% along with UO_2(Mixed Oxide, MOX) is used to increase burn up in some PWRs.

Light-water-cooled and moderated reactors (LWRs) are most common, accounting for nearly 85% of the power reactors. The LWRs consists of ~65% pressurized water reactor (PWR) and their Russian version VVER and ~20% boiling water reactors (BWRs). The PWRs are popular in USA, France, Russia, Ukraine, China, and RO Korea. The BWRs are common in USA, Japan, Sweden, and Germany. Pressurized heavy-water-cooled and moderated reactors (PHWRs) contribute to some 10% of the reactors and are the backbone of the nuclear power program in Canada and India. Graphite-moderated advanced CO_2 gas-cooled reactors (AGRs) are in operation only in UK. Very few graphite-moderated water-cooled reactors (RBMK) are in operation in Russia and Ukraine. AGR and RBMK reactors are being phased out.

3.3.3 Fast Neutron Reactors

Fast neutrons are used to sustain fission chain reaction in fast reactors. Fuel for fast reactors has high enrichment to facilitate interaction of high-energy neutrons with fissile atoms. The core size of fast reactor is compact, and power density is high. The extraction of heat is accomplished by highly conducting coolants like Na, Pb, Pb–Bi, etc. When fast reactors are designed to generate more fissile material than they consume, they are known as fast breeder reactors (FBRs). This is achieved by arranging a blanket of fuel pins consisting of depleted fuel pellets around the main core. The depleted uranium discharged from thermal reactors mostly contains ^{238}U which can be used by fast breeder reactors to produce fissile ^{239}Pu, leading to enhanced utilization of uranium. The term doubling time associated with fast reactors is the period in which the fissile material in a fast reactor is produced to an extent that it is enough for recharging the reactor in use and also for another fast reactor of equivalent capacity. The generation of radioactive waste is reduced in fast reactors because minor actinides are produced at lower rate due to lower capture to fission and also consumed through fission. The plutonium produced in the blanket material has high Pu-239 content because of lesser burn up.

Most commonly used fuel for fast reactors is (U, Pu)O_2 (MOX) fuel. PuO_2 (20–25%) is mixed with natural or depleted UO_2 in MOX fuel [4]. The fuel has good compatibility with Na and Pb coolant. The fabrication flow sheet and reprocessing experiences of MOX fuel is well established. However, the problem of this type of fuel is its lower thermal conductivity and lower density. The breeding ratio of the fuel is low, and it has high doubling time.

The other types of fuels considered for fast reactor application are metallic, mixed carbide, and mixed nitride fuel. Metallic fuel U–Pu–Zr has been used in experimental reactors. The advantage of this fuel is its high thermal conductivity and high breeding ratio. But, the performance of the fuel is hampered due to its high swelling characteristics. Mixed carbides (U, Pu) C and nitrides (U, Pu) N are better fuels than oxide with reference to thermal conductivity, high metal density, and higher breeding ratio.

3.4 Types of Fuels

As described above, a number of different types of fuels are used in nuclear reactors, and technologies have been developed for fabrication of the fuels in a safe and economic manner. The fuels are broadly classified into ceramic (oxide and non-oxide), dispersion and metallic fuels. The following paragraphs briefly describe the fabrication of the fuels.

3.4.1 Ceramic Fuels

Ceramic fuels which have been used in different types of reactors maybe further classified as follows:

1. Oxide Fuel: (i) UO_2, (ii) Mixed Oxide Fuel- $(U, Pu) O_2$, $(Th, {}^{233}U)O_2$, $(Th, Pu)O_2$.
2. Non-Oxide Fuels: (i) Mixed Carbide $(U, Pu) C$, (ii) Mixed Nitride $(U, Pu) N$.

3.4.1.1 Oxide Fuels

a. UO_2 Fuel

UO_2 is the most commonly used fuel in thermal reactors. The production of UO_2, starting from mining of uranium ores, has been dealt with in Chap. 2. UO_2 powder is manufactured by the following methods.

1. Ammonium di-uranate (ADU) route.
2. Ammonium uranium-carbonate (AUC) route.
3. Integrated dry or wet route (IDR).
4. Uranyl nitrate denitration route.

Ammonium Di-Uranate (ADU) route

ADU powder is produced from uranyl nitrate solution by passing NH_3 gas in to the solution which results in precipitation of the powder [5]. ADU is subjected to calcination at temperature between 873–973 K and reduction at 923 K to get sinterable grade powder. The freshly prepared UO_2 powder is pyrophoric and difficult to handle. It is therefore stabilized at room temperature by controlled oxidation in CO_2. The stabilized UO_{2+x} powder is easy to handle. The powder produced by ADU route is non-free flowing.

Ammonium Uranium Carbonate (AUC) route

AUC process starts with either UF_6 or uranium nitrate hexahydrate (UNH) as the starting material [6]. Cylinders containing UF_6 are heated with steam for evaporation. CO_2 and NH_3 are injected into the system containing UF_6. This results in the formation of uranyl ammonium carbonate which precipitates as yellow crystal in slurry form. The slurry is dried to produce AUC powder. The powder is relatively coarse and free flowing having size in the range of 10–20 μm compared to 3–4 μm obtained from ADU route. Because of the coarse nature of the powder, steps like pre-compaction and granulation are not required during the fabrication of fuel pellets.

Integrated Dry or Wet Route (IDR)

Enriched UO_2 $({}^{235}U > 0.7$ wt%) powder is produced from UF_6 obtained from the enrichment plant either by dry route, e.g., integrated dry route (IDR) or wet route. In IDR route, UF_6 gas from the enrichment plant is first treated with steam and then followed by hydrogen treatment at 923–973 K to produce UO_2 powder [7]. In the wet route, UF_6 is hydrolyzed to oxyfluoride and then treated with ammonia to get ADU followed by calcination, reduction, and stabilization to get UO_2 powder.

Denitration of Uranyl Nitrate

In this method, uranyl nitrate solution is thermally decomposed, and oxides of uranium are obtained in a single step unlike ADU which involves many steps to get the final product. This method uses fluidized bed reactor for denitration. To begin with, U_3O_8 particles in a particular range are used as bed material. The bed temperature is set around 673 K. Uranyl nitrate solution is sprayed over the bed material, and due to the heat of the fluidized bed, the solution decomposes releasing moisture and volatile gases. UO_3 is obtained as the product. It is reduced to UO_2 using NH_3 in static bed reduction furnace at around 1023 K temperature. Thus, several steps such as precipitation, filtration, drying, and calcinations involved in ADU route are avoided in this process, and also, it offers savings in cost.

Production of sintered pellets and fuel bundles

The production of sintered UO_2 pellets is carried out following the conventional powder pellet route. The flow sheet for fabrication of sintered pellets and fuel elements is presented in Fig. 3.1. The different process steps followed for this route of production are pre-compaction of fine powder, granulation, final compaction followed by sintering at high temperature [5]. The green UO_2 powder is first pre-compacted and granulated through a sieve to increase the bulk density and flowability of the powder. The granulated powder is admixed with lubricant and binder in a blender. The addition of lubricant eliminates interparticle friction and friction between die and powder particles during final compaction. Binder addition provides handling strength to the pellets. The green density is maintained at 55% TD (TD—theoretical density of UO_2 is 10.96 gm/cm³ {10,960 kg/m³}). Sintering of the green pellets is carried out at 1923 K. During sintering, intermediate soak of pellets is maintained. First soak is set at 473 K for 2 h to remove moisture from the pellets. The subsequent soaks are provided at 1073 K for 2 h and 1923 K for 4 h.

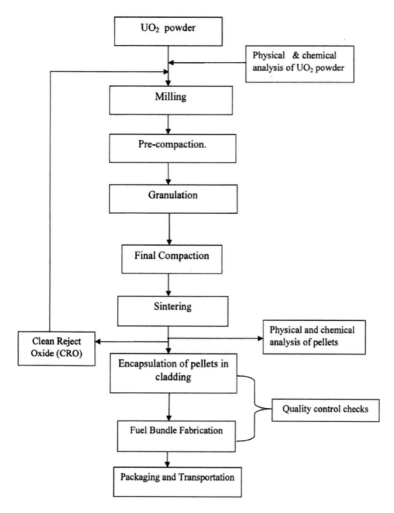

Fig. 3.1 Flow sheet for fabrication of UO₂ fuel bundle

The soak at 1073 K is provided for obtaining O/M ratio of fuel near 2. The sintered pellets of density ranging between 94 and 98% TD are required for different reactors depending upon the burn up. The accepted pellets after quality control checks are encapsulated in clad tubes to fabricate the fuel pins. The clean reject oxide (CRO) obtained after quality control checks on pellets are recycled. The pins are assembled to manufacture fuel bundles which are of different designs depending on the type of the reactors.

b. Mixed Oxide (MOX) Fuel

Mixed oxide (MOX) fuel contains more than one oxide material, e.g., fissile PuO_2 is blended with natural UO_2 or ThO_2 to produce Urania or thoria-based MOX fuel.

MOX fuel is a good alternative to low-enriched uranium oxide (LEU) used in light or boiling water reactors. The development of mixed oxide (MOX) fuel fabrication technology for recycling of reprocessed fissile PuO_2, $^{233}UO_2$ material in uranium or thorium matrix is important for the implementation of the closed fuel cycle. The fuel fabrication methodology depends upon fuel characteristics that include radioactivity, radiotoxicity, reactivity of the material with the environment and fuel specifications [8]. The worldwide acceptance of oxide fuel is well established due to its ease of fabrication, good irradiation resistance in reactor, and easy reprocessing.

The fabrication of Pu-bearing MOX fuel demands ventilated leak-tight glove boxes, as plutonium is an alpha-active material having high specific activity and high biological half-life. Handling of aged or high % Pu-bearing MOX fuel necessitates shielding of the glovebox (Fig. 3.2). The aged plutonium contains a significant amount of ^{241}Am which is a gamma emitter. The spontaneous fission in the even isotopes of plutonium and (α, n) reaction with a light element like O_2 and N_2 provides sources for emission of neutrons. The most important aspect of Pu handling is the possibility of criticality which is avoided by regulating the mass of the material handled, geometry of the container, and administrative regulation. Handling of ^{233}U-bearing MOX poses challenge to fuel fabricators, as reprocessed ^{233}U is always associated with ^{232}U whose daughter product emits strong gamma radiation [9, 10]. The fabrication of $(Th, U)O_2$ MOX fuel using conventional powder metallurgy process is difficult. It requires heavy shielding in all steps of fuel fabrication. These fine powders also have poor flow ability, which makes automation and remote fuel fabrication difficult. This has necessitated development of other processes like coated agglomerate pelletization (CAP), sol–gel microsphere pelletization (SGMP), impregnation, etc. The various processes available for the $(U, Pu)O_2$ and $(Th, U)O_2$ MOX fuel fabrication are described in the following paragraphs:

1. Powder oxide pelletization (POP).
2. Co-precipitation technique.
3. Sol–gel microsphere pelletization (SGMP).
4. Impregnation.
5. Coated agglomerate pelletization (CAP).
6. Impregnated agglomerate pelletization (IAP).
7. Vibratory compaction (Vi-Pac) process.

Powder oxide pelletization (POP)

This route is followed for commercial fabrication of MOX fuel. Figure 3.3 presents the flow sheet mentioning different steps followed during the fabrication of MOX fuel by POP route.

Mixing and milling

The powders are weighed in proper proportion and subjected to mixing and milling in a high stirred ball mill known as attritor. The mixing and milling of the powders result in homogenization, reduction in the particle size, and increase in the specific

Fig. 3.2 Glove box train for fabrication of plutonium-bearing fuels

surface area, apparent and tap density resulting in uniform die fill. Homogeneous mixing of the powders avoids formation of fissile-rich agglomerate, which may act as a hot spot and affect the performance of the fuel in the reactor [11, 12]. Lubricants and binders are added in the powder to reduce friction during compaction and provide strength to pellets [13]. Suitable quality control/process control checks are employed to ensure correct proportion of PuO_2 in the mixed powder [14].

Pre-compaction and granulation

The pre-compaction and granulation steps are needed to improve the flow ability of MOX powder into the die during final compaction. It helps in achieving uniform density of the pellets in the final compaction step. Hydraulic or mechanical press is used for pre-compaction. The granulator produces granules by breaking pre-compacted green pellets.

Final Compaction

The hydraulic or mechanical rotary press is employed to compact the granules into the pellets (Fig. 3.4). The pressure applied during final compaction is nearly three times the pre-compaction pressure. It ensures breaking of hard granules and better contact between the powder particles.

Sintering

The sintering of MOX fuel pellet is carried out in resistance heating furnace under a controlled reducing atmosphere ($N_2 + 7$ Vol% H_2) at approximately 1923 K. This

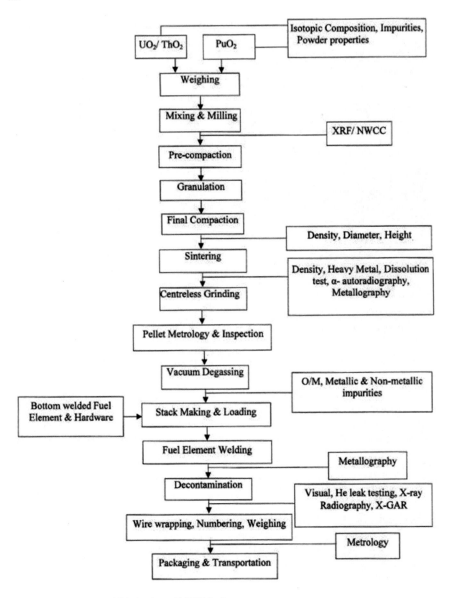

Fig. 3.3 Flow sheet of fabrication of MOX fuel

step provides enough strength and density to the pellets. Oxygen to metal ratio (O/M) is an important parameter of the fuel pellet, and it is achieved by adjusting the oxygen potential, i.e., (H_2O/H_2) ratio of the sintering atmosphere. The gross in-homogeneity in the sintered MOX pellets arising due to improper mixing of powders can lead to problem in the dissolution during reprocessing and hot spot problem during reactor operation. Hence, the evaluation of the macro- and micro-homogeneity of the sintered

Fig. 3.4 Rotary press used for pelletization

MOX pellet is ensured by the dissolutions and α-autoradiography, respectively. Figure 3.5 presents a photograph of annular MOX pellets for fast reactors.

Centerless Grinding

Oversize pellets are centerlessly ground to get the allowed range of diameter. The system consists of two wheels known as regulating and grinding wheels. It differs from normal grinding operation as no spindle or fixture is used to locate and secure pellets. The pellets are placed between the two wheels, and the speed of their rotation relative to each other determines the rate at which material is to be removed from the pellet. The pellets are fed into the system from one end, and after grinding, they come out from the other end.

Inspection of Pellets

The sintered MOX pellets are inspected for density, dimensions, and visual defects. The visual defects include lamination cracks, end chipping, end capping, pits, and

Fig. 3.5 Annular MOX
pellets for fast reactors

blister. They are subjected to both destructive and nondestructive tests as explained in Chap. 4.

Degassing

The degassing of fuel pellets and hardware is carried out to eliminate moisture content. The degassing is carried out at a temperature of 373 K and at a vacuum of 10^{-2} torr. The degassed sintered MOX pellets are analyzed for O/M, metallic and non-metallic impurities.

Stack making

The degassed pellets both fissile and insulation are arranged to form a linear stack meeting the requirement of stack length and weight. Equipment like vibratory bowl and linear feeders are used for stack making and loading of the pellets. The degassed pellets are loaded to vibratory bowl feeder, which arranges the pellets on the linear feeder. The stack of the pellets is prepared on the V-block fitted with a linear feeder. The linear feeder pushes the pellets into one end welded fuel tubes. The hardware such as spring support and plenum spring, etc., is loaded into the fuel tube.

End Plug welding

Gas tungsten arc welding (GTAW) is commonly used for end plug welding of the fuel elements. It is autogenous welding, as the thickness of the tube is less. Argon and helium gases are used as a shielding gas for the bottom and top-end plug welding, respectively. The loaded fuel elements are aligned inside the leak-tight welding chamber and gripped using the collet. The welding chamber is evacuated and pressurized with inert gas for 2–3 times subjected to 2–3 cycles of evacuation and pressurization. The arc generated on tungsten electrode performs welding (Fig. 3.6). During top-end plug welding, helium as bond gas fills the clad at 1–2 kg/cm^2 pressure. Laser welding is also being adopted for end plug welding in some of the manufacturing plants.

Decontamination

The process of decontamination uses ultrasonic cleaning method with demineralized water as medium. This step ensures that there is no loose contamination on the fuel elements and that the fixed contamination if any is below the allowed limit. Laser decontamination technique is also adopted in a few manufacturing plants.

Inspection of Fuel elements

The fuel elements are subjected to various quality control checks like leak testing, X radiography, visual examination, and metrology checks which are covered in detail in Chap. 4.

Co-precipitation Technique

This technique involves mixing of uranyl and plutonium nitrate solution in the required ratio. Addition of hydrazine to the solution reduces the oxidation state of uranium and plutonium, which facilitates co-precipitation. U(VI) is reduced to U(IV) and Pu(IV) to Pu(III). The addition of oxalic acid results in co-precipitation [15]. The precipitate thus obtained is then dried and calcined in air. Plutonium is re-oxidized from Pu(III) to Pu(IV) in the calcination process. The homogeneous MOX powder becomes the starting material for POP route of fuel fabrication. The

Fig. 3.6 Welding of fuel pins inside glove box

MOX pellets obtained from the co-precipitated powder show a high degree of homogeneity. However, the commercial adoption of co-precipitation route is not attractive as it involves handling of the large amount plutonium bearing liquid waste.

Sol–gel microsphere pelletization (SGMP) process

SGMP process uses a dust-free, free-flowing sol–gel-derived microsphere as a starting material for fabrication of MOX fuel pellets [16]. In the internal gelation process, the nitrate solutions of the fuel (U, Pu, Th) metal are mixed with hexamethylenetetramine (HMTA) at 273 K along with urea to from the feed solution. The droplets of the feed solution are gelled into spherical particles after coming in contact with silicon oil maintained at 378–383 K. HMTA on hydrolysis releases ammonia at 363 K, which causes precipitation of the metal ion as hydrated oxide gel. CCl_4 washing removes the silicon oil. The addition of carbon in the initial nitrate solution results in production of porous oxide microspheres. The process of calcination and H_2 reduction produce the soft, porous oxide microspheres.

The advantages of SGMP process are its amenability for automation and remote handling, excellent micro-homogeneity of fuel, and minimal powder handling. However, disposal of a large amount of Pu/^{233}U-bearing active liquid waste generated during the process is a matter of concern.

Impregnation process

This is an attractive technique for fabrication of ^{233}U or Pu-bearing thoria-based MOX fuel pellets. The flow sheet of fabrication is as shown in Fig. 3.7 [10, 17]. In this process, low density, ThO_2 pellets (<65% TD), or porous ThO_2 microspheres prepared by the sol–gel process are used as a starting material [17, 18]. The use of pore former produces low-density porous microspheres.

The low-density ThO_2 pellets or microspheres are subjected to uranyl nitrate (^{233}U) or plutonium nitrate solution under vacuum. The solution enters the pellets/microspheres because of capillary forces. The impregnated pellets or microspheres are subjected to calcination in the air to convert the nitrate solution in the desired chemical form. U/Pu loading in the pellets/microspheres is dependent upon the density of the pre-sintered pellets, the concentration of solution and duration of impregnation. This process avoids handling of fine ^{233}U/Pu-bearing powders and restricts handling of ^{233}U/ Pu to only few steps. However, multiple impregnation and calcination are required for achieving higher percentage loading of fissile content. The fissile concentration in the peripheral region of the pellet remains higher than that of the interior portion [17].

Coated agglomerate pelletization (CAP) Process

This process has been developed for the fabrication of thoria-based MOX fuel containing $^{233}UO_2$ or PuO_2. The performance of different steps of the process flow sheet needs shielded or unshielded facility as shown in Fig. 3.8. Attritor machine mills the virgin ThO_2 powder, to break its platelet morphology. The mixing of ThO_2 powder with organic emulsion (binder, lubricant, and water) in the desired proportion yields a paste. The roller extruder uses the paste for extrusion. The extrudes made

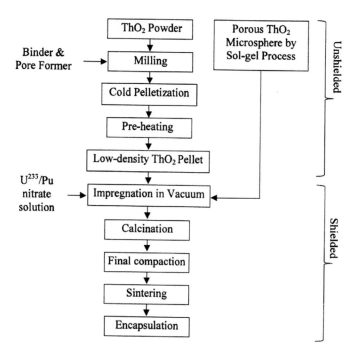

Fig. 3.7 Flow sheet based on impregnation process for manufacturing (Th, U)O$_2$/(Th, Pu)O$_2$ fuel

out of the paste from the roller extruder are dried in an oven. The dried extrudes are broken and spheriodized by a checkered plate rotating at an optimum speed in spherodizer. The tablet coating pan rotating at a suitable rpm does desired coating of calcined ^{233}UO$_2$/PuO$_2$ powder over ThO$_2$ spheroids. The coated spheroids are compacted and sintered in reducing (N$_2$+ 7 vol % H$_2$) atmosphere. The CAP process has advantages over POP process like less powder handling, reduction of the number of steps for shielded operation, and reduction of man-rem to the operators [19].

Impregnated agglomerate pelletization (IAP) process

This process involves spray coating of uranyl nitrate solution over ThO$_2$ spheroids obtained through the extrusion spherodization route [19]. The coated spheroids are dried, final compacted, calcined, and then sintered in reducing atmosphere to obtain high-density pellets. The problems associated with the CAP process like sticking of ^{233}UO$_2$ powder on equipment and glove boxes during coating, non-uniform microstructure, and inhomogeneous fissile concentration in the thoria matrix are resolved in this process [20]. Use of Uranyl nitrate in solution form takes care of the issues mentioned earlier. The solution easily penetrates into ThO$_2$ spheroids during coating in contrast to UO$_2$ powder, which mostly sticks to the external surface of spheroids. This results in uniform microstructure and good homogeneity of the fuel pellets fabricated by the IAP route.

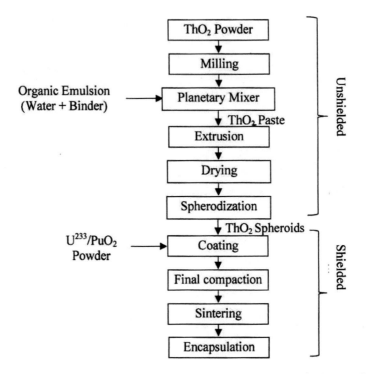

Fig. 3.8 Flow sheet based on coated agglomerate pelletization (CAP) process for manufacturing MOX fuel

Vibratory compaction process

In this process, different-sized sol–gel-derived microspheres produced by SGMP route are used. The microspheres are placed in the clad tubes by vibratory compaction method [10]. The maximum achievable smear density by this process is 90% TD. This process is amenable to automation and remote handling. The problems associated with this process are segregation of the fine powder in the clad tubes, high center-line temperature of the fuel pin, and fuel wash out in case of any clad breach.

c. **Non-Oxide Fuel**

Non-oxide fuels include mainly carbide and nitride of uranium and solid solution of uranium with plutonium. These fuels are regarded as advance and alternative fuel being developed for FBRs because of their high heavy metal density and high thermal conductivity. The experience of operation and irradiation with these fuels is very less worldwide. India is the only country in the world which has used uranium–plutonium mixed carbide fuel as driver fuel [21] for Fast Breeder Test Reactor (FBTR) at Kalpakkam though carbide fuel has been used in experimental pins in some of the reactors in other part of the world.

Nitride fuel is considered better as compared to carbide fuel. It is not as pyrophoric and reactive as carbide. Hence, it does not need ultra-high purity inert gas like N_2,

Ar, or He. A nitride fuel core gives better neutron economy than an oxide fuel core because of harder neutron spectrum. The problem associated with reprocessing of this fuel is radioactive ^{14}C formed by (n, p) reaction in ^{14}N [22]. The production of ^{14}C can be avoided provided ^{15}N is used instead of ^{14}N during fuel fabrication. This would certainly add to fuel economy because of enrichment process involved in production of ^{15}N. The fabrication of carbide and nitride fuel involves a greater number of process steps and a cover of inert gas inside glovebox to take care of pyrophoricity issue involved with these fuels.

Mixed (U, Pu) C and (U, Pu) N fuel

The fabrication of MC (M stands for solid solution of U and Pu) and MN fuel involves more steps as compared to oxide fuel besides the requirement of inert gas cover, as they are highly susceptible to hydrolysis and oxidation. The fabrication of mononitride is easier as the higher nitride phases get dissociated at higher temperatures, unlike carbide fuel where the higher carbide phases are more stable.

MC and MN are synthesized by (a) melting of uranium, plutonium and carbon, (b) production of metal powder from bulk metal by hydriding and de-hydriding followed by treatment with methane/propane and nitrogen to produce carbide and nitride, and (c) carbothermic reduction of oxide powder of metal and graphite [22–24].

Fabrication of MC and MN from the oxides is the most convenient route for mass production. Carbothermic reduction of UO_2, PuO_2, and graphite is carried out in static bed. The overall reaction followed during the process is as follows:

$$MO_2 + 3C = MC + 2CO \uparrow \tag{3.1}$$

UO_2, PuO_2, and graphite are weighed in stoichiometric proportion, mixed homogeneously and then subjected to heating at high temperature in vacuum atmosphere (Fig. 3.9).

In single-step synthesis, the end product is always associated with M_2C_3, and it is difficult to avoid the formation of this product [25]. It is practically not possible to fabricate relatively oxygen free MC. Presence of oxygen and nitrogen in MC acts as carbon equivalent. These atoms replace carbon from the MC lattice to form higher carbide phase. The two-step synthesis is an improvement over single-step synthesis. It aims at fabricating MC with low nitrogen and oxygen and less Pu volatilization losses. In two-step synthesis, carbon is taken more than stoichiometric value so that only M_2C_3 is formed. In the second stage, M_2C_3 formed is reduced by H_2 to yield relatively oxygen and nitrogen-free MC. The disadvantage of this process is that time taken to complete the process is more and is associated with explosion hazards of hydrogen.

The formation of MN takes place as per the following reaction:

$$MO_2 + 2C + \frac{1}{2}N_2 = MN + 2CO \tag{3.2}$$

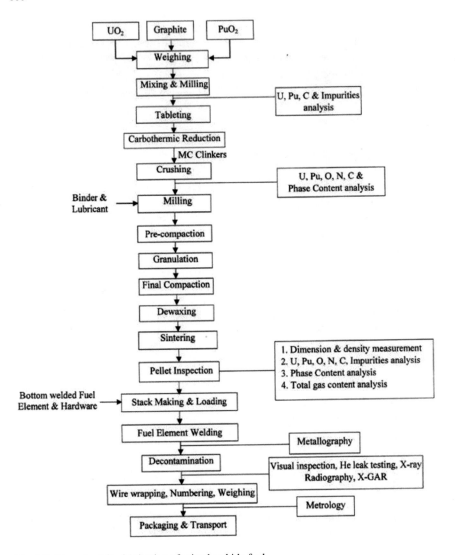

Fig. 3.9 Flow sheet for fabrication of mixed carbide fuel

The three powders are mixed homogeneously, and then, tablets made of the powder are loaded in a furnace for the nitriding process to high temperature in flowing nitrogen and hydrogen gas. After the completion of nitriding process, the product is cooled in the same atmosphere. High purity of nitrogen is maintained during reaction. The product is then analyzed for oxygen and carbon.

After preparation of MC and MN, the clinkers are subjected to crushing and milling in stirred ball mill for 2 to 3 h. Zinc behenate and naphthalene are mixed with the milled powder as binder and lubricant, respectively. The powder is then subjected to pre-compaction at 150 to 200 MPa followed by granulation through 30 mesh sieve. The granules so obtained are compacted at 250 to 300 MPa. In the case of MC, the green pellets are sintered at 1923 K for 4 h in $Ar+H_2$ atmosphere. Sintering is carried out in vacuum or $Ar+N_2$ atmosphere for fabrication of MN pellets.

3.4.2 Dispersion Fuel

Dispersion fuels consist of metallic or ceramic fuel bearing particulates dispersed in either metallic or ceramic matrix. The dispersion fuel isolates the fuel from the substantial volume of the matrix. The localized fission in the dispersed fuel particle minimizes irradiation-induced damages without affecting matrix properties. The selection of dispersion fuel and matrix is important for achieving the desired properties of dispersion fuel and its irradiation behavior in the reactor. Ceramic fuels are commonly used in thermal and fast reactors due to its numerous advantages like ease of fabrication, better irradiation resistance, and easy reprocessing. However, it has some shortcomings such as low thermal conductivity, low heavy atom density, higher fission gas release, and so on. The dispersion fuel can be designed to have optimal favorable properties of the ceramic, metallic, or alloy fuel. Though dispersion fuel is commonly used in the research reactors, it is a potential candidate for thermal, fast, and small-size reactors. The dispersion fuels are classified as (i) METMET dispersion fuels (ii) CERMET dispersion fuels (iii) CERCER dispersion fuels and (iv) Coated particle dispersion fuels.

METMET Dispersion Fuel

METMET fuel refers to dispersion of the metallic fuel in a metallic matrix. This type of fuel has been considered as alternative fuel for light-water and research reactors. U-9Mo alloy in Mg matrix is an example of such fuel. The other examples of this type of fuels are UAl_x, U–5Zr–5Nb, U_3Si_2, U–9Mo dispersed either in Al or Zr matrix. Uranium and its alloys under dispersion condition show better corrosion and swelling resistance than that of the bulk fuel materials [25]. A protective matrix and electrochemical passivation of the fuel particle gives the improved corrosion resistance. The swelling of fuel particles is restrained because of the matrix.

Highly enriched UAl_x dispersed in Al matrix has been extensively used in research reactor cores for production of high neutron flux. Plutonium–aluminum alloy has also been used as fuel in some research reactors. The fuel is operated at low temperature compared to power reactors. The fuel has plate geometry, and it is fabricated by "Picture-frame" technique. Figure 3.10 presents the flow sheet followed to fabricate plate fuel by this technique. In order to comply with framework of international Reduced Enrichment for Research and Test Reactor (RERTR), development of low-enriched high-density fuel has been pursued in different countries. The main objective of the program is to develop the technology needed to replace use high-enrichment uranium (HEU) with low-enrichment uranium (LEU) instead of in research and test reactors, and to without significantly affecting the performance, economic, or safety aspects of the reactors. The objective has been achieved by using high fuel volume fraction and high-density fuel alloy such as U_3Si_2, U_3Si, and U_6Fe instead of UAl_x. The fabrication of U_3Si_2 by powder metallurgy route has also been carried out.

Vibrocompaction followed by impregnation is another method used for fabrication of METMET fuel. U–5Zr–5Nb, U_3Si, and U–9Mo alloy powders are prepared from their alloy via centrifugal atomization process. The various sizes of atomized alloy powder particles are filled in the cladding tube by vibrocompaction process, and then, it is impregnated with a molten matrix material like eutectic Zr alloy or Al. The controlled cooling of the fuel rod is carried out after impregnation.

CERMET dispersion fuel

CERMET is dispersion of ceramic in metal matrix. This fuel combines the favorable properties of both ceramics and metals. The refractory and brittle ceramic fuel particulates, such as UO_2, PuO_2, and PuN, are uniformly dispersed in highly conducting and ductile metal matrix such as Al, Zr, and Mo. CERMET possesses higher thermal

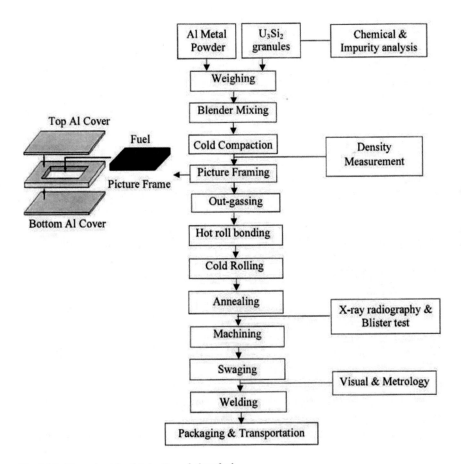

Fig. 3.10 Flow sheet for fabrication of plate fuel

conductivity compared to the ceramic fuel. This facilitates the fuel to operate at low center line temperature which subsequently amounts to reduced stored energy of the reactor core. Al and Zr matrix is the choice for LWRs due to their low thermal neutron absorption cross-section. LWR-type experimental fuel pins with (UO_2–Zr) and (UO_2–Al). CERMET fuels have been successfully tested in Russia. Dispersed fuel of PuO_2-M type (M: Al, Zr) and PuN-M type (M: Mo, W, Cr) are suitable for light-water and fast reactors, respectively, for plutonium burning. Different methods for fabrication of CERMET fuel element are as described below.

Vibropacking of fuel

The first step for preparation of fuel pins by this method is production of granules of desired size distribution. This is followed by blending of the particles of various size fractions. Simultaneous loading and compaction of blended particles in one end plug welded clad tube is carried out. The packing density can vary from 60 to 90% of TD by controlling the granule particles in one, two, or three sizes. The second step is filling of the pin by molten matrix material till saturation in a special furnace. The final step is to weld the top end of the fuel cladding.

Hot pressing of fuel pellets

In hot pressing, pressure and temperature are applied simultaneously on the compacts. The starting materials needed in this method of fabrication are fuel particles, inert matrix material, and the fuel cladding. The fuel particles are coated with inert matrix material by powder metallurgy technique. The fuel particles are then subjected to hot pressing to fabricate fuel pellets. The pellets are then loaded in the cladding tube, followed by vacuuming, filling of inert gas, and top-end welding.

Fabrication by the zone melting process using powder mixture

In this method of fabrication, two types of the powder are mixed and filled inside cladding tube and vibropacked. The cladding is subjected to degassing followed by zone melting from bottom to top under vibration. The tube is finally welded at the top.

Fabrication by the zone melting of prepressed fuel pellets

The two powders are mixed uniformly followed by prepressing of the powders at room temperature. The prepressed pellets are placed in the clad tube and exposed to zone melting arrangement from bottom to top with simultaneous column pressing. This is followed by upper-end plug welding.

Fabrication by powder metallurgy together with extrusion

This fabrication method involves steps like production of fuel granules, fabrication of compacts, fuel meat, and finally fuel elements. The sintered compacts are put in a barrel (Al alloy or zirconium-based alloy), and the plug is inserted through one of its ends. The assembly is then deformed by 2–5%. Subsequently, the open butt end of the barrel is rolled, and the assembly is extruded at temperature at 773–873 K for 80–100% deformation. The fuel meat is degassed and inserted into a clad tube. The joint rolling of fuel meat and clad tube is performed. The end plugs are inserted at both the ends of the tube, and butt welding is carried out.

Fabrication by powder metallurgy together with impregnation

In impregnation method, filling of cladding tube is done with fuel and matrix granules followed by vacuum annealing at temperature higher than the melting point of the matrix material. It coats the fuel granules and the cladding under the action of capillary forces forming connection between fuel granules itself and between the fuel granules and cladding. The connecting bridges increase the heat conductivity of the fuel meat.

Cold compaction, Sintering, and Encapsulation

This is the conventional powder metallurgical technique to produce green pellets followed by sintering [26]. The two types of powder are first blended to achieve the homogeneity. The green compacts are then fabricated following steps like precompaction, granulation, and final compaction. The green pellets are then sintered and encapsulated in cladding material by suitable welding process. Figure 3.11 presents the flow sheet for fabrication of CERMET by this method.

CERCER Dispersion Fuels

It consists of macroscopic or microscopic dispersion of ceramic fuel particulates such as PuO_2 and PuN in a ceramic matrix such as MgO, Al_2O_3, $MgAl_2O_4$, and ZrO_2. These types of composite fuels with inert matrix have been developed for disposal of excess plutonium coming from reprocessing as well as nuclear weapons and transmutation of minor actinides (MA: Am, Np, and Cm) in the waste management stream. CERCER fuel system is of two types, namely homogeneous and heterogeneous fuel. In the homogeneous fuel, single-phase solid solution of the fissile fuel is possible with a matrix material, for example, dispersion of PuO_2 in CeO_2 or stabilized ZrO_2

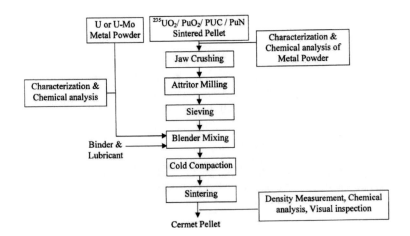

Fig. 3.11 Flow sheet for fabrication of CERMET fuel

matrix. The heterogeneous fuel has a two-phase structure in which the dispersion of fissile material is done in an inert matrix. The dispersion of PuO_2 or AmO_2 particles in MgO, Al_2O_3, or spinel $MgAl_2O_4$ is an example of a heterogeneous system.

Coated Particle Dispersion Fuels

It consists of uniform dispersion of multi-layered coated fuel kernel generally termed as TRI-isotropic (TRISO) particle in the graphite matrix. Coated particle dispersion fuel is a potential candidate for high-temperature gas-cooled reactor (HTGR). The spherical or prismatic block-type (pellet/pin) fuel elements containing TRISO particles are used in HTGR. The fabrication of HTGR fuel element need to be carried out in 3 steps, as given below:

1. Preparation of kernels containing ^{235}U or ^{233}U.
2. Multi-layer coating of kernels to form TRISO particles.
3. Fabrication of spherical or prismatic block-type fuel element.

The spherical fuel kernels are prepared using sol–gel method. The gelled product is reduced and sintered at high temperature for obtaining high-density microspheres. The different layers of the TRISO particle are as shown in Fig. 3.12. The fuel kernel is coated with silicon carbide (SiC) layer sandwiched between two pyrolytic carbon layers. Each layer acts as a tiny pressure vessel for retention of fission products and fuel at elevated temperature. The first porous buffer layer of carbon provides volume to accommodate fission products and kernel swelling. The high-density inner pyrolytic carbon layer acts as a barrier to gross diffusion of fission product and actinides. The intermediate SiC layer gives a superior diffusional barrier for metallic fission products than pyrolytic carbon layers. The outermost pyrolytic carbon layer gives chemical and corrosion protection to the kernel from the external environment. The coating of the fuel kernel is carried out in a fluidized bed reactor using chemical vapor deposition (CVD) process. The gases flowing over through the fluidized bed undergo thermo-chemical decomposition, and their constituents deposit onto the surfaces of fuel kernels. Propylene (C_3H_6) along with Ar carrier gas is used to form the pyrolytic carbon layer. SiC layer is coated using a mixture of methyl trichlorosilane (MTS: CH_3SiCl_3), hydrogen, and argon carrier gas.

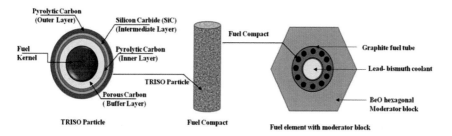

Fig. 3.12 Schematic of TRISO fuel particle, fuel compact, and moderator block of CHTR

The spherical fuel elements have been used in the pebble bed-type HTGR of Germany, Russia, and China. The flowsheet for the manufacturing of the spherical TRISO fuel element is shown in Fig. 3.13. The graphite powder along with phenolic resin binder is mixed in a conical mixer, and the mixture is then fed to a kneading machine. The paste-like mixture is converted into granules through extrusion followed by drying operations. The granules are milled in a hammer mill, homogenized, and kept ready for pressing. Over coating of TRISO particles with graphite matrix powder is carried out in a rotating drum. The homogenized graphite matrix powder and over-coated TRISO particle are fed to the prepressing machine. The prepressed molds along with additional fuel matrix material are final compacted. The machined green fuel spheres are subjected to carbonization and annealing heat treatments. The carbonizing heat treatment is carried out at 1073 K to provide strength and to remove all organic material. The annealing process is carried out under vacuum at approximately 2173-2223 K to eliminate residual impurities in the graphite matrix and to improve the strength and corrosion resistance.

The prismatic block or compact-type fuel elements have been used in Japanese and US design of HTGR. The fuel element in the US design is a hexagonal graphite block, which is filled with TRISO fuel compacts and sealed. The "pin-in block"-type design is used in the Japanese HTGR. The fuel pellets encapsulated in graphite sleeves which sit in bores made in a hexagonal block are used. In Indian Compact High Temperature Reactor (CHTR) design, the graphite fuel tube carrying fuel compact is centrally located in a prismatic BeO moderator block as shown in Fig. 3.12. The blended mixture of graphite powder, TRISO particles, and phenol binder is compacted in the press to obtain cylindrical pellets containing a dispersion of TRISO particles in the

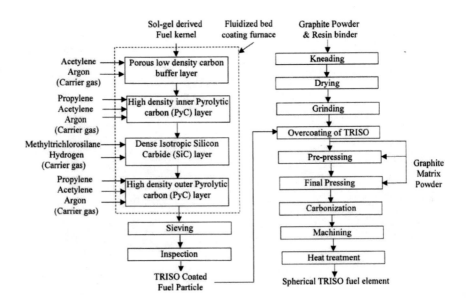

Fig. 3.13 Flow sheet for the fabrication of TRISO fuel and spherical fuel element

graphite matrix. These fuel pellets after heat treatments are used in fuel elements as per design. An advanced version of HTGR called very high temperature gas-cooled reactors (VHTGRs) is also under consideration as one of the GENIV reactor systems.

3.4.3 Metallic Fuels

3.4.3.1 Metallic Uranium as Fuel

Metallic uranium has been used as a fuel for non-power reactors with large cores. These reactors use heavy water as moderator and light or heavy water as coolant. Natural uranium metal is used in the form of metal rod or cluster cladded in aluminum. Figure 3.14 presents the flowsheet of fabrication of metallic uranium fuel with major quality control checks. Uranium ingots of nuclear grade obtained by magnesio-thermic reduction of UF_4 melted in an induction furnace and cast into billets. The billets are hot rolled at 893 K. Alternatively, extrusion of shorter billets is also adopted. The texture developed during hot rolling/extrusion process is removed by beta quenching. Beta heat-treated rods are straightened. Beta treatment removes the texture and avoids irradiation growth because of the anisotropy of uranium. The heat-treated rods are straightened and machined to remove surface oxidation. The accepted uranium rods are then canned in 1S aluminum with good mechanical bonding between the clad and uranium metal. The end closure welding is done by TIG. In the case of cluster, the pins are assembled in a fixture and welded. The cluster/rods are dispatched to reactor site reactor site after all the required quality control checks.

3.4.3.2 Uranium–Plutonium-Based Alloys as Fuel

Metallic fuels are considered as candidate fuels for fast breeder reactors (FBRs) due to high breeding potential, high thermal conductivity, high fissile atom density, low doubling time, and ease of fabrication as compared to other ceramic fuels. Irradiation experiments with uranium-based fuel pins were carried out in various experimental fast reactors, namely, Experimental Breeder Reactor I (EBR-I), EBR-II, the Enrico Fermi Reactor, and Dounreay Fast Reactor (DFR). A few shortcomings of metallic fuels like low solidus temperature, high swelling rate, high fission gas release, and susceptibility to chemical and mechanical interaction with cladding materials suppress their full potential, and thus, their burn up was limited to a few atom percent (at.%) only. Before the full potential of metallic fuel was revealed, the global trend of development of fast reactor fuel was directed toward oxide fuels. However, continuous effort and attention were given to increase the burn up limit of the driver fuels of those existing experimental fast reactors (EBR-II) at Argonne National Laboratory (ANL) in United States. It was observed that reducing the fuel "smear density" (the cross-sectional area ratio of the fuel slug to the cladding material inside) to ~75% was effective in promoting fission gas release before the fuel slug

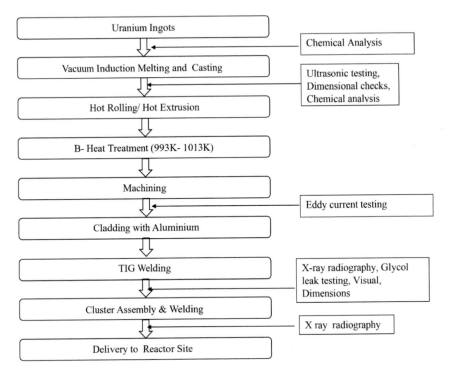

Fig. 3.14 Flow sheet of fabrication of metallic uranium fuel with major quality control checks

comes in contact with cladding material. This suppresses the fuel–clad mechanical interaction (FCMI) to a great extent at an early stage of irradiation. The design burn up limit of the Mk-II driver fuel was then increased to 8 at.% with the smear density adjustments.

Optimization of driver fuel composition was another challenge in those initial years. The Mk-I and Mk-II driver fuels of EBR-II were uranium-fissium alloy, i.e., U–5 wt% Fs alloy, where fissium (Fs) is an equilibrium composition of fission products: 2.46Mo, 1.96Ru, 0.28Rh, 0.19Pd, 0.1Zr, and 0.01Nb (in wt%), left in the melt-refining process. The U–Pu–Fs alloys did not show satisfactory compatibility with the cladding materials, necessitating development of various other U–Pu-based alloy fuels. These fuels were explored in terms of having optimum physical properties, irradiation performance, and compatibility with cladding materials. Researchers considered U–Pu–Zr-based ternary alloys would be the best suitable candidate as a driver fuel for fast reactors because of their higher solidus temperature and compatibility with stainless steels as cladding material [27]. Thus, fuel composition and fuel-cladding gap, the two key emerging metal fuel design features were incorporated in the integral fast reactor (IFR) program [28, 29] initiated at ANL in 1984 with U–Pu–10 wt% Zr as fuel slug having ~75% smear density. A larger number of U–Pu–Zr fuel pins were irradiated in EBR-II and FFTF to high burn up (20 at.%) [30].

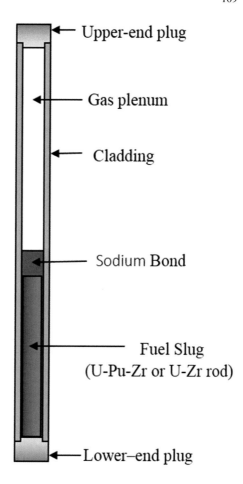

Fig. 3.15 Schematic view of a metallic fuel pin

Upper-end plug

Gas plenum

Cladding

Sodium Bond

Fuel Slug
(U-Pu-Zr or U-Zr rod)

Lower–end plug

The pyrochemical route was chosen for the reprocessing of metallic fuels. The development of pyro-processing technologies was based on electro-refining methods for separating actinide elements from fission products, and these developmental activities started in the 1980s. Pyro-metallurgical and electrochemical processing were two key reprocessing elements in the integral fast reactor (IFR) program. As the fuel coming out from the reactors will be highly radioactive, the reprocessing operations are to be carried out in hot-cell facility.

A schematic view of a metallic fuel pin is presented in Fig. 3.15. The cylindrical fuel alloy rod, called fuel slug, is inserted inside the fuel pin in such a way that a plenum area above the slug is maintained. A relatively large plenum volume is provided above the fuel slug to accommodate the pressure of the fission gas accumulating during irradiation. Because sodium does not react with U–Pu–Zr alloys, the annular gap between the fuel slug and the cladding can be filled with sodium (Na-bonded fuel) to facilitate efficient thermal conduction from the fuel to the coolant.

Both mechanically bonded fuel pins with U–Pu alloys and sodium bonded pins with U–Pu–Zr ternary alloys are being explored for FBRs [31].

Injection casting has been applied in the fabrication of the EBR-II driver and test fuel pins since the 1960s [32]. The other alternative processes of fuel slug fabrication include centrifugal casting, continuous casting, and gravity casting. Unlike in injection casting, the furnace containing the fuel melt is not evacuated in gravity casting, centrifugal casting, and continuous casting. Since there is no evacuation, the condition is favorable for suppressing evaporation of Am . The following paragraph describes fuel slug fabrication methods, focusing on the injection casting process. The fabrication flow sheet of the fuel slug is shown in Fig. 3.16.

An outline/flow sheet of an injection casting process is illustrated in Fig. 3.17. The starting materials (uranium, plutonium, and zirconium metals) are charged into the graphite crucible in the injection casting furnace. The silica tube molds with the top ends closed are set above this crucible. In order to avoid any unwanted reaction of both crucible and mold with the molten alloy, the interior surface of the graphite crucible and silica mold is coated with yttria and zirconia, respectively. The crucible is inductively heated up to ~1833 K, which is sufficiently higher than the liquidus temperature of the fuel alloy (e.g., 1656 K for U–10 wt% Zr). The homogeneity of the alloy is ensured by keeping the melt at high temperature and at the same time by electromagnetic stirring with full power to the crucible [32, 33]. After the vessel is evacuated, the molds are lowered through the drive shaft, and their bottom ends are immersed in the melt. At this point, the furnace atmosphere is filled with argon gas. The pressure difference between the inner region of the mold (vacuum) and the

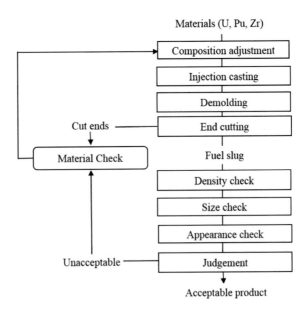

Fig. 3.16 Flow sheet of fabrication of metallic fuel slug

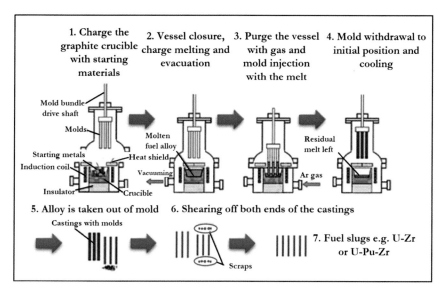

Fig. 3.17 Outline of injection casting process

furnace (Ar pressure) facilitates the melt in the graphite vessel to enter the mold (injection). The injected melt inside the mold is solidified quickly from the top to the bottom end. Once completely cooled, the fuel alloy castings are taken out of the silica molds. The mold must be broken when the castings are to be taken out. As the molds are not reusable, the mold shards add up to the radioactive waste load. The final acceptable fuel slugs are obtained by shearing off both ends of the castings. The fuel slug diameter is controlled by the inner diameter of the mold, and hence, no grinding step is required for the fuel slug surface. The important casting parameters such as mold preheat temperature, molten alloy temperature, pressurization rate in injection, and cooling rate after injection should be determined carefully based on the mold dimensions and alloy composition. The various casting defects such as shrinkage pipes, micro-shrinkage, and hot tears may arise from inappropriate casting parameters.

3.4.3.3   Uranium Supported Thorium Alloys as Fuel

As thorium does not have any fissile isotope, it has to be alloyed with uranium or plutonium to be used as a fuel. The details of thorium metal production methods are described in Chap. 2.

Thorium alloy development program started in the early 50 s in the US nuclear laboratories, viz., BMI, ORNL, and Ames. However, due to the abundant resources of uranium, the enthusiasm and interest in thorium fuel program could not sustain worldwide. Th–U alloy is generally prepared by melting the constituents under inert atmosphere in an arc melting setup. The as-cast microstructure of Th–U alloys varies from single phase in the high Th alloys (<3 wt% U) to eutectic at 96 wt%U. With higher uranium content, the uranium forms a separate phase in the thorium-rich matrix. In Th-30wt%U alloys, the uranium-rich phase forms layers surrounding the grains. This type of microstructure is not desirable from irradiation stability point of view. In Th-52wt%U alloy, a well-developed interconnected network of uranium is seen. For enhancing the irradiation stability of the alloys with higher uranium content, the uranium should be stabilized in isotropic gamma phase. This could be done by alloying it with gamma stabilizing elements, viz., Mo, Nb, Zr, etc. With still higher uranium content, the thorium-rich phase is present in the form of dendrites, and uranium becomes the continuous phase. Th-96wt%U is a eutectic structure where fibrous eutectic products form colonies inside the microstructure. Thermal conductivity of high thorium Th-U alloys is found to be quite superior compared to that of uranium and other uranium alloys that are being used as metallic fuel. Thermal expansion coefficient values of high thorium Th–U alloys are also comparable with that of prevailing oxide fuels.

3.5 Accident Tolerant Fuel (ATF) Concepts

A lot of work has been carried out in different countries to design a safe and economical design for fuels and also safe operation of the reactors [34]. In view of the vulnerability of the fuel designs to extreme accident conditions as shown by Three Mile Island and Fukushima Daiichi accidents, research has been initiated to develop accident tolerant fuel (ATF) which would be more resistant to fuel failure. The strategy is to develop nuclear fuels and claddings with enhanced accident tolerance characteristics for use in the current fleet of commercial reactors, primarily LWRs or in new reactor concepts (GEN-III+). Accident tolerant fuel for LWRs got greater attention because of the recent accidents mentioned above. ATF technology will enhance the safety and performance of the fuel. The use of coated claddings, doped pellets, new cladding materials (FeCrAl, SiC, Mo), higher enrichment of the fuel, new fuels (UN, U_3Si_2), and double clad tubes are being tried out specially for LWRs. The use of coated cladding will enhance the protection of fuel rods against debris fretting and provide better material behavior and oxidation resistance. Doped pellets will have reduced rigidity thereby reducing the risk of damage of the clad and allowing more flexible reactor operations. The larger grain size of the doped pellets helps in higher retention of fission gas within the pellets reducing the quantity of radioactive gases released to the environment. The proposed FeCrAl-based alloy instead of zirconium alloy for the clad will result in improved high temperature steam oxidation, strength at normal and high temperature accident conditions. It will improve corrosion performance also. Increased enrichment of ^{235}U up to 10% will have the advantages of increased operational flexibility, a smaller number of fuel assemblies and higher burn up. Higher burn up of the fuel will give longer reactor cycles and less fuel assemblies. The use of UN pellets, SiC cladding, and extruded metallic fuel are some of the long-term accident tolerant fuel technologies being investigated. Benefits of UN pellets are increased in uranium, high melting point and better thermal conductivity. SiC cladding has the advantages of structural integrity at very high temperature and better high temperature steam oxidation resulting in less hydrogen generation under design basis accident and severe accident conditions. Extruded metallic fuel will have increased thermal conductivity and complete retention of fission gas. The potential challenges of the new concepts must be overcome before implementing the new technology. Irradiation of a number of test fuel assemblies using ATF technology is in progress.

3.6 Automation in Nuclear Fuel Manufacture

Automation is being introduced at various stages of fabrication in nuclear fuel fabrication facilities in different countries. Reduction of dose received by the personnel, repeatability, increased production without compromising on quality and reduction of manpower are some of the advantages of automation. A number of fabrication plants have implemented complete automation in fuel assembly manufacture, quality control, and material handling. Processes like SGMP and vibrocompaction are more amenable to automation and remotization. It is possible to automate the movement of fuel powder both outside and inside the gloveboxes (for plutonium oxide). Transfer systems are available for transport of compacted and sintered pellets from one workstation to another. The sintering furnaces are available within built automation for movement of the pellets. The automated pellet feeder systems are being used for loading the pellets into the clad tubes. Automated systems for stacking of pellets, rotation of the fuel rods for X radiography helium leak testing, and ultrasonic testing are a few examples of automation in the nuclear industry. Pick and place systems are in use wherever feasible and convenient. Innovative automation systems have been developed for use in fuel production line including fuel assemblies. Advanced image vision system and laser-based systems are available for online checking of dimensions and measurement of density of the pellets. The innovative automation systems being used in fuel manufacturing facilities have resulted in higher productivity, better quality assurance and reliability of the final product, and reduced radiation exposure to the workers.

3.7 Summary

Fabrication of nuclear fuels of various types for nuclear reactors has been presented in this chapter. The types of reactors and design philosophy of nuclear fuels have been described before fabrication technology is introduced. The fuels used for different types of reactors are given in a tabular form also. The steps followed in the manufacture of ceramic fuels, both oxide [UO_2, Mixed Oxide–(U, Pu)O_2, (Th, ^{233}U)O_2, (Th,

Pu)O_2 and non-oxide [Mixed Carbide (U, Pu)C, Mixed Nitride (U, Pu)N] are given in detail along with the flowsheets. The various processes for pelletization followed in nuclear fuel fabrication facilities are described. Widely used powder oxide pelletization (POP) technique is covered in more detail. Advanced processes like sol–gel microsphere pelletization (SGMP), impregnation, and coated agglomeration process (CAP) are also described. Fabrication of dispersion fuels containing metallic or ceramic particulates dispersed in metallic or ceramic matrix is also explained in this chapter. Dispersion fuels are classified into different types based on the nature of fuel particulates and dispersing matrices and the fabrication techniques used are described. Fabrication technology for advanced fuels like coated particle dispersion fuels (TRISO) for high temperature gas-cooled reactor (HTGR) is also presented in this chapter. Manufacture of metallic fuels for research reactors and fast breeder reactors (FBRs) is briefly described in this chapter. New trends in nuclear fuel technology, viz., Reduced Enrichment for Research and Test Reactor (RERTR) program, accident tolerant fuel (ATF) and automation in nuclear fuel fabrication facilities are also briefly covered. In brief, an attempt has been made in this chapter to present the technology for fabrication of both conventional and advanced nuclear fuels for reactors.

Questions

1. What are the factors to be considered while designing fuel elements?
2. Describe different types of thermal neutron power reactors and fuel elements used in them?
3. What are fast reactors? Mention the fuels used in fast reactors?
4. Explain the fabrication of UO_2 bundles for thermal reactors with a flowsheet.
5. What are the processes available for manufacture of mixed oxide (MOX) fuel?
6. Explain in detail powder oxide pelletization (POP) for MOX fuel.
7. Compare the different processes for manufacture of MOX fuels.
8. Describe impregnation process for MOX fuel.
9. What is CAP process?
10. Describe the fabrication of mixed carbide and mixed nitride fuels.
11. What are dispersion fuels? Briefly describe the fabrication of different types of dispersion fuels.
12. Describe the fabrication of plate-type fuel.
13. How is CERMET fuel fabricated?
14. Describe coated particle dispersion fuel.
15. How are metallic fuels fabricated?
16. How is metallic uranium fuel for non-power reactors fabricated?
17. Explain accident tolerant fuel (ATF) concepts.
18. Explain the role of automation in nuclear manufacture.

References

1. Power reactor information system, IAEA (2019)
2. Nuclear power and research reactors. https://www.world-nuclear.org/information-library/nuclear-fuel-cycle/nuclear-power-reactors/nuclear-power-reactors.aspx
3. Experiences and trends of manufacturing technology of advanced nuclear fuels, IAEA-TECDOC-1686, International Atomic Energy Agency, Vienna (2012)
4. S.C. Chetal, V. Balasubramaniyan, P. Chellapandi, P. Mohanakrishnan, P. Puthiyavinayagam, C.P. Pillai, S. Raghupathy, T.K. Shanmugham, C.S. Pillai, The design of the prototype fast breeder reactor. Nucl. Eng. Design **236**(7–8), 852–860 (2006)
5. P. Balakrishna, C.K. Asnani, R.M. Kartha, K. Ramachandran, K.S. Babu, V. Ravichandran, B.N. Murty, C. Ganguly, Uranium dioxide powder preparation, pressing and sintering for optimum yield. J. Nucl. Techn. **127**(3), 375–381 (1999)
6. M. Becker, *Manufacture of Uranium Dioxide Powder* (Siemens, AG, 1976)
7. S.G. Brandberg, The conversion of uranium hexafluoride to uranium dioxide. Nucl. Tech. **18**(2), 177–184 (1973)
8. H.S. Kamath, A. Kumar, Fast reactor fuel fabrication. Fast React. Technol., IANCAS Bull. **1**(4), 26–34 (2002)
9. Thorium fuel cycle-potential benefits and challenges, IAEA-TECDOC-1450, 49–64 (2005)
10. BNM-CEA/DTA/LPRI, Nuclide 2000, Version 2, Nuclear and Atomic Decay Data, BNM-CEA/DTA/LPRI, 30th June 2004
11. Status and Advances in MOX Technology, IAEA Technical Report Series No.415, International Atomic Energy Agency, Vienna, pp. 37–41 (2003)
12. .H. Lee, Y.H. Koo, D.S. Sohn, Nuclear fuel behavior modelling at high burnup and its experimental support, IAEA-TECHDOC-1233, IAEA, Vienna, p. 247 (2001)
13. S. Mishra, S.N. Rahul, I.D. Godbole, A.K. Mishra, A. Kumar, H.S. Kamath, Use of polyethylene glycol and oleic acid in fabrication of nuclear fuel pellets, in *Powder Metallurgy Association of India, PMAI Conference* (1999)
14. A. Karande, A. Fulzele, A. Prakash, M. Afzal, J.P. Panakkal, H.S. Kamath, Determination of PuO_2% in power reactor mixed oxide fuel blends (0.4–44% PuO_2) by neutron well co-incidence counting technique, J. Radioanal. Nucl. Chem. **284**, 451–455 (2010)
15. E.D. Collins, S.L. Voit, R.J. Vedder, *Evaluation of Co-Precipitation Processes for the Synthesis of Mixed-Oxide Fuel Feedstock Materials* (Oak Ridge National Laboratory, United States, N. p., 2011). Web. https://doi.org/10.2172/1024695
16. N. Kumar, R.V. Pai, J. Joshi, S. Mukerjee, V. Vaidya, V. Venugopal, Preparation of (U, Pu)O_2 pellets through sol–gel microsphere pelletisation technique. J. Nucl. Mater. **359**(1–2), 69–79 (2006)
17. T.R.G. Kutty, M.R. Nair, P. Sengupta, U. Basak, A. Kumar, H.S. Kamath, Characterization of (Th, U)O_2 fuel pellets made by impregnation technique. J. Nucl. Mater. **374**(1–2), 9–19 (2008)
18. R.V. Pai, J. Dehadraya, S. Bhattacharya, S. Gupta, S. Mukerjee, Fabrication of dense (Th, U)O_2 pellets through microspheres impregnation technique. J. Nucl. Mater. **381**(3), 249–258 (2008)
19. T.R.G. Kutty, K. Khan, P. Somayajulu, A. Sengupta, J.P. Panakkal, A. Kumar, H.S. Kamath, Development of CAP process for fabrication of ThO_2–UO_2 fuels, Part I, fabrication and densification behavior. J. Nucl. Mater. **373**(1–3), 299–308 (2008)
20. P. Khot, Y. Nehete, A. Fulzele, C. Baghra, A. Mishra, M. Afzal, J.P. Panakkal, H.S. Kamath, Development of Impregnated Agglomerate Pelletisation (IAP) process for fabrication of (Th, U)O_2 mixed oxide pellets. J. Nucl. Mater. **420**(1–3), 1–8 (2012)
21. C. Ganguly, P.V. Hegde, G.C. Jain, U. Basak, R.S. Mehrotra, S. Majumdar, P.R. Roy, Development and fabrication of 70% PuC-30% UC fuel for the fast breeder test reactor in India. Nucl. Tech. **72**(1), 59–69 (1986)
22. H. Matzke, *Science of Advanced LMFBR Fuels* (Physics Publishing, North Holland, Amsterdam, 1986)
23. H. Blank, K. Richter, M. Coquerelle, H. Matzke, M. Campana, C. Sari, I. Ray, Dense fuels in Europe. J. Nucl. Mater. **166**(1–2), 95–104 (1989)

24. Development status of metallic, dispersion and non-oxide advanced and alternative fuels for power and research reactors, IAEA-TECDOC-1374, IAEA, Vienna (2003)
25. S. Mishra, P.S. Kutty, A. Kumar, G.J. Prasad, *International Conference on Fast Reactors and Related Fuel Cycles: Next Generation Nuclear Systems for Sustainable Development—FR17* (Yekaterinburg, Russian Federation, June 26–29, 2017)
26. S. Mishra, P.S. Kutty, T.R.G. Kutty, S. Das, G.K. Dey, A. Kumar, Cermet fuel for fast Reactor-Fabrication and characterization. J. Nucl. Mater. **442**(1), 400–407 (2013)
27. C.E. Stevenson, *The EBR-II Fuel Cycle Story* (American Nuclear Society, La Grange Park, IL, 1987)
28. G.L. Hofman, L.C. Walters, in *Nuclear Materials*, Part 1, B.R.T. Frost (ed.) *Material Science and Technology, A Comprehensive Treatment*, R.W. Cahn, P. Haasen, E.J. Kramer, (eds.), vol. 10A, pp. 1–43 (VCH Verlagsgesellschaft, Weinheim, Chapter 11994).
29. C.E. Till, Y.I. Chang, Progress and status of the integral fast reactor (IFR) fuel cycle development, in *Proceedings of the International Conference on Fast Reactor and Related Fuel Cycles* (Kyoto, Japan, October 28–November 1, 1991)
30. M.T. Simnad, Nuclear reactor materials and fuels. Encycl. Phys. Sci. Technol. Elsevier **10**, 775–815 (2002)
31. K. Devan, A. Bachchan, A. Riyas, T. Sathiyasheela, P. Mohanakrishnan, S.C. Chetal, Physics design of experimental metal fueled fast reactor cores for full scale demonstration. Nucl. Eng. Des. **241**, 3058–3067 (2011)
32. D.E. Burkes, R.S. Fielding, D.L. Porter, D.C. Crawford, M.K. Meyer, A US perspective on fast reactor fuel fabrication technology and experience part I: metal fuels and assembly design. J. Nucl. Mater. **389**, 458–469 (2009)
33. T. Ogata, T. Tsukada, Engineering-scale development of injection casting technology for metal fuel cycle, in *Proceedings of 7th International Conference on Advanced Nuclear Fuel Cycles and systems*, Global 2007, Boise, ID, USA, September 9–13 (2007)
34. Accident tolerant fuel concepts for light water reactors, IAEA-TECDOC-1797, Vienna (2016)

Further Reading

1. B.R.T. Frost, *Nuclear Fuel Elements: Design, Fabrication and Performance*, 1st ed. (Elsevier, 1982)
2. S. Glasstone, A. Sesonske, *Nuclear Reactor Engineering: Reactor Systems Engineering* (Springer Science & Business Media, 1994)
3. IAEA Reference Data Series No. 2 Nuclear Power Reactors in The World International Atomic Energy Agency, Vienna (2019)
4. K. Baur, E. Von Collani, *Nuclear Fuel Quality Management Handbook Volume I: Fabrication, Operation, Disposal and Transport of Nuclear Fuel*, Technical Editor Peter Rudling, Advanced Nuclear Technology International, Skultuna, Sweden, March 2010
5. D. Olander, Nuclear fuels—present and future. J. Nucl. Mater. **389** (2009)

Chapter 4
Quality Control of Nuclear Fuels

D. B. Sathe, Amrit Prakash, and Jose P. Panakkal

4.1 Introduction

Quality of nuclear fuel loaded in the reactors is an important factor which ensures safe and economic operation of nuclear reactors and useful life of the fuel elements/assemblies. A variety of techniques are being used for assuring the quality of the fuel elements fabricated in a fuel manufacturing facility. The techniques are broadly classified into physical and chemical quality control. This chapter gives an introduction to quality control and describes basic aspects of various techniques used and their application in the quality control of nuclear fuels. Since the most common type of fuel used is ceramic fuels, techniques used for ceramic fuels are covered in more detail. The safety aspects and issues (chemical and radiological) while carrying out quality control functions in a nuclear fuel fabrication facility are described in detail in Chap. 11. It is felt that definitions of a few common terms used in the field of quality control will be useful to the beginners and are given below [1, 2].

D. B. Sathe
Fuel Fabrication-Integrated Nuclear Recycle Plant Operation, Nuclear Recycle Board, Bhabha Atomic Research Centre, Tarapur 401504, India
e-mail: dbsathe@barctara.gov.in

A. Prakash
Radiometallurgy Division, Nuclear Fuels Group, Bhabha Atomic Research Centre, Mumbai 400085, India
e-mail: amritp@barc.gov.in

J. P. Panakkal (✉)
Formerly Advanced Fuel Fabrication Facility, Nuclear Fuels Group, Bhabha Atomic Research Centre, Mumbai 400085, India
e-mail: jpanakkal@yahoo.com

© The Author(s), under exclusive license to Springer Nature Singapore Pte Ltd. 2023   117
B. S. Tomar et al. (eds.), *Nuclear Fuel Cycle*,
https://doi.org/10.1007/978-981-99-0949-0_4

Quality: It is a measure of fitness for use (from the point of view of the user) and conformance to the specifications (from the point of view of the manufacturer). It is the totality of features and characteristics of a product or service that has bearing on its ability to satisfy a given need.

Quality Control: It is the regulatory process through which we measure the actual quality performance and compare with standards and act on the difference. It is the operational techniques and activities that sustain the product or service quality to the specified requirement.

Quality Assurance: It is the activity of providing to all concerned, the evidence needed to establish confidence that the quality function is being performed adequately.

Inspection: It is the evaluation of the quality of some characteristic in relation to a standard, and the main purpose is to determine whether the products conform to the specifications.

Reliability: The probability of a product performing a required function without failure under given conditions for a specified period of time.

Statistical Quality Control: It uses statistical methods for monitoring the quality of products and services.

Quality control of nuclear fuels starts from the raw material stage itself. The quality control points are indicated in the flow sheets of fabrication of fuels described in Chap. 3 (Figs. 3.1, 3.3, 3.7, 3.8, 3.9, 3.10, 3.11, 3.13, 3.14 and 3.17). The techniques used for checking the feed materials, intermediate products, and the final product are discussed in the following sections.

4.2 Physical Quality Control Techniques

A number of physical quality control techniques are available to fuel manufacturers to assure the quality of fuel at different stages of fabrication. The tests are non-destructive and destructive. Non-destructive evaluation techniques which do not adversely affect the usefulness of the products can be carried out on a 100% basis. Destructive tests can be done only on statistical basis.

4.2.1 Powder Characteristics

The physical properties of powders used for fabrication of nuclear fuels are important for getting the desired density and other characteristics of sintered pellets. UO_2 and PuO_2 powder possess different particle size, particle morphology, and specific surface area and hence exhibit different behavior during pelletization. The properties of the powder change during the processing of the powder and hence the resultant powder behavior. In order to obtain homogenous fuel pellets, powder properties like tap density, bulk density, specific surface area, and powder morphology are checked for their conformance to the specification [3]. Bulk and tap density are a measure of packing of powder particles, and the properties are affected by parameters such as particle size, shape, and agglomeration. Agglomeration and flaky or irregular particle size make it difficult to achieve high packing density and can affect the pelletization process. Bulk density is determined by measuring the volume of a known mass of material (100g) in a tared measuring cylinder. Tap density is determined by mechanically tapping a graduated cylinder until the powder volume remained constant, and this volume is recorded. The density values are calculated by dividing the mass by the volume. Specific surface area of UO_2 and PuO_2 powder is measured using a gas adsorption technique, and particle size is determined by optical microscopy/laser-based particle size analyzer. The powder morphology is examined at high magnification using a scanning electron microscope (SEM) and is used as a research tool during the development of process.

4.2.2   Physical Inspection of Fuel Pellets

Since nuclear fuel in the form of ceramic pellets is used in most of the reactors, methods used for physical inspection of fuel pellets are described in this section.

The performance of the fuel may be affected by a variety of physical quality factors such as density, grain size, diameter, surface roughness, surface chipping, and cracking of the pellets apart from their chemical characteristics. High density is generally specified for the pellets so that the pellets have good strength and required fissile content. The outside diameter of the pellets is specified to maintain the specified pellet to clad gap. The physical integrity of the pellets is very important to have good performance during irradiation. Physical inspection involves the inspection of pellets to evaluate physical characteristics such as density, linear mass, dimensions, and surface defects. Density and linear mass are measured on statistical samples representative of the batches. Density of the pellets is measured using geometrical or immersion technique. As per sampling result, the batch is further processed for visual inspection and diameter sorting of pellets. Dimensional and visual inspection of all the fuel pellets is generally carried out using semi-automated systems to reduce dose to the operators. Attempts are being made in different fuel manufacturing plants to develop automated techniques using laser and digital processing techniques [4]. The pellets of acceptable dimensions are subjected to visual inspection under proper illumination for detecting the presence of any surface defects such as chips, pits, metallic or non-metallic inclusions, and cracks. Oversize pellets are ground to correct size and again subjected to inspection. The accepted pellets are sent further for loading, and the pellets rejected due to various reasons are sent for re-pelletization after treatment such as oxidation. Necessary safety precautions should be taken in handling the pellets containing plutonium as described in Chap. 3.

4.2.3 Metallography

Metallography is the study of microstructure of engineering materials and is considered as an integral aspect of material testing. Metallography is a critical quality control step for checking the correct processing of materials and the reliability of the products and helps in failure analysis. Metallographic procedure involves sectioning of the specimen at the location of interest followed by grinding and polishing to obtain mirror finish surface. Polished surface is etched to reveal the microstructure. The most common technique for etching is selective chemical etching. Other techniques such as electrolytic, thermal, laser, ion beam, and plasma etching have also found specialized applications in various industries depending upon their suitability and applications.

Metallographic procedure for welds and ceramics is almost same except the choice of etchant which varies depending on the composition and the type of material. Low-speed cut-off wheels are utilized for sectioning of samples. Ample coolant and proper speed control are essential in all sectioning operations. The entire process is designed to produce a scratch free surface by employing a series of successively finer abrasives in steps from 240, 320, and 400 to 600 grit SiC. Polishing involves the use of abrasives, suspended in a water solution, on a cloth covered with electrically powered wheel to achieve mirror finish surface on the sample.

Microscopic examination of a properly polished and un-etched specimen reveals structural features such as porosity, inclusions, cracks, and other physical imperfections. Etching is used to highlight and sometimes identify microstructural features or phases present. Etching occurs when the acid or base is applied on the surface of the specimen. The sample gets etched because of the difference in the rate of attack of the various phases present and their orientation. The etched sample is viewed under metallurgical microscope to observe the grain structure and allied defects.

A metallurgical microscope has a magnification varying from 25 to 1000X. Scanning electron microscopes (SEMs) (magnifications up to 20,000X) and transmission electron microscopes (TEMs) (magnifications up to 100,000X) are used for detailed microstructural study.

During the fabrication of nuclear fuel pins, metallography is a destructive test used for checking the quality of the end plug welds on a statistical basis during qualification and production of end plug welding. Resistance welding, tungsten inert gas (TIG) welding, and laser welding are used for encapsulation of the pellets. Quality of resistance end plug welds is primarily checked by metallography. Metallography is used for monitoring the quality of TIG welds and laser welds during qualification and on production welds. Non-destructive techniques like radiography and ultrasonic testing are used to check the quality of the welds. Metallographic examination gives information about the depth of penetration, cracks, porosities, and other defects in the weld (Fig. 4.1).

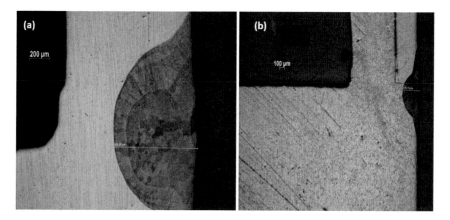

Fig. 4.1 Metallography of tube-end plug TIG weld: (**a**) good penetration of the weld and (**b**) lack of penetration of the weld

Apart from weld metallography, ceramography of fuel pellets is carried out to evaluate grain size, porosity distribution, cracks, tear outs, and inclusions. Grain size affects fuel plasticity (creep) and fission gas release which in turn determines fuel life/performance in the reactor. Hence, ceramography is carried out to evaluate whether the fuel pellets have optimum grain size (Fig. 4.2). Metallography is also used for checking the bonds of fuel plates.

Fig. 4.2 Ceramography of fuel pellet showing the grain structure

4.2.4 Radiography Testing (RT)

Radiography is used in a very wide range of fields like medicine, engineering, nuclear, and security. In general, radiography is used for inspecting materials to detect hidden flaws by using the ability of short wavelength electromagnetic radiation to penetrate various materials. The intensity of the radiation that penetrates and passes through the material is either captured by a radiation sensitive film (film radiography) or by a planar/line array of radiation sensitive sensors (real-time radiography).

In radiographic testing, the specimen to be inspected is placed between the radiation source and radiation sensitive film. The radiation source can either be X-ray machine or radioactive source (Co-60, Cs-137, Ir-192). The attenuation of the radiation in the material depends on the thickness and density of the object and is manifested as the variation of intensity of the radiation coming out recorded by the radiation sensitive film. As shown in Fig. 4.3, the variation in the degree of darkening of film is used to detect discontinuities and also to determine the thickness or variation in composition of material [5].

The energy of the radiation used depends on the thickness, density, and atomic number of the specimen, i.e., higher voltage/energy is needed for penetrating larger thickness and highly attenuating materials. The radiation coming out of the object will cause darkening of the processed film due to the interaction of the radiation falling on the film. The variation of the thickness of the object due to discontinuities, such as porosity or cracks, will appear as dark outlines on the film. Inclusions of low density, such as slag, will appear as dark areas on the film while inclusions of high density, such as tungsten, will appear as light spots. The time taken for radiography (exposure) for a given object is decided by the strength or intensity of the radiation source to get optimum density on the film. The sensitivity of the radiographic technique is measured using image quality indicators (IQIs) or penetrameters. Sensitivity S is defined as

$$S = \frac{\Delta x}{x} \times 100 \tag{4.1}$$

where Δx = Change in thickness detectable and x = Thickness of object.

Sensitivity of the radiographic technique can be improved by decreasing the energy of the radiation. The geometric unsharpness, (U_g) which plays a major role in the definition of the image decreases by increasing the distance between the radiation source and the film, decreasing the distance between the film/detector and the object and the size of the radiation source. It is expressed as

$$Ug = \frac{Fd}{D} \tag{4.2}$$

where Ug = Geometric unsharpness, F = Source size,

d = Distance from source side of object to the film,

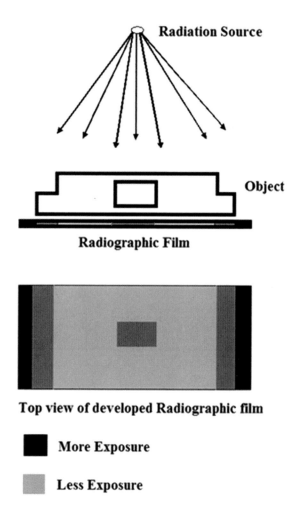

Fig. 4.3 Principle of radiography

D = Distance from source of radiation to object.

Radiography plays a major role in the inspection of end plug welds used for sealed fuel elements, fuel plates also for intermediate inspection of cast fuel billets.

The integrity of the end plug welds is crucial from the point of maintaining the structural integrity of the fuel pins in the reactor. The integrity of the welds is checked by various NDT techniques like visual inspection, leak testing, and X-ray radiography and ultrasonic testing.

Nuclear fuel pins are inspected using X-ray radiography because gamma radiography does not achieve the required sensitivity. Both the top and bottom end plug welds are checked by X-ray radiography to assess the integrity of the welds. Being cylindrical in shape, the weld profile should be compensated by using shape correction block to obtain uniform thickness and hence uniform optical density across the image. Simultaneously 5 to 10 pins are inserted in the shape correction blocks and are exposed to X-rays. Generally, exposure is taken in three orientations to cover the full circumference. Exposed films are chemically processed for converting latent image into permanent image. The films are viewed on high intensity illuminator. Defects encountered are lack of penetration (LOP), wall thinning, tungsten inclusion, porosity, etc. Typical joint and welded tubes placed in a shape correction block and X-ray radiographs are shown in Fig. 4.4.

Digital radiography is an advanced technology based on digital detector systems in which the x-ray image is displayed directly on a computer screen without the need for developing chemicals or intermediate scanning and is being adopted in a number of fuel manufacturing facilities [6]. As the image quality required for inspection of end plug welds is very high, use of micro-focal (less than 100 μm) X-ray radiography unit is recommended. Advanced micro-focus units have a focal spot size as low as 5 μm which is achieved by focusing the electron beam on the target in combination with sufficient cooling to avoid overheating. Micro-focus radiography facilitates

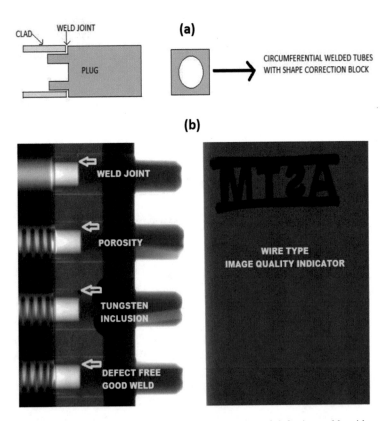

Fig. 4.4 (**a**) Radiograph of typical weld joint, (**b**) radiograph of defective welds with porosity, tungsten inclusion, and defect free good weld (left side) and wire type image quality indicator (right side)

observation of minute details of the object through magnified X-ray images, which in turn enhances the flaw detection capability improving the reliability in comparison with conventional radiography.

Radiography is also used for inspection of nuclear fuel plates for the location of the fuel core in the rolled plate and checking gross homogeneity.

It is essential that qualified and certified personnel should carry out radiographic testing and interpretation since false interpretation of radiographs can be expensive and interfere seriously with productivity. Relevant safety regulations should be followed during radiographic testing. X-ray and gamma radiation are invisible to the naked eye and can have serious health and safety implications.

4.2.5 Ultrasonic Inspection

Ultrasonic testing is being used extensively in the nuclear field for non-destructive evaluation of components, intermediate products, and final products. Ultrasonic testing uses high-frequency sound waves to perform examinations and make measurements. Ultrasonic testing is used for dimensional measurements, thickness, material characterization, flaw detection, and so on.

High frequency sound waves (frequency greater than 0.1 MHz), that is, ultrasonic waves are sent to the material/specimen by means of a transducer coupled to the material using a couplant. The ultrasound waves reflected from the interface of the discontinuity or the back wall are detected by the same transducer in pulse echo method (Fig. 4.5). The time taken by the ultrasound in the material helps in determining the location of the discontinuity. In transmission method, the ultrasound waves transmitted though the specimen are detected by a transducer kept at the other end (Fig. 4.5). The testing is done by longitudinal and shear waves depending on the applications. Other types of waves are also used for some special applications. Low-frequency ultrasound waves are used for testing highly attenuating materials like castings. Ultrasound waves of higher frequency are used for increasing the sensitivity of detection. Angle transducers are used when the application demands the waves are to be sent at an angle to the specimen like welds to cover the region of interest. Immersion ultrasonic testing is also carried out with both the specimen and the probe immersed in a liquid like water in some cases (Fig. 4.5). Ultrasound is also used to measure thickness of the specimen using the velocity in the material and the time taken for travel in the material.

Immersion ultrasonic testing is used for inspecting fuel pin end plug welds for thermal and fast reactors. High-frequency pulse echo shear wave technique is used for this purpose [7]. The fuel billets are also checked using ultrasonic testing for any defect in the cast material. Bond testing of fuel plates is also being carried out using ultrasound waves.

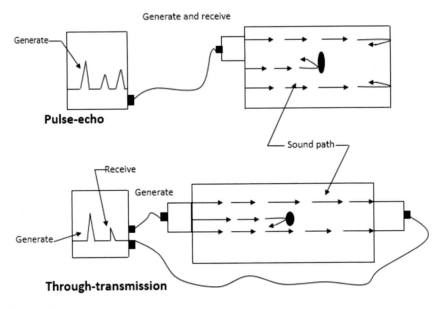

Fig. 4.5 Ultrasonic testing

4.2.6 *Decontamination Check*

The fabricated fuel elements/assemblies have to be checked for any traces of radioactive materials on the surface, and they have to be decontaminated by suitable cleaning techniques like ultrasonic cleaning, laser, and manual decontamination. Any loose contamination on the external surface of the fuel pin/assembly is not allowed. The loose contamination on the pins is checked by taking swipe samples on entire pin surface and counting of these swipe samples in a ZnS(Ag)-based counting system. The fixed contamination on the surface is checked by putting the ALSCIN (ZnS(Ag)) monitor close to the pin surface. After the decontamination of the fuel elements, they are checked by the health physics personnel for both loose and fixed contamination.

4.2.7 *Leak Testing*

Leak testing is a quality control check used to detect manufacturing defects which help in verifying the integrity of products and improving consumer safety. Leak testing techniques detect or measure the leakage of a gas or liquid from the product.

A number of leak testing techniques are available, spanning from very simple approach to complex systems. The most commonly used leak test methods are underwater bubble test, bubble soap paint, pressure and vacuum decay, and tracer gas detectors (halogen, helium and hydrogen). The first three techniques, due to their characteristics and sensitivity, can be used only for gross leak detection. Tracer gas leak testing methods are more sensitive.

Leak testing of fuel elements is very important since it reduces the possibility of contaminating the reactor systems by fission gas leaking out of the fuel elements under irradiation. In nuclear field, mainly high sensitivity techniques like mass spectrometer-based helium leak testing methods are employed. Glycol tests are also employed for testing fuel plates. Helium leak detector, also known as a mass spectrometer leak detector (MSLD), is used to detect leaks/leak rates into or out of the product.

The tracer gas, helium, is introduced to a test part (e.g., first end plug welded clad tube) that is connected to the leak detector (Fig. 4.6a). Helium leaking through the test part enters the detector system of the mass spectrometer tuned for helium, its partial pressure is measured, and the results are displayed on a meter. A welded fuel pin containing helium is kept in a chamber which is evacuated, and helium leaking out of the fuel pin is detected by a mass spectrometer leak detector (Fig. 4.6b). The leak rate allowed for nuclear fuel pin is less than 1×10^{-15} MPa m^3/sec.

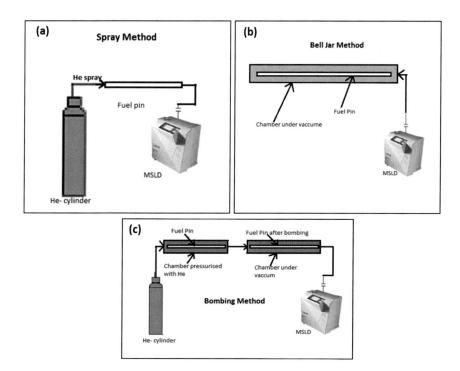

Fig. 4.6 Leak testing of fuel elements

Leak testing of sealed fuel elements in which helium is not used as a cover gas is carried out immediately after bombing (pressurizing) the product with helium (Fig. 4.6c). Care should be taken to complete the test without delay so as to prevent the loss of helium which has penetrated into the product during bombing.

4.2.8 Radiation-Based Techniques for Fissile Distribution and Composition

A number of radiation-based methods are used in nuclear industry to determine fissile distribution and composition of the fuel elements. Concentration of fissile isotopes in the fuel may lead to hot spot conditions affecting the safety of the reactor. During the manufacture of nuclear fuel, homogeneity of the fuel is checked at different stages.

(a) **Alpha autoradiography of fuel pellets**

Micro-homogeneity of fissile material in fuel pellets is checked using alpha autoradiography by recording alpha radiation emitted by the fissile isotopes like ^{235}U or ^{239}Pu on a film sensitive to alpha radiation [8].

Concentration of alpha tracks at a particular portion on the film indicates the agglomeration of alpha emitting nuclei at the corresponding location on the surface of the pellet. The processed film is viewed under optical microscope at 100x to observe distribution of alpha tracks across the film as shown in Fig. 4.7.

Neutron radiography and electron probe micro-analysis (EPMA) are other techniques available for evaluating micro-homogeneity.

(b) **Gamma scanning**

In passive gamma scanning (PGS), the intensity of the radiation coming out of the fuel pins is recorded by a detector. It is a non-destructive and quantitative testing technique for quality control of nuclear fuel pins. The technique is utilized for detection of cross mixing of pellets of different composition and to ascertain the correct composition in fuel pins. PGS can also estimate the fissile content and its spatial distribution across

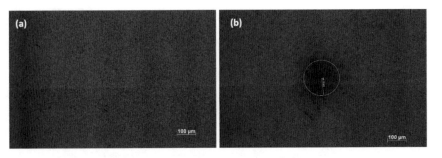

Fig. 4.7 Alpha autoradiograph of MOX pellet (**a**) uniform distribution and (**b**) non-uniform distribution

the length of the fuel pins and stack length [9]. Thallium activated sodium iodide detector, NaI(Tl), is used for gross gamma scanning. High purity germanium (HPGe) is used for isotopic analysis.

Passive gamma scanning records only the natural gamma rays coming mainly from surface or subsurface layer of the nuclear fuel pellets since the gamma rays from the central region of the pellets may be self-shielded depending on the diameter of the pellets. Active gamma scanning of nuclear material is used to obtain information about the fissile distribution inside the fuel pellets. This technique uses thermalized neutrons from a neutron source to induce fission in the fuel pellets and detects fission gamma rays. The fission gamma being of higher energy gives information about the fissile distribution inside the fuel pellets.

(c)   **Other techniques**

Gamma attenuation technique is used for determining fissile distribution in fuel plates by using an external gamma source like ^{241}Am and counting the attenuated gamma rays using a detector.

Alternatively, gamma autoradiography (GAR) (mapping of gamma rays emanating from the fuel) gives information about the distribution/concentration of fissile material in the fuel elements. A combination of X-radiography and gamma autoradiography (XGAR) gives additional information about the hardware [10].

Passive gamma scanning and XGAR of the fuel pins are used as final check for loading of pellets of specified composition and sequential loading of the fuel pellets and hardware. Radiography of the fabricated fuel elements using neutrons (neutron radiography) also gives more quantitative information about the fissile distribution since neutrons can penetrate the fuel pellets and give internal details also. But neutron radiography of all the fuel elements is not a practical proposition.

4.2.9   *Metrological Inspection of Hardware and Fuel Elements/Assemblies*

Metrology is an important aspect of quality control of nuclear fuel elements/assemblies. Metrology is carried out at all the stages of fabrication right from the hardware components such as tubes, bottom end plugs, top end plugs, springs, spring supports, and components required for assemblies. The finished fuel elements and assemblies are also subjected to metrological inspection using specially designed gages before final dispatch to the reactor site. The standard tools used for metrology are Vernier Caliper, micrometer, height gage, optical metrology system, and specially designed gages. Semi-automatic/automatic systems are used for dimensional inspection of components, fuel elements, and assemblies, and the data are stored.

4.3   Chemical Quality Control of Nuclear Fuels

Chemical quality control assures that the quality of the fabricated fuel conforms to the chemical specifications laid down by the fuel designers [11] in order to deliver designed output of the fuel in a nuclear reactor. These specifications include major, minor, and trace constituents of the fuel, isotopic content, and other characteristics which in turn affect the properties and performance of the fuel during in-pile performance. The specifications of the constituents in the final fuel product are therefore very stringent, and their quantification is of paramount importance. In this context, quality control exercise provides the scope for the evaluation of physical and chemical properties of the fabricated fuel from each batch before encapsulation in the fuel pins. It must be mentioned that apart from stringent quality control, significant emphasis has also been given to process control to maintain the robustness of process, thereby increasing the acceptance rate as well as improving quality of the fabricated fuel.

Metallic impurities like B, Cd, and rare earths, viz. Dy, Eu, Gd, and Sm being neutron absorbers affect neutron economy, and hence, their control is very important. Non-metallic impurities such as H, C, F, Cl, and moisture play a crucial role in corrosion of the clad and various other aspects. Moisture also alters oxygen to metal ratio (O/M) of the fuel and releases hydrogen which causes buildup of pressure. Similarly, pellets containing higher carbon can cause pressure buildup due to release of gases like carbon monoxide (CO).

Higher content of metallic impurities in general may dilute the fissile content as well as indicate the poor quality of hardware and equipment used in the fabrication. Based on their severity, they are divided as critical, major, and minor impurities. Similarly, uranium and plutonium content as well as their isotopic composition are ensured in order to have the required fissile content. Oxygen to metal ratio is one of the very critical specifications which affects properties such as thermal conductivity, melting point, number of phases, chemical reactivity, and mechanical strength.

It is well known that different kinds of fuels have been used in reactors, and the specification for a specific element in fuels varies depending on the type of the reactor. The specifications of the fuels, its effect on fuel properties and performance, and the analytical techniques for their determination have been discussed briefly in this section. The analytical methodology for the various types of fuel, viz. metal, alloys, oxides, and carbides is almost same, and there may be a variation only in sample preparation due to their nature in handling.

4.3.1   Heavy Metal Content

Uranium and plutonium are the major heavy elements present in fuels, although some metallic and non-metallic elements are present invariably in the fuel as impurities. Since these elements are fissile, accurate quantification and strict quality check are necessary to meet the specifications assigned to the fuel for their optimum performance in the reactor. A large number of destructive and non-destructive analytical techniques are available for determination of U and Pu in nuclear fuel samples, but redox titrimetric methods offer several advantages over other techniques in terms of cost, time, precision, accuracy, space, and the effort. Most of the other methods involve expensive equipment, take more time, and are not capable of giving results with an accuracy and precision better than 1%. Redox titrimetry is a low-cost technique with high precision and accuracy of the order of 0.2% (RSD).

The first step in any redox titrimetric method is dissolution of the sample in a suitable medium. UO_2 usually dissolves in HNO_3, whereas dissolution of PuO_2 is relatively difficult. PuO_2 which is not sintered dissolves in boiling concentrated HNO_3, but sintered PuO_2 pellets dissolve in the presence of HF. The fluoride ions attack solid PuO_2, and the process of displacement and dissociation continues [12]. Fuels containing both plutonium and uranium are dissolved by refluxing in a 1:1 mixture of concentrated H_2SO_4 and HNO_3 at elevated temperature. The resulting solution after dissolution is then adjusted to suitable acidity for further analysis.

4.3.1.1 Uranium

Uranium exists in the solution as U(VI) which is determined by Davis and Gray method [13]. In this process, reduction of U(VI) is carried out homogenously with excess of Fe(II) in strong H_3PO_4 solution containing HNO_3 and sulfamic acid (NH_2SO_3H). The excess unreacted Fe(II) is selectively oxidized to Fe(III) by reaction with HNO_3 in presence of molybdate (Mo^{+6}) catalyst. It is then diluted with H_2SO_4 containing VO^{+2} and then titrated with standard solution of $K_2Cr_2O_7$. The end point of titration is detected either by potentiometry or by bi-amperometry or by using barium diphenyl sulfonate indicator which turns to red-violet from colorless beyond the end point of titration. Plutonium does not interfere in the analysis of uranium as it is first reduced to Pu(III) by Fe(II) and later back oxidized to Pu(IV) by HNO_3 in presence of Mo(VI). A major drawback of the method is the use of highly complexing H_3PO_4 from which recovery of Pu after the analysis is cumbersome. With the objective of eliminating the use of H_3PO_4, the method is modified where Ti(III) is used for reducing U(VI) to U(IV) in H_2SO_4 medium. It is followed by destruction of the excess Ti(III) by HNO_3. Nitrous acid (HNO_2) generated during this process is destroyed by adding sulfamic acid. The solution is then diluted to bring down the acidity, and Fe(III) ions are added to produce Fe(II) by reaction with U(IV). The Fe(II) is titrated with standard $K_2Cr_2O_7$ using bi-amperometry to detect the end point.

Although bi-amperometry is the most preferred redox titrimetric method for the quantitative determination of U and Pu in nuclear fuel in laboratories, recovery of precious Pu and U from analytical waste is a cumbersome process. It requires the separation of various metallic impurities (Fe, Cr, Ti, Ag, K, etc.) present in the redox titrants added during the analysis. Also, bi-amperometry requires separate solutions for the determination of U and Pu in the same sample. Controlled potential coulometry (CPC) is another well-established technique for precise and accurate determination of U and Pu, and it generates analytical waste without any metallic impurities [14]. It does not need any standard reference material as the calculation is based on physical constants. Two different working electrodes (a Hg-pool electrode for U and a Pt electrode for Pu) are required for the determination of U and Pu. Uranium determination is carried out by applying controlled potential to a mercury pool electrode (vs. SCE). Plutonium interferes in the determination by reacting with mercury pool. Mercury is oxidized to Hg_2^{+2} by Pu(IV). Mercury cannot therefore be used as working electrode for samples containing plutonium. Platinum can serve as working electrode for this purpose, but it involves hydrogen evolution reaction that reduces current efficiency for the reduction of UO_2^{+2}. It is therefore challenging to determine U directly in presence of Pu by controlled potential coulometry. Alternatively, U(VI) can be reduced to U(IV) by a suitable reducing agent followed by oxidation with excess Fe(III). The amount of Fe(II) produced during the oxidation is equivalent to the amount of U(VI) present in the solution and can be determined by controlled potential coulometry.

4.3.1.2   Plutonium

Due to radiotoxicity associated with plutonium, strict safety norms must be followed to protect the analysts and lab environment. The detailed radiological safety requirements while handling plutonium are discussed in Chap. 11. Drummond and Grant method [15] is routinely used for the determination of plutonium in nuclear fuel samples. In this method, plutonium present in the solution is oxidized quantitatively to Pu(VI) by using excess AgO. Remaining part of excess AgO is neutralized with sulfamic acid. After that Pu(VI) is reduced to Pu(IV) using known excess of Fe(II). The amount of Fe(II) remaining after the reduction is determined by titration with standard $K_2Cr_2O_7$ solution. The end point of the titration is determined either by potentiometry or bi-amperometry. Since uranium exists as U(VI) in solution, it remains unaffected during the determination of plutonium. However, addition of large excess of AgO or Fe(II) will invoke positive bias in measured plutonium concentration as Ag^+ produced in the oxidation of plutonium can oxidize Fe(II).

In controlled potential coulometry, Pu is quantitatively reduced electrolytically to trivalent state in 1 M H_2SO_4 solution at a platinum electrode whose potential is controlled at $+0.3$ V versus SCE. It is then oxidized to tetra valent state at a controlled potential of $+0.7$ versus SCE. The accumulated coulombs are equivalent to Pu content in the solution. The interference of iron can be nullified by incorporation of chemical oxidation step by which Pu is oxidized to hexavalent state prior to performing the coulometric determination. Although there are several ways to detect the end point of titration, potentiometric and bi-amperometric methods are preferred due to the simple, convenient, and relatively maintenance free instrumentation.

X-ray fluorescence (XRF) spectrometry is being used as a process control check for determination of Pu/U ratio. In XRF, intensity of characteristic X-rays emitted by the elements on irradiation of the sample by X-rays is determined in the spectrometer [16]. The intensity of X-rays emitted depends on the concentration of the elements, and it forms the basis of quantitative analysis by XRF. XRF is used regularly as a process control check during the fabrication of carbide fuel. Attempts have been made to determine Pu content in final sintered MOX pellet using XRF as an alternative technique to redox titrimetric methods [17].

4.3.2 Isotopic Composition

It is necessary to determine the isotopic composition of the actinide element used as the nuclear fuel to estimate the fissile content of the fuel being loaded in the reactor. Mass spectrometry is the most commonly used method for the determination of isotopic composition of different actinide elements [18]. The important factors that govern the choice of mass spectrometric technique to be used include the type of matrix, amount of material available, expected concentration of the actinide material, and the desired precision and accuracy. Thermal ionization mass spectrometer (TIMS) is commonly used for determining the isotopic composition and concentration of actinide elements like uranium, plutonium, americium, and so on. In TIMS, the spectrometer consists of a thermal ionization source, a magnetic sector analyzer with an extended geometry, secondary electron multiplier as well as multi-Faraday cup detector system (Fig. 4.8). A motorized turret assembly loaded with number of samples is inserted in the ion source, and these samples can be analyzed sequentially without disturbing the vacuum in the mass spectrometer. The multi-Faraday cup detector system allows simultaneous acquisition of data of all the isotopes. This is particularly useful because the samples involved contain small amount of actinide element. This also eliminates the time dependent fluctuations involved during determination by sequential analysis of various isotopes with a single detector. Multi-Faraday cup detector system also allows online correction due to the isobaric interference. Isobaric interferences must be resolved which occur due to the presence of isotopes of same mass number (e.g., ^{238}U in ^{238}Pu or vice versa). To obtain good

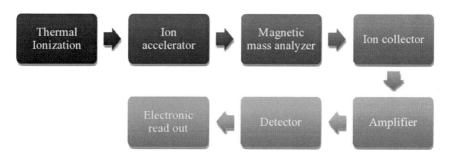

Fig. 4.8 Schematic of TIMS

results by TIMS, it is essential that the actinide element should be present in highly pure form free from other elements which otherwise may lead to a variety of effects such as suppression of ion current of the actinide element and instability in the ion beam or isobaric interference.

Different mass spectrometric techniques used are inductively coupled plasma mass spectrometry (ICPMS), secondary ions mass spectrometry (SIMS), glow discharge mass spectrometry (GDMS), and the advanced techniques like resonance ionization mass spectrometry (RIMS) and accelerator-based mass spectrometry (AMS) which provide ultrahigh sensitivity.

Mass spectrometry is regularly being used in reprocessing as well as fuel fabrication facilities to determine the abundance of various isotopes of actinide elements.

4.3.3 Phase Analysis

Powder X-ray diffraction (XRD) has been used extensively as a powerful technique for qualitative and quantitative analysis of polycrystalline materials [19]. Qualitative phase analysis deals with identification of different crystalline phases present in a mixture while quantitative phase analysis may be employed to determine their relative concentration in the mixture. X-ray diffraction pattern of the powdered sample is recorded using monochromatic X-rays. The diffracted X-ray intensity is measured as a function of scattering angle (2θ). The inter-planar spaces are calculated from Bragg's equation:

$$n\lambda = 2d \sin\theta \tag{4.3}$$

where λ is the wavelength of monochromatic X-rays, d is inter-planar spacing, and θ is the angle of diffraction.

The experimentally determined values of inter-planar spacings and their relative intensities are compared to that of standard materials available in the International Powder Diffraction File (PDF) database which is compiled and maintained by Joint Committee for Powder Diffraction Standards (JCPDS). Since each crystalline phase has its own characteristic diffraction pattern and acts as a unique "fingerprint," comparison of XRD pattern with PDF database can identify phases of crystalline materials even in a mixture also. Apart from identification of phases, XRD may be used to confirm the formation of solid solution and determine the precise lattice parameters of the unit cell. The other applications include determination of particle size and micro-strains, textural characteristics, etc. High-temperature XRD may be used to probe phase transitions and investigate high/low-temperature phases.

Quantitative phase analysis aims to determine phase compositions (fraction of phases) in multiphase polycrystalline materials. It is based on the correlation between the diffraction intensities of a given phase and its volume fraction in the multiphase sample. Several methods are known for quantitative phase analysis. Rietveld refinement has been widely used for quantitative phase analysis [20]. In this method, a theoretical diffraction pattern is generated based on models of the crystal structures of the phases present in the sample and compared to the observed pattern. Least-squares procedure is then used to minimize the difference between the calculated diffraction pattern and the observed diffraction pattern by adjusting the parameters of the model of the crystal structure. That procedure gives the fractions of the phases present in the sample, lattice parameters, and refinement of their crystal structures (atomic coordinates, site occupation factors).

Application of X-ray diffraction in the context of quality control of nuclear fuel is to confirm the formation of solid solution, detection and identification of phases, and their quantitative determination. X-ray diffraction is used for determined the texture of metallic fuels after hot processing like rolling and extrusion during the fabrication of metallic fuels.

4.3.4 Oxygen to Metal Ratio (O/M)

The oxygen to metal ratio (O/M), where M is U + Pu which is one of the important specifications of oxide nuclear fuel (U, Pu) O_{2+X} or UO_{2+X}. O/M affects certain physical and chemical properties of nuclear fuel affecting its performance in the reactor [21]. Change of O/M affects the thermal conductivity and melting point of the fuel. It controls the chemical state of the fission products and their interaction with the fuel. It also governs the potential for the interaction of the liquid sodium coolant with the fuel matrix to form low density sodium uranate in the event of minor breaches in the cladding. The O/M ratio also affects the number of phases in mixed oxide for a given U/Pu ratio and the extent of deviation from the stoichiometric composition.

O/M is more important in the case of fast reactor fuel than thermal reactor fuel. O/U ratio of uranium oxide in the beginning of irradiation is close to stoichiometric composition, i.e., O/U is 2.00. The fission of ^{235}U is less oxidizing than that of ^{239}Pu as it produces a smaller number of noble metal fission products (Ru, Rh, and Pd). The released oxygen from UO_2-based fuels during fission in thermal reactors is consumed mainly by rare earths fission products forming respective oxides with no significant change in oxygen potential. But in the case of mixed oxide (U, Pu)O_2 fuel in fast reactors, the fission is mainly due to Pu which forms small amount of noble metal fission products leading to increase in oxygen potential with burn-np. Some of the fission products such as Cs, Te, and iodine are suspected to cause increased internal attack of stainless-steel cladding material in presence of higher oxygen at the clad surface.

Methods of measurement

Being a critical specification, the method of measurement should be very precise and accurate. The common method used is gas equilibration method. This method involves measuring the change of weight that occurs when an oxide sample is heated in an atmosphere in which oxygen potential is maintained at approximately -100 kcal/mole [22]. An atmosphere with oxygen potential close to -100 kcal/mole produces an O/M ratio of 2.00 from the starting oxide which is either hypo or hyper stoichiometric and contains 0–100 mol% of PuO_2 in UO_2. To achieve this, sintered mixed oxide is heated at 1073 K for 3 h in flowing Ar and 8% v/v H_2 containing water at a partial pressure of 4 torr obtained by passing the gas mixture over water trap maintained at 273 K. The sample is then cooled up to 573 K in the same flowing gas (moist argon/8% H_2) and then in dry gas of 92%Ar and 8% H_2 up to room temperature. Thus, the thermal equilibrium of sintered U–Pu mixed oxide or uranium oxide in moist argon hydrogen atmosphere produces $MO_{2.000}$. The O/M values of the equilibrated samples are calculated from the change in weight with respect to the reference state (stoichiometric composition) using the expression:

$$O/M = 2 \pm x = 2 \pm \left\{ \frac{\Delta W}{W} \right\} \times \left\{ \frac{M}{16} \right\} \qquad (4.4)$$

where W is the weight of sample at O/M $= 2.00$, ΔW is the change in weight relative to the stoichiometric sample weight W, and M is the molecular weight of stoichiometric mixed oxide.

The oxygen potential of ~100 kcal/mole can also be achieved by CO/CO_2 gas mixture in ratio of 10/1. Considering the accuracy of the balance and precision of the measurement of the weight ~10 μg, the total error in the measurement of the weight is ±100 μg. This is equivalent to maximum error of ±0.002 in O/M for 1 g of the initial weight of the sample.

O/M can also be determined by dissolution methods, galvanic cell technique, solid state coulometric titrations, and X-ray diffraction method.

4.3.5   Trace Metal Assay

The thermo-physical and metallurgical properties are highly dependent on the presence of the metallic constituents even at trace level. Neutron being the primary particle responsible for nuclear fission, the economy of neutron is of prime importance. In view of this, the elements with high neutron absorption cross section (B, Cd, Dy, Eu, Gd, and Sm) have stringent specification limits. Some of the rare earth elements, i.e., Ce, Nd, Tb, etc., and transition elements like Mn, Co, and Ni can form activation products. Metals like Zn having low melting point are responsible for liquid metal embrittlement altering the fuel structure and consequent failure whereas the refractory metals (W, Mo, Ta) do not allow the fuel to creep, thereby increasing fuel clad mechanical interaction resulting in the damage of the clad.

The process pick up and the condition of the process equipment is monitored by analyzing common elements like Fe, Cr, and Ni. Additionally, the presence of higher concentration of Fe and Ni causes sintering problems. The presence of light elements

Fig. 4.9 General Schematic of a Spectrometer

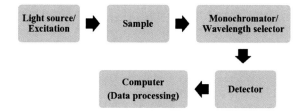

leads to decrease in density and fissile content of the fuel materials. In general, the specification limits for fast reactor fuel are more relaxed as compared to thermal reactor fuel from the point of view of fissile content.

Spectrometric techniques

Atomic spectroscopy consists of three techniques, namely atomic absorption, atomic emission, and atomic fluorescence. Among these, atomic absorption spectroscopy (AAS) and atomic emission spectroscopy (AES) are most commonly being used.

Atomic absorption is the process that occurs when a ground state atom absorbs energy in the form of light of a specific wavelength and is elevated to an excited state. The amount of light energy absorbed at this wavelength will increase as the number of atoms of the selected element in the light path increases. The relationship between the amount of light absorbed and the concentration of the analyte present in known standards can be used to determine unknown concentrations by measuring the amount of light they absorb.

Atomic emission spectroscopy is a process in which the light emitted by the excited atoms or ions is measured. The emission occurs when sufficient thermal or electrical energy is available to excite a free atom or ion to an unstable energy state. Light is emitted when the atom or ion returns to a more stable configuration or the ground state. The wave lengths of light emitted are specific to the elements which are present in the sample. The schematic of the components of a spectrometer is shown in Fig. 4.9.

Spectrometric techniques are regularly being used for the determination of trace metals in nuclear fuels, viz. UO_2, PuO_2, (U–Pu)O_2, ThO_2 as well as metal-based fuels. Techniques based on atomic emission (AES) or absorption spectrometry (AAS) are chosen according to the application, i.e., nature of the metallic impurity. DC Arc carrier distillation technique based on AES has largely been used for common metallic impurities like B, Cd, Fe, Cr, Ni, Co, V, Zn, Mo, Si, W, etc., while inductively coupled plasma-atomic emission spectrometry (ICP-AES) is used for the determination of rare earths. This is primarily due to the refractory nature of the rare earths which makes the determination difficult with the desired sensitivity by DC Arc. DC Arc as an excitation source is advantageous for the analysis of refractory matrices because of the possibility of using carrier distillation technique. Carrier distillation technique [23] provides a method for physical separation of analytes of intermediate volatility from refractory samples. In order to achieve this, a carrier with intermediate volatility is added at a few percentage levels to the sample. Addition of carrier results in maintaining sample temperature nearly constant around the boiling point of the carrier. The halide carriers also facilitate volatilization of analytes by conversion of their relatively refractory oxides to volatile halide forms due to thermos-chemical reactions occurring in the electrode crater. A number of carriers and their combinations have been reported in the literature. Carriers commonly used are silver, alkali, and alkaline earth halides, Ga_2O_3, ZnO, Sb_2O_3, etc. The sensitivity of the method largely depends upon the choice and proportion of the carrier, while the precision of determination is governed by the excitation source like arc/plasma and detection device. Though DC Arc has been proved to be very useful for the analysis of nuclear fuels in solid form it suffers from poor precision.

The techniques like laser ablation inductively coupled plasma mass spectrometry (LA ICP MS) [24], glow discharge mass spectrometry (GDMS), etc., meant for direct solid sample analysis cannot be used for these types of samples due to the unavailability of the matrix matched certified reference materials. The other option is to bring the matrix into solution and then analyze it by ICP-MS, ICP atomic emission spectroscopy (ICP-AES), or ion chromatography (IC). Among these, ICP-AES is regularly being used owing to its versatility of analyzing major, minor, and trace elements like rare earths particularly Dy, Eu, Gd, and Sm as well as other transition elements. It is suitable for glove box adaptation and easy to operate under glove box condition. It has better sensitivity and precision of determination. However, due to the high excitation temperature in ICP, tolerance of major matrix in the aqueous phase is extremely low as compared to that in the DC Arc source. The elements need to be chemically separated due to matrix interference and then analyzed. Due to the complexity of different fuel materials, separate solvent extraction systems have been optimized to achieve efficient performance. The extent of recovery of the elements from the major matrices is verified by radiotracer technique. Hence, additional number of liquid–liquid contacts have to be carried out before the analysis by ICP-AES technique. 30% TBP/Xylene/4 M HNO_3 system is the obvious choice for the separation of UO_2, PuO_2, and $(U–Pu)O_2$-based matrices. The various solvent extraction systems adopted in the analysis of nuclear materials are listed in Table 4.1.

Table 4.1 Solvent extraction systems adopted in the analysis of nuclear materials

Matrix	Dissolution	Solvent
$U_3O_8/U_3O_8–PuO_2$	6M HCl	20% TnOA/Xylene/6M HCl
PuO_2	$HNO_3 + HF$	30% TBP/Xylene/4M HNO_3
U_3Si_2	$HNO_3 + HF$	40% TBP
ThO_2	$HNO_3 + HF$ (microwave)	30% Cyanex 923/Xylene/4M HNO_3
Al–U and Al–Pu	HCl	TnOA

Atomic absorption spectrometry (AAS) [25] is based on the atomization of the samples followed by absorption of characteristic radiation by the ground state/low-level excited atoms. AAS instrument requires characteristic radiation source, high-temperature source for generation of atomic vapor (atomization) from sample, and detector system capable of measuring absorption signal. Hollow cathode lamp (HCL) is the preferred source of characteristic radiation as it provides a narrow line emission. Two modes of atomization prevalent in AAS technique are flame and electro thermal atomization (ETA). ETA AAS has several advantages like requirement of small volume/mass of the sample, analysis in solid form, improved sensitivity, and lower detection limits. The elements that are refractory in nature (W, Si, rare earths, etc.) as well as those elements that form stable carbides have poor atomization efficiency in AAS and hence difficult to analyze routinely with required detection limits. ETA AAS also has poor precision of determination as compared to flame AAS.

Spectrometric techniques therefore offer a comprehensive solution for trace metal assay of nuclear materials. Conventional methods like DC Arc carrier distillation technique and ICP AES are regularly being used for this purpose.

4.3.6   Non-metallic Impurities

Apart from affecting the neutron economy, non-metallic impurities may react with the cladding and cause failure. If gaseous, they may cause of clad ballooning due to increased internal pressure and even rupture of clad. There is also a chance that they may react with some of the fission products and influence their mobility.

4.3.6.1   Hydrogen

Hydrogen is one of the very critical specifications in thermal reactors, and the allowable limit is less than one ppm mainly because zircaloy is used as the clad material, and the presence of hydrogen may lead to embrittlement. The solubility of hydrogen in zirconium is 60 ppm at 573 K and 1 ppm at room temperature.

In fast reactors, the specification of hydrogen is relaxed as compared to thermal reactors since hydrogen can diffuse out easily from stainless-steel clad under the operating temperature of the reactor.

The most common method adopted for the determination of hydrogen is inert gas fusion (IGF) technique [26]. The fusion is carried out at approximately 2273 K without using any flux. The released gases are passed through Schutze (I_2O_5) reagent to selectively oxidize CO to CO_2 without affecting hydrogen. CO_2 formed is trapped by ascarite (NaOH) and anhydrone (anhydrous magnesium perchlorate). H_2 and N_2 are then separated using gas chromatographic column. Hydrogen comes out first and is detected by thermal conductivity detector.

4.3.6.2   Chlorine and Fluorine

Various types of acids are used during the mining of uranium, thorium, and reprocessing of the spent fuel from the reactors to get plutonium. These operations may lead to pick up of chloride and fluoride as impurities in the fuel pellet. The chloride and fluoride impurities can cause de-passivation of the oxide film on the internal surface of the clad tube. Presence of moisture further accelerates the chances of corrosion due to the formation of respective acids. The specification is less stringent in carbide and nitride fuels as compared to oxide fuels because of less moisture content.

Chlorine and fluorine are determined by pyro hydrolyzing [27] the sample at 1173 K under moist N_2/Ar condition. The liberated Cl and F as HCl and HF are collected in an acetate or carbonate buffer depending on the detection method employed for Cl and F. They are determined by spectrophotometry, ion selective electrode, or by ion chromatography [28].

4.3.6.3   Carbon

Carbon is one of the critical impurities in both fast and thermal reactor fuel. The source of carbon is mainly the binder and lubricant used during the fabrication of the fuel pellets. Though these materials evaporate during the dewaxing stage or sintering, there is always a probability of trace material remaining in the pellet. In case of UO_2/MOX sintered pellets, it may be retained in CO or carbide form. In case of thermal reactors, this uranium carbide reacts with coolant water and leads to fuel fragmentation and dispersal in presence of zircaloy clad defects. In the case of fast reactors, uranium hyper carbides may react with stainless-steel clad and cause clad carburization.

Carbon is the major constituent of carbide fuel, and C/M is an important parameter to be measured to control the hyper carbide (M_2C_3) at the specified optimum level. In nitride fuels also, carbon as impurity is to be controlled to avoid SS clad carburization which results in embrittlement.

Determination of carbon involves the combustion of carbon present in the sample to carbon dioxide (CO_2), purification from the accompanying gases, and detection by infra-red detector [29] or thermal conductivity detector. In thermal conductivity detection technique, a known amount of sample in the form of powder is subjected to combustion at more than 1273 K in a quartz lined graphite crucible or alumina crucible containing copper or tungsten as accelerator in flowing purified oxygen atmosphere by induction heating. The purified and concentrated CO_2 is detected by the thermal conductivity detector maintained at 318 K. In case of infra-red absorption method, CO_2 is separated and measured. Commercial analyzers are available to measure the infra-red absorption of CO_2 for the determination of carbon. In combustion cum electrical conductivity detection technique, sample is heated in flowing purified oxygen using resistance furnace at 1173 K. Purified CO_2 is absorbed in baryta solution. Barium hydroxide reacts with carbon dioxide and forms sparingly soluble barium carbonate. This results in change in the electrical conductivity of the baryta solution. The concentration of carbon content can be evaluated from the calibration plot drawn between magnitudes of change of electrical conductivity and the concentration of carbon content.

Free carbon and combined carbon in the fuel are evaluated by subjecting the independent aliquots of samples to combustion at two different temperatures or by dissolution method.

4.3.6.4   Nitrogen

The main source of nitrogen in oxide fuels is the raw material itself, i.e., uranium and plutonium oxides used for the fabrication of the fuel may get contaminated with nitrogen picked up during the processing steps from mining to purification. The other source is the nitric acid used in almost all the processing steps due to its non-corrosive behavior unlike hydrochloric acid. The specification of nitrogen is very stringent for thermal reactor fuels. But there is no specification for nitrogen for oxide fuel for fast reactors. In the case of carbide fuels, it may favor release of CO, which leads to carburization. Nitrogen may also interact with the cladding material to form zirconium oxynitrides. These nitrides lead to poor corrosion resistance of zircaloy. In the case of steel, chromium reacts with nitrogen and gets precipitated at temperatures ranging from 773 to 1323 K. This leads to embrittlement reducing the toughness of steel.

Nitrogen is picked up during the fabrication of carbide fuel in inert gas atmosphere. Nitrogen impurity present in the carbide fuel is completely soluble in MC phase and is present as MCNO. Nitrogen acts like oxygen and favors the formation of CO leading to carburization of clad.

Nitrogen is determined by i) inert gas fusion (IGF) technique [30] and Kjeldhal's technique [31]. In inert gas fusion technique, the sample is fused in a graphite crucible at high temperature (2773–3273 K), and filtered and purified nitrogen gas is detected by thermal conductivity detector. In Kjeldhal's technique, the sample is dissolved in acid to convert nitride nitrogen into ammonium salt, and the liberated ammonia after distillation is detected by spectrophotometry or ion chromatography. The demerit of this technique is that trace amount of free nitrogen will not be detected.

4.3.6.5   Oxygen

Determination of oxygen in carbide fuels is by inert gas fusion (IGF) technique described earlier. CO formed as a result of reaction of oxygen with graphite is detected, and oxygen is estimated.

4.3.6.6 Sulfur

During the fabrication of nuclear fuels, sintering of the pellets is carried out in argon/nitrogen-hydrogen atmosphere to obtain the required density and other characteristics. During sintering, sulfur, if present, will react with hydrogen to release H_2S which may lead to disintegration of the pellets. Sulfur is usually determined by combustion cum IR detection method. During the combustion of sample in oxygen atmosphere, SO_2 is formed. After purification of the gases released, SO_2 is be detected by IR at 7400 or 8700 nm. It can also be determined by combustion cum mass spectrometry.

4.3.7 Total Gas Analysis

The total gas content is defined as the amount of non-condensable gases released from the material when it is heated at the specified temperature for the specified time under static vacuum. The temperature and duration up to which the pellet is to be heated depend upon the center-line temperature of the fuel expected in the reactor based on the type of reactor. Stringent specifications are laid down for sintered fuel pellets for total gas content. This is because the occluded gases apart from pressurizing the clad have a detrimental effect on the thermal conductivity of the cover gas.

Hot vacuum extraction (HVE) technique is employed to measure total gas content. Generally, hydrogen (H_2), methane (CH_4), nitrogen (N_2), carbon monoxide (CO), and carbon dioxide (CO_2) are present in the gases released from the nuclear fuel pellets. The source for these gases in fuel pellets may be due to entrapping of gases during fabrication, sintering, and reaction products at reactor operating conditions. The composition of the total gas also needs to be determined to find out the source of the gas as well as its implications. The gas composition is determined by coupling with quadrupole mass spectrometry (QMS) technique and is called HVE-QMS. Instead of QMS, gas chromatograph (GC) also can be used known as HVE-GC.

The sintered pellet is heated, under static vacuum conditions, for the specified time at the specified temperature, and the released gases are extracted into a pre-calibrated (volume) chamber. The pressure of the system is measured, and finally, the volume of the gases is calculated as given below.

$$V_T = (PS - PB)V_S T_{STP}/(T_S P_{STP} 1000 w_s) \tag{4.5}$$

where V_T is the volume of total gas (cc/g), P_S is the pressure of the gases released during sample analysis (atm), P_B is the pressure of the gases released during blank analysis (atm), V_S is the volume of the collection chamber, T_{STP} is the standard temperature (273 K), T_R is the room temperature (298 K), P_{STP} is the pressure at STP (1 atm), W_S is the mass of the sample (g), and 1000 is for conversion from liter to cm^3.

4.3.8   Dissolution Test

The mechanical mixing route used in the fabrication of MOX pellets can lead to homogeneity problems if the mixing or milling techniques and sintering cycles are not proper. The inhomogeneity can be classified into (i) macro-inhomogeneity and (ii) micro-inhomogeneity. Though it is possible to avoid macro-inhomogeneity by adopting good milling and sintering parameters, it is not possible to eliminate micro-inhomogeneity of Pu distribution. The gross inhomogeneity can lead to serious problems during reprocessing of the spent fuel due to difficulty in dissolution of Pu rich phase in nitric acid, as addition of HF is not also desirable in the reprocessing process. It may also create hot spot condition during the operation of the reactor and affect Doppler coefficient. The dissolution method involves refluxing of sintered pellet in 10 M nitric acid, and the residue (if any) is examined for its plutonium content. If the pellet dissolves completely without having any plutonium residue, it is generally presumed to be homogeneous.

4.3.9   Weld Chemistry Test

The weld chemistry test is specified to ensure that impurities like hydrogen, nitrogen, and oxygen which affect the mechanical and corrosion properties of zircaloy end plug welds are well within specification limits. While hydrogen is important due to hydriding problem, the presence of nitrogen and oxygen above specified limits may cause increased corrosion. The analysis for the impurities is carried out by commercial determinators.

4.3.10   Cover Gas Analysis

Helium finds extensive applications in nuclear industry due to its chemical inertness, radiation stability, and thermal conductivity. It is used as a blanket gas in pressurized heavy water reactor (PHWR), as bonding material between fuel and clad in most of the thermal and fast reactors and as a coolant in gas cooled reactors. The specification of helium purity varies with application. It is particularly stringent for its use as a bonding material in LWR/fast breeder reactor oxide or carbide fuels. Helium in the fuel pin is analyzed by gas chromatographic technique. The gas chromatograph is equipped with a thermal conductivity detector, a molecular sieve, sub-atmospheric gas sample transfer facility, and a data transfer system. Procedures have been standardized for the trace analysis of H_2, O_2, and N_2 in small quantities of He filled in stainless-steel tubes employing sub-atmospheric injection technique.

The specifications, both physical and chemical of nuclear fuels for different types of reactors, vary depending on the design. A compilation of important specifications of fuel pellets and fuel pins and the methods used for determining them for thermal and fast power reactors are given in Tables 4.2, 4.3 and 4.4. This compilation gives the readers an understanding of the elements of specifications and methods (conventional and advanced) in a nutshell. It should be understood that the exact values should be taken from the approved specification document for the reactor under consideration.

4.4 Summary

An attempt has been made in this chapter to present the importance of quality, and how it is achieved in the manufacture of nuclear fuels. Important terms used in the field of quality control are defined in the beginning of the chapter before describing the quality control techniques. Quality control procedures at different stages of manufacture, viz. powder, pellet/fuel core, fuel element, and fuel assembly are given in detail. Initially, physical quality control techniques are presented. The non-destructive testing techniques used are clearly explained. A number of photographs/illustrations are also given to make the readers understand the techniques. Inspection of fuel pellets, end plug welds of fuel pins, and inspection of final fuel elements and assemblies are covered. The section on chemical quality control describes the techniques for chemical characterization of the fuels using various techniques. The techniques are described explaining the basic theory involved in the determination of the chemical characteristics. Techniques for determining heavy metal content, metallic and non-metallic impurities, isotopic analysis and oxygen to metal ratio, and others are explained in the chapter. Quality control techniques both conventional and advanced techniques are presented. Additionally, typical specifications of fuels for thermal and fast reactor and the techniques used are tabulated for the benefit of the readers.

Questions

1. What are the physical characteristics of the fuel powder to be checked?
2. Describe the physical inspection of fuel pellets.
3. Discuss the role of metallography during quality control of nuclear fuels,
4. Explain the process of radiography testing and its application in quality control of fuel elements.
5. Explain alpha autoradiography and its significance.
6. Describe the various tests to be carried out to monitor the quality of fuel pin welds.
7. Explain the principle of ultrasonic inspection.
8. Describe the leak testing methods for checking the integrity of fuel elements.
9. Explain gamma scanning and its use in the quality control of nuclear fuel elements.
10. What are the chemical methods used for determination of heavy metal content in nuclear fuels?
11. How is isotopic composition is determined?
12. List the different mass spectrometric techniques used.

Table 4.2 Important physical characteristics of fuel pellets and fuel pins and test methods

	Test methods	
	Conventional	Advanced
A. Fuel pellets		
1. Dimensions	Micrometer, Go-no go Gauges	Automated pellet inspection system, Image vision system, Laser based system
2. Density	Geometric or immersion method	Automated pellet, inspection system, image vision system
3. Physical integrity	Visual Examination, Magnifier	Automated pellet Inspection system, Image vision system
4. Grain size	Microstructural evaluation	Image analysis
5. Micro-homogeneity (MOX pellets)	Alpha autoradiography	Image analysis
B. Fuel pins		
1. Dimensions	Metrological inspection, go-no go gages	LVDT, Laser-based system
2. Leak testing	Mass spectrometer-based Helium leak Detector	
3. End plug welds		
i. TIG/LASER welds	Metallography(destructive), X-radiography	Micro-focal radiography, Flat panel detector system, Ultrasonic testing
ii. Resistance welds	Metallography (destructive)	Ultrasonic testing
4. Internal components	X-radiography, Gamma scanning, Gamma autoradiography	
5. Decontamination check	Counting system	

Table 4.3 Important chemical specifications of nuclear fuel pellets and test methods

Impurity	Thermal reactors		Fast reactors	Analytical technique (Conventional)	Advanced analytical technique
	Natural UO$_2$	Enriched UO$_2$	Mixed oxide (U, Pu) O$_2$ (MOX)		
Metallic impurities (in ppm)					
Ag	1	25	20	DC Arc-AES for: Al, Be, B, Ca, Cd, Co, Cr, Cu, Fe, Li, Mg, Mn, Mo, Na, Ni, Pb, Si, Sn, V, W, and Zn	ICP-OES, ICP-MS ETA-AAS, HPLC, ion chromatograph, ETA
Al	250	400	250		ICP
B	*	1	3	ICP-AES: other refractory and non-volatile elements, e.g., Dy, Eu, Gd, Sm, etc	
Be			10		
Ca	See specifications for Ca + Mg		See specifications for Ca + Mg		
Cd	*	1	See specifications for Cd + Dy + Sm + Gd		
Co	100	75	100		
Cr	250	400	500		
Cu	20	400	100		
Eu	*		3		
Fe	500	400	1500		
Gd	*	1	See specifications for Cd + Dy + Sm + Gd		
Mg	See specifications for Ca + Mg	200	See specifications for Ca + Mg		
Mn	20		200		
Mo		400	300		

(continued)

Table 4.3 (continued)

		Thermal reactors		Fast reactors	Analytical technique (Conventional)	Advanced analytical technique
	Impurity	Natural UO$_2$	Enriched UO$_2$	Mixed oxide (U, Pu) O$_2$ (MOX)		
	Na		400			
	Ni	250	400	200		
	Pb		400	100		
	Si	500	200	500		
	Sm	*		See specifications for Cd + Dy + Sm + Gd		
	Sn			25		
	V		400			
	W		100	200		
	Zn		400	500		
	Ca + Mg	200		800		
	Cd + Dy + Sm + Gd	*		20		
Non-metallic impurities (ppm)	C	100	200	300	Inert gas fusion technique using IR detector	
	N	75	100	–	Kjeldahl's method	Inert gas fusion technique using TC detector
	Equivalent hydrogen content	1	Moisture content in the rod is specified	3	Inert gas fusion technique	

(continued)

Table 4.3 (continued)

	Impurity	Thermal reactors		Fast reactors	Analytical technique (Conventional)	Advanced analytical technique
		Natural UO$_2$	Enriched UO$_2$	Mixed oxide (U, Pu) O$_2$ (MOX)		
	Cl	25	15		Pyro-hydrolysis	Electro chemical method using ion selective electrode, spectrophotometry, IC
	F	15	25			
	Cl + F			50		
Equivalent boron content (EBC) (ppm)		<1.5 (*to be included in EBC)	<2.5			
Total impurities (ppm)		<1500	<2500	<5000		
Heavy metal content (minimum)	U	U 87.7%	U 87.7%	U + Pu 87.7	Davis–Gray method	Ti(III) method, WDXRF-based method
	Pu				Drummond and Grant method	Ti(III) method, WDXRF-based method same
Gas content in the pellet		<0.04 cc/gm inclusive of water vapor		<0.1 cc/g at 1650 °C (Excluding Hydrogen)	Hot vacuum quadrupole mass spectrometer (HVQMS)	Same
O/M		1.99–2.02	1.99–2.03	1.96 to 2.00	Thermogravimetric method involving gas equilibration	XRD-based method involving evaluation of cell parameter, thermogravimetric method

(continued)

Table 4.3 (continued)

| | | Thermal reactors | | Fast reactors | | |
	Impurity	Natural UO_2	Enriched UO_2	Mixed oxide (U, Pu) O_2 (MOX)	Analytical technique (Conventional)	Advanced analytical technique
Macro-homogeneity				Residue after dissolution shall not exceed 1.0 wt.%	Dissolution test by refluxing with 12M HNO_3 for 8 h	

Table 4.4 Other tests

Characteristic	Technique used
1. Isotopic analysis	Thermal ionization mass spectrometer
2. Weld corrosion test for zircaloy welds	Autoclave test
3. Weld chemistry of zircaloy welds	Commercial determinators for individual elements
4. Cover gas analysis of the fuel pin	Gas chromatograph

13. How is X-ray diffraction and X-ray fluorescence are useful in the quality control of nuclear fuels?
14. How is oxygen to metal (O/M) ratio is determined?
15. What are the various test methods for trace metal assay?
16. Describe the techniques used for determining non-metallic impurities, hydrogen, chlorine, fluorine, carbon, nitrogen, oxygen, and Sulfur.
17. What is dissolution test and explain its significance.
18. What is weld chemistry test?
19. Explain cover gas analysis and total gas analysis.

References

1. J.M. Juran, *Quality Control Handbook* (McGraw Hill Book Co., New York, 1974)
2. Quality Assurance, BSI Handbook 22, 1983.
3. Guide to powder characterization, American Laboratory, media 20/document, 2019.
4. D. Mukherjee, K. Majeesh, C. Baghra, T. Soreng, J.P. Panakkal, H.S. Kamath, Online integrated visual inspection and sorting system for fast reactor fuels. Nucl. Eng. and Desi. **240**, 1392–1396 (2010)
5. R. Halmshaw, *Physics of industrial radiology* (Elsevier publishing company Inc., New York, 1966)
6. D. B. Sathe, Nagendra Kumar, N. Walinjkar, A. K. Hinge, Amrit Prakash, Mohd. Afzal, J.P. Panakkal, Evaluation of end plug welds of nuclear fuel pins with X-ray real time radiography system, Second International Conference on Advances in Nuclear Materials (ANM 2011), BARC, Mumbai, Feb 9–11, 2011.
7. D. Mukherjee, M. Saxena, D.B. Sathe, J.P. Panakkal, H.S. Kamath, Development of an ultrasonic testing technique for the NDT of Breeder Reactor end cap weld. Mater. Eval. **64**, 1097–1101 (2006)
8. O.J. Wick, *Plutonium handbook* (American Nuclear Society, Volume-I, 1980)
9. K.V. Vrindadevi, T. Soreng, D. Mukherjee, J.P. Panakkal, H.S. Kamath, Non-destructive determination of PuO_2 content in MOX fuel pins for fast reactors using Passive Gamma scanning. J. Nucl. Mater. **399**, 122–127 (2010)
10. J.P. Panakkal, D. Mukherjee, H.S. Kamath, Nondestructive Evaluation of Uranium-Plutonium Mixed Oxide (MOX) fuel elements by Gamma Autoradiography, Proceed. 17th WCNDT, Shanghai, Oct 25–28, 2008.
11. M.V. Ramaniah, Analytical chemistry of fast reactor fuels-A review. Pure & Appl. Chem. **4**(4), 889–908 (1982)

12. J.L. Ryan, L.A. Bray, E.J. Wheelbright, G.H. Bryan, Transuranium elements, a Half century, L.R. Morss, and J. Fuger (Eds.), American Chemical Society, Washington DC, 1992, 288.
13. W. Davies, W.A. Gray, Rapid and specific titrimetric method for the precise determination of uranium using iron(II) sulphate as reductant. Talanta **11**, 1203 (1964)
14. N. Gopinath, J.V. Kamat, H.S. Sharma, S.G. Marathe, H.C. Jain, Coulomatric determination of uranium by successive addition method. Bull. Electrochem. **5**, 805 (1989)
15. J.A. Drummond, R.A. Grant, Potentiometric determination of plutonium by argentic oxidation, ferrous reduction and dichromate titration. Talanta **13**, 477–488 (1966)
16. E.P. Bartin, Introduction to X Ray spectrometric analysis, Springer Science & Business Media, 2013.
17. A. Pandey, F. Khan, A. Kelkar, P. Purohit, P. Kumar, V. Kumar, D. Sathe, R.B. Bhatt, P.G. Behere, Non-destructive determination of uranium and plutonium in annular (U, Pu) O_2 mixed oxide sintered pellets by wavelength dispersive X-ray fluorescence spectrometry. J. Radioanal. Nucl. Chem. **326**, 423–433 (2020)
18. J.S. Beckerand, H.J. Dietze, Inorganic Mass spectrometric methods for trace, ultratrace, isotope and surface analysis. Int. J. Mass Spectr. **97**, 1–35 (2001)
19. B.D. Cullity, Elements of X-ray Diffraction, Addison-Wesley Publishing, 1956.
20. D.L. Bish, S. Howard, Quantitative phase analysis using the Rietveld method. J. Appl. Cryst. **21**, 86–91 (1988)
21. C.E. Johnson, I. Johnson, P.E. Blackburn, C.E. Crouthamel, Effect of oxygen concentration on properties of fast reactor mixed oxide fuel. React. Tech. **15**(4), 303 (1972)
22. C.E. Mcneilly, T.D. Chikalla, Determination of oxygen/metal ratios for uranium, plutonium and (U, Pu) mixed oxides. J. Nucl. Mater. **3**, 77 (1971)
23. B. F. Scribner, H. R. Mullin. J. Res. Nat. Bur. Stan., RP 1753, (1946) 37, 379–389.
24. B. Fernandez, F. Claverie, C. Pecheyran, O.F.X. Donard, Direct Analysis of solid samples by fs-LA-ICP-MS). Trends Anal. Chem. **26**, 951–966 (2007)
25. Bernhard Weltz. *Atomic Absorption Spectrometry*. Michael Sparling, Wiley VCH, Third Edition (1999).
26. K.V. Chetty, J. Radhakrishna, Y.S. Sai, N. Balachander, P. Venkatramana, P.R. Natarajan, Inert gas fusion for the determination of hydrogen. Nitrogen and Oxygen in UO_2, UC and UN, Radiochem. Radioanal. Lett. 58 (1983)161.
27. J.C. Warf, Analytical chemistry of the Manhattan Project National Nuclear Energy Series, Div III, Volume I, Mc Graw Hill, NewYork, (1950).
28. A. Pandey, A. Kelkar, R.K. Singhal, C. Baghra, A. Prakash, M. Afzal, J.P. Panakkal, Effect of accelerators on thoria based nuclear fuels for rapid and quantitative pyrohydrolytic extraction of F^- and Cl^- and their simultaneous determination by Ion Chromatography, J. Radioanal. Nucl. Chem. February 2012, 2012.
29. G.C. Swanson, M.C. Burt, M.C. Lambert, R.W. Stromatt, D.L. Shawell, S.A. Hedl 1580 FP (1978).
30. B.A. Taylor, Graphite resistance furnace for use in the inert gas fusion method of determining gases in solids, A E R E –R 5763 (1968).
31. J.E. Rein, G.M. Matlack, G.R. Waterbury, R.T. Phelps, C.F. Metz, LA-4622 (1971).

Further Reading

1. Nuclear, Solar and Geothermal Energy, Annual Book of ASTM Standards, Section 12, vol. 12.01
2. Nondestructive Testing Handbooks on Radiography and Radiation Testing (RT), Leak Testing (LT), Ultrasonic Testing (UT), Visual Testing (VT) and Overview, ASNT
3. *Proceedings of International Conference on Characterization and Quality control of Nuclear Fuels (CQCNF 2012)*, Hyderabad, 27–29 February 2012 (IAEA, 2012)

4. *ASM Handbook of Metallography and Microstructures*, vol. 9 (ASM, Ohio, 2004)
5. M. Fayed et al. (eds.), *Handbook of Powder Science and Technology* (Chapman and Hall, New York, NY, 1997)
6. IAEA-TECDOC-1166 Advanced methods of process/quality control in nuclear reactor fuel manufacture, in *Proceedings of a Technical Committee Meeting held in Lingen, Germany, 18–22 October 1999* (1999)
7. M.T. Kelley, W.L. Belew, G.V. Pierce, W.D. Shults, H.C. Jones, D.J. Fisher, Controlled-potential polarography and coulometry as microanalytical techniques. Microchem. J. **19**, 315–333 (1966)
8. H. Bailly, D. Menessier, C. Prunier, *The Nuclear Fuel of Pressurized Water Reactors and Fast Neutron Reactors Design and Behavior* (Lavoisier Publishing, 1990)
9. G.L. Booman, W.B. Holbrook, J.B. Rein, Coulometric determination of uranium (VI) at controlled potential. Anal. Chem. **29**, 219 (1957)
10. K. Klukkola, High temperature Electrochemical Study of uranium Oxide in the UO_2–U_3O_8 region. Acta. Chemica. Scandinavica, 16–0327 (1962)
11. K. Klukkola, C.J. Wagner, EMF measurements on Galvanic cells with solid electrolytes. Electrochem. Soc. **104**, 379 (1957)
12. J.L. Drummond, V.M. Sinclair, Some aspects of the measurement of the oxygen to metal ratio in solid solutions of uranium and plutonium dioxides, in *Sixth Conference on Analytical Chemistry in Nuclear Reactor Technology*, Gatlinburg, Tennessee, October 1962 (1962)
13. A.L. De Souza, M.E.B. Cotrim, M.A.F. Pires, An overview of spectrometric techniques and sample preparation for the determination of impurities in uranium nuclear fuel grade. Microchem. J. **106**, 194–201 (2013)
14. R.K. Malhotra, K. Satyanarayana, Estimation of impurities in reactor grade uranium using ICP-AES. Talanta **50**, 601–608 (1999)
15. American Standards for testing of materials, C696 Standard Test methods for chemical, Mass spectrometric, and spectrochemical Analysis of Nuclear Grade uranium dioxide powders and pellets, ASTM 27.120.30 (2019)
16. American Standards for testing of materials, C698- Standard Test methods for chemical, Mass spectrometric, and spectrochemical Analysis of Nuclear Grade Mixed Oxides (U, Pu)O_2 powders and pellets, ASTM 27.12030 (2016)
17. A. Saha, S.B. Deb, B.K. Nagar, M.K. Saxena, Determination of trace rare earth elements in gadolinium aluminate by inductively coupled plasma time of flight mass spectrometry. Spectrochim. Acta. Part B **94–95**, 14–21 (2014)
18. G.V. Ramanaiah, Determination of yttrium, scandium and other rare earth elements in uranium-rich geological materials by ICP-AES. Talanta **46**, 533–540 (1998)
19. V.C. Adya, M. Kumar, A. Sengupta, V. Natrajan, Inductively coupled plasma atomic emission spectrometric determination of indium (In) and gallium (Ga) in thorium matrix after chemical separation using cyanex 923 extractant. At Spectros. **36**, 261–265 (2015)
20. P. Verma, K.L. Ramakumar, Determination of alkali and alkaline earth elements along with nitrogen in uranium based nuclear fuel materials by ion chromatography (IC). Anal. Chim. Acta. **601**, 125–129 (2007)
21. C.E. Crouthamel, C.E. Johnson, Thiocyanate spectrophotometric determination of Molybdenum and tungsten. Anal. Chem. **24**(1780) (1952)
22. D. Srivastava, S.P. Garg, G.L. Goswami, Thermodynamic analysis of mixed carbide, carbonitride and nitride fuels for fast breeder reactors. J. Nucl. Mater. **161**, 44 (1989)
23. S.G. Kulkarni, G.A. Ramarao, V.K. Manchanda, P.R. Natrajan, Gas Chromatographic analysis of high purity helium at ambient pressure. J. Chrom. Sci. **23**, 68–74 (1985)
24. A. Skoog Douglas, H.F. James, R. Crouch Stanley, *Principles of Instrumental Analysis*, 6th edn. (Thomson Brookes, 2007)
25. S.K. Aggarwal, H.C. Jain, *Introduction to Mass Spectrometry*, Indian society for mass spectrometry, Perfect Prints (1997)

Chapter 5
Thermophysical and Thermochemical Properties of Nuclear Fuels

S. Anthonysamy and D. Das

This chapter deals with the phase equilibria, thermophysical and thermochemical properties of advanced nuclear fuel materials. The main focus is on actinide oxides, carbides and nitrides, in addition to metal alloys. These fuel materials are refractory in nature and can be used at high temperatures as fuels in fast reactors. For the successful exploitation of these potential fuel materials and to understand and predict their behavior during reactor operation, it is desirable to have a detailed knowledge of the appropriate regions of the phase systems along with accurate data on their thermophysical and thermochemical properties.

The behavior of nuclear fuels during irradiation is largely dependent on their physical and chemical properties which change with temperature and burnup. Under sustained irradiation over a long period inside the reactor, the thermophysical properties evolve with the variations in fuel composition, grain morphology porosity distribution and distribution of fission products. As the burnup increases, these parameters undergo significant changes from their values at virgin state. In fast reactor fuels, the changes in thermophysical and thermochemical properties are significant due to very high linear power rating and high burnup, and these will be discussed later (refer the chapter describing PIE studies). The fuel in its virgin state has uniform composition and microstructure with controlled porosity uniformly distributed in the phase. The state is achieved following rigorous fabrication process. Irradiation produces scores of fission products (fps) according to their relative yields for a given fuel, and these products get redistributed in the fuel matrix and settle down in different chemical states. Typical yields of fps in thermal and fast reactor fuels are presented

S. Anthonysamy (✉)
Formerly Metal Fuel Recycle Group, Indira Gandhi Centre for Atomic Research, Kalpakkam 603120, India
e-mail: savarianthonysamy@gmail.com

D. Das
Formerly Chemistry Group, Bhabha Atomic Research Centre, Mumbai 400085, India
e-mail: dasd1951@gmail.com

Table 5.1 Typical fission product yields (atom %) from different reactor fuels

FPs	^{235}U PWR (45 GWD/T)	MOX PWR (7.8% PuO$_2$) (45 GWD/T)	^{233}U-AHWR (20 GWD/T)	^{239}Pu FBR (100 GWD/T)
Xe + Kr	12.8	12.8	16.2	12.6
Pd + Ru + Tc + Rh	16.0	23.8	7.1	22.9
Mo	11.7	11.0	10.3	10.7
Zr	13.3	9.5	16.7	9.8
Y + RE	25.3	23.1	26.6	23.4
Ba + Sr	7.2	5.3	9.4	5.3
Cs + Rb	11.0	10.8	9.8	11.0
Metalloids + Halogens	1.8	2.2	2.8	2.4
Ag + Sn + Cd	0.7	1.4	0.3	1.2

in Table 5.1. There is progressive buildup of fps in the cladded fuel with burnup. The fission gases settle mostly as dispersed microbubbles inside fuel matrix, and a part of the gases is released at the fuel–clad gap and in the plenum space. For metallic fuels, the released part is significant. There is change in the fuel morphology and stoichiometry, development of heterogeneity due to insoluble phase formation with fps, and increased porosity. In the process, the fuel parameters progressively change with irradiation inside the reactor. Thermophysical properties such as melting point, density, heat capacity, elasticity, thermal conductivity and thermal expansion behavior are essentially considered in the design and performance modeling, and in safety analysis of the fuel. Thermochemical properties of fuel, fission products and clad are also considered to evaluate the long-term fuel containment within clad over the burnup period. The clad without significant distortion or any breach is the most important requirement for fuel containment. For evaluating the containment, besides thermal and thermochemical properties, the data on transport behavior of gaseous and volatile fps are required. Gaseous and volatiles undergo transport to the cooler periphery of fuel–clad interface causing concern to the containment. Reactive volatiles corrode the clad wall. The evaluations are to be carried at normal irradiation as well as in power ramp situation. Power ramp can occur because of occasional lag in the feedback control over the fission chain. In the case of breach of clad under off-normal situation, the fuel–clad–coolant interaction can lead to spread of radioactive fuel and fission products, and particularly, the release of the volatile radioactive components. These aspects are also considered in the performance evaluation of fuels.

Table 5.2 Characteristic properties of allotropes of uranium

Phase	Stability temperature range, K	Crystal structure, cell dimension, Å	Density, 10^{-3} kg/m^3
α-U	<942	Orthorhombic, a = 2.853, b = 5.865, c = 4.955	19.05
β-U	942–1049	Tetragonal a = 10.795, b = 5.865	18.37
γ-U	1049–1408	Face-centered, cubic a = 3.525	18.06
α-Pu	Up to 390	Simple monoclinic	19.86
β-Pu	390–486	bc-monoclinic	17.70
γ-Pu	486–588	fc-orthorhombic	17.14
δ-Pu	588–729	Face-centered cubic	15.92
δ'-Pu	729–748	bc-tetragonal	16.00
ε-Pu	748–913	Body-centered cubic	16.51
α-Zr	Up to 1135	Hexagonal close packed	6.49 at 293 K
β-Zr	1135–2128	Body-centered cubic	

5.1 Metallic Fuels

5.1.1 Uranium

Uranium has three allotropic modifications (α, β and γ); their characteristics are presented in Table 5.2. Uranium crystals are characterized by anisotropy along the symmetry axes of crystals.

5.1.2 Plutonium

Plutonium is a man-made transuranium element of the actinide series. Plutonium is formed by neutron capture by the isotope ^{238}U and subsequent two-stage β$^-$ decay of intermediate products. The mechanism of its formation is the following:

$$\underset{92}{^{238}}U \xrightarrow[(n,\gamma)]{} \underset{92}{^{239}}U \xrightarrow[(-\beta)]{} \underset{93}{^{239}}Np \xrightarrow[(-\beta)]{} \underset{94}{^{239}}Pu \tag{5.1}$$

Plutonium has six crystal modifications (α, β, γ, δ, δ' and ε). Their characteristics are given in Table 5.2.

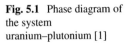

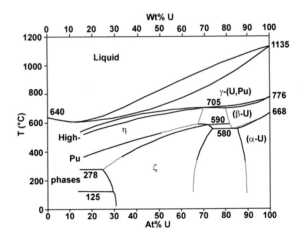

5.1.3 Phase Equilibria in the System U–Pu

Phase diagram of the system uranium–plutonium [1] is shown in Fig. 5.1. The solidus and liquidus both have minima at the same intermediate composition and temperature (~12 at.% U and 883 K). There are two intermediate U–Pu phases: ζ-(U,Pu) with ~25–72 at.% U and is stable at room temperature, and η-(Pu–U), which has between ~3 and 70 at.% U and is stable at temperatures between 550 and 978 K. With the exception of a continuous solid solution between γ-U and ε-Pu, no Pu phase can dissolve more than ~2 at.% U, and no U phase can dissolve more than ~20 at.% Pu.

5.1.4 Phase Equilibria in the System U–Zr

Phase diagram of U–Zr [2] is shown in Fig. 5.2. The system is characterized by a high-temperature bcc solid solution between γ-U and β-Zr that separates into two body-centered cubic phases ($\gamma1$ and $\gamma2$) at lower temperatures, the existence of a single intermediate phase (δ-UZr$_2$) that is stable at room temperature, and limited solubility of U in α-Zr and of Zr in α-U and β-U.

5.1.5 Phase Equilibria in the System Pu–Zr

Key features: (a) continuous solubility between the bcc phases β-Zr and ε-Pu, (b) extensive solubility of Zr in δ-Pu and limited solubility in α-Pu, β-Pu, γ-Pu, and δ'-Pu and (c) the existence of θ-(Pu–Zr) as an intermediate phase that is stable at room temperature.

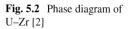

Fig. 5.2 Phase diagram of
U–Zr [2]

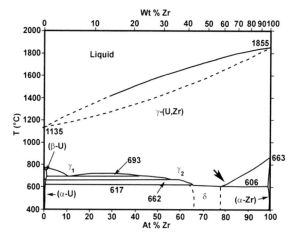

5.1.6   Phase Equilibria in the System U–Pu–Zr

All of the phases in the U–Pu–Zr system at temperatures above ~773 K are phases
in the U–Pu, U–Zr and Pu–Zr systems. Some of the binary compounds are modified
by dissolution of ternary elements. Allotropes γ–U, ε–Pu and β–Zr form solid solution over complete composition range, whereas other allotropes have reasonable or
limited solubility for other elements.

The first phase to crystallize for all U–Pu–Zr ternary compositions is a body-
centered cubic solid solution between γ–U, ε–Pu and β–Zr, which transforms to a
number of lower-temperature phases by solid-state reactions.

5.1.7   Phase Equilibria in the System U–Al

There are three uranium aluminide compounds in this system: UAl_2 (congruent
melting point: 3070 K, density: 8.14 g/cm³, structure: face-centered cubic), UAl_3
(decomposes peritectically; density: 6.8 g/ cm³, structure: simple cubic) and UAl_4
(decomposes peritectically at 1350 K, density: 6.06 g/ cm³, structure: orthorhombic).
There is no significant solid solubility of uranium in aluminum. Consequently, in the
80 wt. % aluminum range, the microstructure of the alloy consists of UAl_4 dispersed
in a matrix of nearly pure aluminum. At higher uranium contents, UAl_3 is retained
by rapid cooling or by the addition of an element such as silicon, which suppresses
the peritectic reaction of UAl_3 to UAl_4. Alloys containing up to 68 wt.% uranium, in
the equilibrium condition, have aluminum and UAl_4 with UAl_4 present as platelets
in the eutectic phase and as primary crystals in hypereutectic alloys. However, due to
nonequilibrium conditions during solidification, primary crystals of UAl_3 and UAl_4
are formed in aluminum matrix containing about 30 wt.% uranium.

5.1.8   Phase Equilibria in the System Pu–Al

Three Pu–Al phases relevant to nuclear technology are $PuAl_2$, $PuAl_3$ and $PuAl_4$. $PuAl_2$ is isostructural with UAl_2. Unalloyed plutonium has never been tried as a nuclear fuel because of its following unfavorable metallurgical properties: (i) plutonium has a low melting point; (ii) there are six allotropic transformations of plutonium between 293 and 913 K. No other metal shows as many phase changes within such a narrow temperature range; and (iii) there is large volume change which accompanies the alpha to beta transformation. It was, therefore, natural to look for a high-temperature phase which could be stabilized down to the room temperature by a proper combination of alloying and heat treatment. δ–Pu has a face-centered cubic structure and is stable between 598 and 729 K. It exhibits the highest ductility and best formability among all the plutonium allotropes. The δ phase can be retained at room temperature by small additions of Al. Up to 14.5 at.% Al can dissolve in Pu (δ).

The δ-stabilized plutonium has many desirable mechanical and metallurgical properties. In addition, these alloys have negligible volume change during working, produce good quality casting and are more resistant to oxidation compared to unalloyed plutonium. However, the density of delta-stabilized plutonium is less than unalloyed plutonium. δ-stabilized plutonium alloys have been used as fuels in the fast reactors. Al-Pu alloys containing 3–20 wt.% plutonium have widespread application in research reactors. Because of its low parasitic thermal neutron absorption cross section (0.22b) and high thermal conductivity, Al is an excellent inert carrier for the fissile plutonium. Since plutonium has negligible solid solubility with aluminum, these alloys essentially comprised of a dispersion of the intermetallic $PuAl_4$ in an aluminum matrix. Such dispersion fuel has the advantage of minimum radiation damage by restricting the fission recoil damage to the area immediately adjacent to the fissile dispersoid particles, i.e., $PuAl_4$.

5.1.9   Phase Equilibria in the System U–Si

U–Si binary system is characterized by seven intermetallic compounds: USi_3, USi_2, $USi_{1.88}$, U_3Si_5, USi, U_3Si_2 and U_3Si. U_3Si_2 has a narrow homogeneity range ($x = 0.03$ in $U_3Si_{2\pm x}$) at temperatures above 1273 K. The compounds USi, U_3Si_5 and $USi_{1.88}$ exhibit narrow homogeneity ranges. There are three different phases for U_3Si_5: the hexagonal defect AlB_2-type as well as two orthorhombically distorted AlB_2-type-related phases. The tetragonal $USi_{1.88}$ (defect $ThSi_2$-type) at its silicon poor phase boundary is in equilibrium with an orthorhombic phase (defect $GdSi_2$-type) fully ordered stoichiometric USi_2.

U_3Si_2 is being considered as an alternative to uranium oxide for fuel in nuclear reactors. Advantages are (i) higher percentage of uranium and (ii) higher thermal conductivity. Direct replacement of UO_2 with U_3Si_2 enables a reactor to generate more energy from a set of fuel rods. U_3Si_2 has five times better thermal conductivity than UO_2. Hence, it is expected to be a fuel rod capable of equal power output.

U_3Si and U_3Si_2 are choices for incorporation into composite plate fuels in research reactors due to their high uranium densities. This increased uranium density compared with uranium dioxide (UO_2) has made them attractive to a new generation of nuclear fuels. A higher uranium density allows incorporation of U–Si phases into composite fuels with the aim of increasing coping time during a cladding breach before fission products and/or actinides are released. In addition, development of fuels containing higher uranium densities than those of UO_2 could facilitate utilization of alternative cladding materials that offer improved high-temperature performance than zirconium alloys.

5.1.10   Phase Equilibria in the System Pu–Si

Five intermediate compounds have been identified in this system: Pu_5Si_3 (ζ), Pu_3Si_2 (η), $PuSi$ (θ), Pu_3Si_5 (γ) and $PuSi_2$ (κ). Very little solubility of Si in α, β, and γ Pu is reported. Although δ Pu is not stabilized at room temperature by the addition of Si, the δ phase can be retained metastable at that temperature if alloys containing at least 0.75 at.% Si are cooled rapidly from the δ phase temperature region. Up to 14.5 at.% Al can dissolve in Pu(δ).

5.2   Oxide Fuels

5.2.1   Phase Equilibria in the System Uranium–Oxygen

A partial phase diagram of the system uranium–oxygen [3] is shown in Fig. 5.3. Uranium can exist in a number of different valence states. Sixteen well-characterized solid uranium oxides are reported. Though the existence of a large number of solid uranium oxide phases has been reported, the following compounds have been well characterized: UO_2 (O/U = 2.00), U_4O_9 (O/U = 2.25), U_3O_7 (O/U = 2.33), U_3O_8 (O/U = 2.67), UO_3 (O/U = 3.00) and U_2O_7 (O/U = 3.5).

The compound uranium dioxide (urania), UO_2, has a cubic fluorite-type crystal structure and exists over a wide range of composition. The uranium dioxide phase accommodates readily interstitial oxygen anions to form hyperstoichiometric UO_{2+x} and also to lose oxygen anions to form hypostoichiometric UO_{2-x}. The composition of uranium dioxide changes from $UO_{2.00}$ at 600 K to $UO_{2.24}$ at 1400 K. The hypostoichiometric oxide UO_{2-x} exists only at elevated temperatures. At low temperatures, material with O/U < 2 is a mixture of $UO_{2.00}$ and metallic uranium. The predominant defects in $UO_{2\pm x}$ are oxygen vacancies/interstitial oxygen ions. Any uranium

Fig. 5.3 Partial phase
diagram of the system
uranium–oxygen [3]

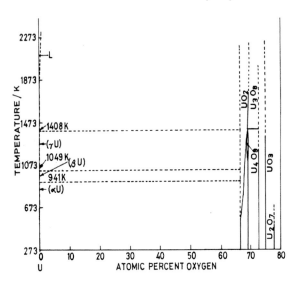

oxide when heated in air at ~900 K results in α-U_3O_8. The theoretical density of this
compound calculated using XRD data is 8.4 g/cc.

5.2.2 Phase Equilibria in the System Plutonium–Oxygen

There are four stable compounds in this system: $PuO_{1.5}$, $PuO_{1.52}$, $PuO_{1.61}$ and PuO_2
[3] (Fig. 5.4). The most stable compound is plutonium dioxide, PuO_2 (plutonia, O/Pu
= 2). Below 2000 K, no oxide of higher oxidation state than PuO_2 is found. Hypo
region of plutonia is very broad. Plutonia has cubic, fluorite-type crystal structure.
The compound melts at 3017 ± 28 K. $PuO_{1.5}$ and $PuO_{1.52}$ are line compounds,
whereas $PuO_{1.61}$ exhibits hyperstoichiometry. The compound is thermodynamically
stable only at temperatures above 573 K.

5.2.3 Phase Equilibria in the System Thorium–Oxygen

Only one solid binary compound, namely thorium dioxide (thoria), ThO_2, exists in
this system. Thoria has cubic, fluorite-type crystal structure. It is a line compound
(stoichiometric with O/Th = 2.00) at least up to 2000 K. The nonstoichiometry of
thoria has not been established without ambiguity. At temperatures above 2000 K,
the compound becomes hypostoichiometric, ThO_{2-x}. The maximum value of x is
0.015 at 2000 K, and the value increases linearly with increasing temperature. At
3013 ± 100 K, the value of x is 0.13 ± 0.04. As the temperature is raised above this

Fig. 5.4 Pu–O phase diagram [3]

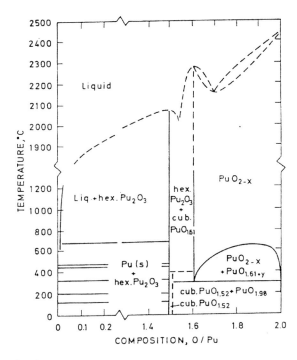

point, the maximum value of x decreases reaching zero at 3663 K. At 3663 K, solid and liquid thoria are in equilibrium.

5.2.4 Phase Equilibria in the System Uranium–Plutonium–Oxygen

A small portion of the U–Pu–O phase diagram at 1073 K [3] is shown in Fig. 5.5. The focus is on dioxide phase of this system as UO_2 and PuO_2 are the most suitable compounds for nuclear fuel applications. UO_2 and PuO_2 have similar crystal structures and other parameters required for making substitutional solid solution as per Hume–Rothery rule. Hence, they make solid solution over the complete range of 'x' $(U_xPu_{1-x})O_2$. Both uranium and plutonium can exist as ions in a number of different valence states. +4, +5 and +6 are the most stable oxidation states in uranium. 3+ and 4+ are the most stable oxidation states in plutonium. Hence, uranium–plutonium oxides show broad ranges of nonstoichiometry, where the oxygen-to-metal ratio (O/M) differs from 2, yet the system consists of only a single phase. The mixed oxide may deviate from stoichiometry in both directions.

Fig. 5.5 Partial phase
diagram of U–Pu–O system
at 1073 K [3]

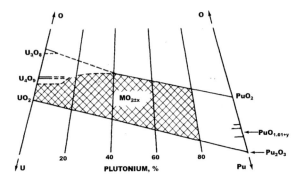

The hyperstoichiometric mixed oxide has thermodynamic properties equivalent to an ideal solution of stoichiometric plutonia and hyperstoichiometric urania. The compound $(U_{1-y},Pu_y)O_{2+x}$ is represented as a mixture of y mole fraction of PuO_2 and (1-y) mole fraction of UO_{2+z}, where $z = x/(1-y)$. Hypostoichiometric mixed oxides consist of U^{4+} and a mixture of Pu^{3+} and Pu^{4+}. The compound with the formula $(U_{1-y},Pu_y)O_{2-x}$ can be treated as an ideal solution of (1-y) mole fraction of UO_2 and y mole fraction of PuO_{2-z}, where $z = x/y$. The extent of nonstoichiometry in the mixed oxide is limited by the supply of cations whose valence can be altered. Hence, the plutonium component of hypostoichiometric mixtures cannot be reduced to a valence less than 3+. The maximum value of z in PuO_{2-z} is 0.5, or the maximum value of x is 0.5y.

5.2.5 Phase Equilibria in the Ternary System Uranium–Thorium–Oxygen

Thoria and urania form a complete series of solid solutions with the fluorite-type of cubic structure. These solid solutions are stable up to the melting point. The unit cell sizes of the solid solutions at room temperature obey Vegard's law over the entire compositional range. However, some experimental results suggest that for the solid solutions with low concentrations (less than 5 mol. %) of thoria, there is atomic ordering. The phase diagram of the system U–Th–O has not yet been satisfactorily established due to experimental problems associated with the measurements of the solidus and liquidus temperatures of ThO_2–UO_2 solution. The value of x in $(U_yTh_{1-y})O_{2+x}$ solid solutions varies with temperature as well as urania content in the solid solutions, i.e., y. In stoichiometric oxides, both uranium and thorium are present in the +4 valence state. In hyperstoichiometric oxides, where oxygen ions are in excess, uranium will exist in +4 and +5 valence states while thorium will remain in the +4 state. The excess O^{-2} ions are accommodated in interstitial sites in the fluorite structure. The cation sublattice remains intact. Uranium and thorium ions are randomly distributed on every available site in the cation sublattice.

5.3 Carbide Fuels

5.3.1 Phase Equilibria in the System Uranium–Carbon

The U–C system is characterized by three compounds: uranium monocarbide (UC), uranium sesquicarbide (U_2C_3) and uranium dicarbide (UC_2) (Fig. 5.6). UC, stoichiometric uranium monocarbide, is a line compound containing 4.80 wt.% carbon. It has a theoretical density of 13.63 g/cm^3 and crystallizes in the fcc-NaCl structure. The compound melts congruently at 2780 K. UC is isomorphous with the actinide metal monocarbides as well as UN, US and UP. There is no evidence of solubility of uranium in UC.

U_2C_3, uranium sesquicarbide, cannot be produced directly by casting or by compaction of powders of uranium metal and graphite. Once formed at elevated temperature, it is stable at room temperature. It decomposes to UC and UC_2 at approximately 2100 K. The thermodynamic stability of U_2C_3 is still controversial. The crystal structure of U_2C_3 is body-centered cubic.

Uranium dicarbide exists in two crystallographic forms. The low-temperature phase is α-body-centered tetragonal CaC_2-type structure. This transforms to the β-fcc cubic KCN-type structure at temperatures above 2000 K. The compound melts congruently at 2720 K. The α-UC_2 phase is retained in a metastable form in material rapidly cooled from the melt. The compound is stabilized by dissolved oxygen. It is reported to transform to U_2C_3 at high temperatures (>1673 K) under stress or after prolonged annealing in vacuum. UC and UC_2 are completely soluble in each other at elevated temperatures, forming a solid solution.

Fig. 5.6 U–C phase diagram [4]

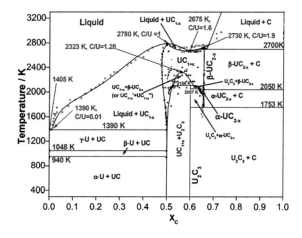

5.3.2 Phase Equilibria in the System Pu–C

There are four stable compounds in the Pu–C system: Pu_3C_2, PuC_{1-x}, Pu_2C_3 and PuC_2 (Fig. 5.7). The plutonium-rich solid region is characterized by the six allotropes of plutonium: α-Pu (simple monoclinic, stable up to 390 K), β-Pu (body-centered monoclinic, stable up to 473 K), γ-Pu (face-centered orthorhombic, stable up to 573 K), δ-Pu (face-centered cubic, stable up to 723 K), δ′-Pu (body-centered tetragonal, stable up to 748 K) and ε-Pu (body-centered cubic, stable up to 910 K). Due to very limited solubility of carbon in Pu metal, the phase transition temperatures of plutonium allotropes are not influenced by carbon impurity. Plutonium melts at 910 ± 5 K. The melting temperature of ε-Pu is reported to decrease by 8 K upon addition of carbon impurity.

Triplutonium Dicarbide, Pu_3C_2(ζ-phase): Pu_3C_2 is a line compound, and its crystal structure is still unknown. Pu_3C_2 forms very slowly with the peritectoidal reaction ε-Pu + 2PuC = Pu_3C_2.

Plutonium Monocarbide, PuC_{1-x}: Plutonium monocarbide, PuC_{1-x}, is always hypostoichiometric with respect to carbon. The monocarbide phase is therefore a defect structure, with some vacant carbon atom lattice sites. This compound decomposes peritectically into Pu_2C_3 and liquid plutonium at 1900 ± 30 K. The phase field of PuC_{1-x} is widest at the eutectic temperature of 848 K (38–48 at. % or C/Pu ranges from 0.74 to 0.94). PuC_{1-x} crystallizes in the NaCl rock-salt fcc structure. The lattice constant of PuC_{1-x} was measured to be $a = 4.954 \pm 0.002$ Å for samples (C/Pu = 0.74) quenched to room temperature from 673 K ≤ T ≤ 908 K. This value remains constant for plutonium-rich compositions, whereas it increases up to 4.973 ± 0.001 Å at the carbon-rich boundary in equilibrium with bcc Pu_2C_3. PuC_{1-x} exists only as a hypostoichiometric compound, possibly because of the smaller size of plutonium atoms compared to uranium atoms, resulting in more vacant sites for carbon interstitials.

Fig. 5.7 Pu–C phase diagram [4]

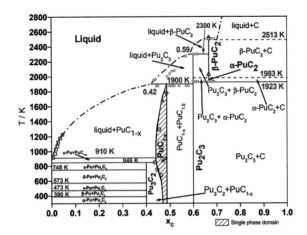

Plutonium Sesquicarbide, Pu_2C_3: The most stable phase in the Pu–C system is bcc Pu_2C_3. Its lattice constant was measured to be $a = 8.129 \pm 0.001$ Å. Photoelectron spectroscopic studies have established that the main difference between the electronic structures of PuC and Pu_2C_3 is the partial occupancy of the $5f^6$ states in Pu_2C_3. Pu_2C_3 melts peritectically into liquid + PuC_2 at 2323 ± 20 K.

Plutonium Dicarbide: Plutonium dicarbide, PuC_2, exists only at temperatures between 1923 and 2513 K. Because the transformation of PuC_2 to $Pu_2C_3 + C$ is extremely rapid, it is difficult to quench the dicarbide to room temperature. The structure of the dicarbide was identified to be tetragonal CaC_2-type, with $a = 3.63$ Å and $c = 6.094$ Å. It undergoes a martensitic phase transition at 1983 K. The high-temperature structure was fcc (KCN) with a lattice parameter $a = 5.70 \pm 0.001$ Å. It is still unclear whether cubic PuC_2 melts peritectically (into liquid + graphite) or congruently.

5.3.3 Phase Equilibria in the System U–Pu–C

Uranium and plutonium monocarbides form a complete range of solid solutions. The (U,Pu)C phase is stoichiometric with respect to carbon over the composition range from 0 to 35 at. % Pu. With further increase in plutonium content, the phase deviates from stoichiometry toward greater metal-to-carbon ratios to form the defective (U,Pu)C phase. A section of the U–Pu–C phase diagram at 843 K is shown in Fig. 5.8.

Uranium and plutonium sesquicarbides form a complete range of solid solutions at temperatures below 2033 K. Cubic UC_2 forms a continuous solid solution with PuC_2 at temperatures above 2273 K. At temperatures below 2273 K, (U,Pu) sesquicarbide is formed in preference to (U,Pu) dicarbide in the hyperstoichiometric phase region. The biphasic field consisting of (U,Pu)C + $(U,Pu)_2C_3$ exists between 50 and 60 atom % C.

In U–Pu–C system, U–Pu metal phase, and the monocarbide and sesquicarbide solid phases possess complete ranges of solutions. In the two-phase region metal + monocarbide, the metal phase will always contain more plutonium than the monocarbide phase. In the case of monocarbide + sesquicarbide two-phase region, the

Fig. 5.8 A section of the U–Pu–C phase diagram at 843 K [5]

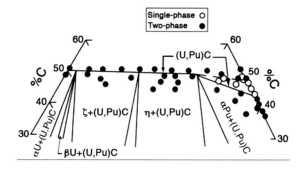

sesquicarbide contains more plutonium than the monocarbide phase. A biphasic mixture containing (U,Pu) monocarbide as the major phase with optimum amount of (U,Pu) sesquicarbide as the second phase is the fuel considered for fast reactors. Single-phase (U,Pu) monocarbide, which is a line compound, cannot be used as the fuel since the free metals formed during burnup would form low-melting eutectics with Fe and Ni of the structural materials. This might cause, in the worst case, a melt-down of the cladding. Since the C/M ratio decreases with burnup, to ensure that the metal phase does not form during irradiation, the fuel to start with is fabricated as a monocarbide with 5–15 vol. % of sesquicarbide.

In case of the carbide fuel, carburization of cladding occurs with the formation of $(Cr,Fe)_{23}C_6$ whenever the carbon potential of the fuel is greater than that of the cladding; the difference in carbon potential serves as the driving force. Carbon potential of plutonium sesquicarbide is much lower than that of uranium sesquicarbide at all temperatures. When $MC + M_2C_3$ biphasic mixture is used as the fuel, since the sesquicarbide is enriched in plutonium due to segregation, the plutonium-rich fuel is more compatible with the stainless steel cladding materials.

5.4 Nitride Fuels

5.4.1 Phase Equilibria in the System U–N

U–N system (Fig. 5.9) is characterized by four compounds, uranium mononitride, UN, (congruent melting at ~3123 K under 2.5 atm nitrogen pressure), α-uranium sesquinitride, U_2N_{3+x} (stable upto1073 K), β-uranium sesquinitride, U_2N_{3-x}, (exists above 1073 K and decomposes into UN and N_2 at 1340 K) and uranium dinitride, UN_2, which is nitrogen deficient. UN has a fcc-NaCl-type structure and exists over a small composition range (N/U ratio: 0.99 ± 0.006 to 1.0 in the temperature range 1873–3073 K). α-uranium sesquinitride is isomorphous with bcc-Mn_2O_3 structure, and if the N/U ratio is greater than 1.7–1.75, then the fcc phase with CaF_2 structure is formed. β-uranium sesquinitride is isomorphous with La_2O_3 (A-type rare earth structure).

5.4.2 Phase Equilibria in the System Pu–N

There is only one compound in the Pu–N system, the stoichiometric mononitride PuN (Fig. 5.10).

The compound undergoes decomposition at 1 bar N_2 and 2843 ± 30 K. The compound melts congruently at high nitrogen pressures and at high temperatures. For example, the compound melts at 3040 K at 25 bar nitrogen. No homogeneity

Fig. 5.9 U–N phase diagram [6]

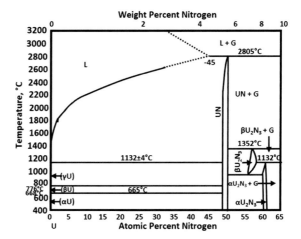

Fig. 5.10 Pu–N phase diagram [7]

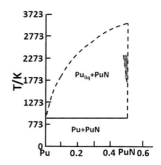

range of PuN was observed at 1 bar N_2. At a nitrogen pressure of 10^{-4} bar, the PuN phase can exist up to N/Pu of 0.968 at 3263 K and N/Pu of 0.954 at 2343 K.

5.4.3 Phase Equilibria in the System U–Pu–N

UN and PuN form a complete range of solid solutions. Figure 5.11 shows a calculated isothermal section of the U–Pu–N phase diagram at 1273 K.

Sesquinitrides do not exist in the Pu–N system. A sesquinitride phase with a Pu/(U + Pu) ratio of 0.15 has been reported in the literature. U–Pu–N phase diagrams with either very small plutonium solubility or up to 15 mol % 'PuN$_{1.5}$' solubility in UN$_{1.5}$ are being considered. Three separate phase regions are considered relevant for the application of nitrides as nuclear fuels: (1) the two-phase region metal + mononitride, (2) the single-phase solid solution and (3) the regions involving the sesquinitrides in the U–Pu–N phase diagram.

Fig. 5.11 A calculated
isothermal section of the
U–Pu–N phase diagram at
1273 K [7]

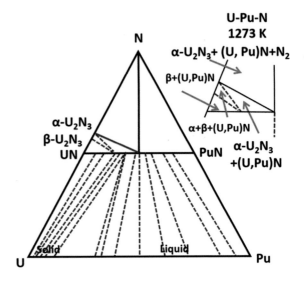

The **Two-Phase Region U–Pu + U–Pu Mononitride**: In the two-phase field $U_{1-x1}Pu_{x1}$(solid or liquid) + $<U_{1-x2}Pu_{x2}N>$, the concentrations of uranium and plutonium in the two phases are different because of the relative thermodynamic stabilities of the phases UN and PuN. The segregation increases with increasing temperature.

The **Single-Phase Mononitride Region**: Because UN exists over a relatively narrow composition range and PuN is almost a stoichiometric compound, the mononitride solid solutions (U,Pu)N also exist over a relatively narrow composition range. Based on the measurement of plutonium pressures using a Knudsen effusion technique over the single-phase mononitride of composition $(U_{0.8}Pu_{0.2})N$, it has been found that the material remained single phase up to 2400 K.

The **Regions Involving the Sesquinitrides**: The nature of the phase diagram in these regions has not yet been established unambiguously. Experimental studies indicate that a sesquinitride phase is formed with a Pu/(U + Pu) ratio of 0.15.

5.5 Oxygen Solubility in Carbides

Oxygen can substitute C in UC to the extent of x = 0.35 in $UC_{1-x}O_x$. $UC_{1-x}O_x$, is regarded as a solid solution between UC and the hypothetical compound 'UO'. The solubility of oxygen in sesquicarbide, U_2C_3, is very low. The dicarbide, UC_2, can accommodate significant amounts of O_2, ~5 mol. % of the C content. The solid solution formed by the dissolution of oxygen in uranium dicarbide can be modeled as an ideal solution of 95 mol % of $UC_{1.91}$ and 5 mol % of UO_2.

Plutonium monocarbide phase requires carbon site vacancies for stability such that the maximum homogeneity range of PuC_{1-x} is $0.64 \leq x \leq 0.94$ at 907 K. Oxygen can substitute C in PuC_{1-x} to the extent of y $= 0.67$ in $PuC_{1-x-y}O_y$ (1273–1673 K). $PuC_{1-x}O_x$ can be regarded as a solid solution between PuC and the hypothetical compound 'PuO'. The solubility of oxygen in plutonium sesquicarbide is very low.

5.6 Oxygen Solubility in Nitrides

Oxygen solubility in UN is not more than 3 mol. % of U. UO-UN solid solution will be ideal or possess rather small deviation from ideality. The solubility of oxygen in ß-U_2N_{3-x} is negligible. At temperatures above 1450 K, α-U_2N_{3-x} and UO_2 are completely miscible. PuN dissolves 13 ± 3 mol % of PuO.

5.7 Nitrogen Solubility in Carbides

UC and UN form a solid solution over the complete composition range. PuC and PuN also form solid solution over the complete composition range. Thermodynamic calculations show that the phase field (U,Pu)CNO + $UN_{1.5}$ is stable over a very narrow composition range. A nitride fuel with (U,Pu)O_2 as second phase is relatively stable over a limited composition but wider temperature range. Therefore, in-pile thermodynamic behavior of the nitride fuel with the (U,Pu)CNO + (U,Pu)O_2 phase field is expected to be superior to (U,Pu)CNO + $UN_{1.5}$.

5.8 Thermophysical and Chemical Properties of Nuclear Fuels

Before elaboration of thermal and thermodynamic properties of nuclear fuels, it is prudent to mention the typical stacking of fuels and thermal states of the fuel pin at a given power rating of nuclear reactor. In fuel pin assembly, ceramic fuels are stacked in columns within cylindrical space of clad. Metallic fuel is usually used in the form of cladded slag of sodium bonded fuel. The slags are fabricated with 20–25% porosity to accommodate gases and fps, minimizing mechanical stress to clad. Liquid sodium is soaked inside porous slag for aiding heat conduction. Ceramic fuel with higher mechanical strength than metal can retain the gases inside fuel matrix that has much less leftover porosity (~3–5%) in their fabrication from powder compaction and sintering route. UO_2 and mixed oxides have been extensively used worldwide in thermal and fast reactors. The performances of urania, plutonia and their solid solutions as reactor fuels are well established. Uranium dioxide (UO_2) is a standard fuel of many nuclear power plants. The main advantages in the use of

Table 5.3 Summary of the different types of fuels

Fuel type	Physical form
Oxide	UO_2, $(U,Pu)O_2$ as pressed and sintered pellets (density >95%)
Metallic (usually alloy)	U–Zr, U–Pu–Zr as Na-bonded alloy slug (smear density $\approx$75% TD)
Carbide/nitride	UC, $(U,Pu)C$, UN, $(U,Pu)N$ as pressed and sintered pellets
Intermetallic compound	Uranium silicide

uranium dioxide are its high melting point, dimensional and radiation stability and its chemical compatibility with other reactor components.

The procedures of fuel fabrication, storage as spent fuel, reprocessing, and waste management are proven for over so many decades. The fabrication and handling of the oxides are easier than the carbide, nitride or metallic fuels. Immense experience gained with these fuels has helped to leap forward for using the alternatives considering their relative merits as will be discussed later in the text. Table 5.3 gives a summary of different fuels that will be considered in the text. It will be important to mention here that this is not a comprehensive list. Many other types of fuel systems such as molten salts, cermet, etc. are also considered for their specific advantages.

The fast reactor fuels generally operate at higher power ratings (400–700 W/cm) as compared to those of thermal reactors (200–400 W/cm). The resultant higher heat flux in fast reactor fuel establishes steeper temperature gradient radially across cladded fuel pin (~8 mm diameter) which is cooled by flowing liquid sodium metal. With the stated power rating of fast reactors, a temperature gradient of about 1200 K is set up at steady state across the radius of ceramic fuel (~2200 K down to 1000 K) and about 300 K for metallic fuel (~1000 K down to 700 K). The stated temperature profile becomes occasionally steeper in power ramp situation. Thermal reactor pins (~12–13 mm diameter) usually cooled by flowing water/heavy water normally establish the temperature profile with central and peripheral temperatures of about 1300 K and 800 K respectively when operating at low power rating (~200 W/cm). At higher power rating (~400–500 W/cm), the central line temperature can be in the range of 1900–2100 K.

Among the mentioned thermophysical properties, thermal conductivity of nuclear fuel is the most important one that influences almost all important processes such as fission gas release, swelling behavior, grain growth, etc. Further, it limits the linear power rating of fuel pin. The heat transfer property undergoes significant deterioration during irradiation because of the progressive development of porosity and grain growth in fuel matrix, formation of microdispersed bubbles containing fission gases, and incorporation of fps in soluble and insoluble forms and change in fuel stoichiometry. The dispersed bubbles as well as the undissolved solid phases impart internal stress to the matrix, by which the fuel pin swells, develops radial microcracks and voids. In the case of helium bonded ceramic fuel, conductivity of fuel–clad gap also deteriorates progressively due to dilution of initially filled helium column by the fission gases. Helium gas at tens of bars pressure is filled in the fuel–clad gap for the heat conduction and for mechanical stability against moderator/coolant pressure existing outside thin walled clad.

The deterioration of fuel–clad gap conductance increases thermal stress because of steeper temperature gradient across fuel. Knowledge of melting point of fuel decides the uppermost limit of its linear power rating. Total melting is strictly avoided as it can severely affect the containment due to intergranular corrosion of clad by the liquid. Thermal conductivity as a function of temperature and simulated fuel composition is used in establishing the temperature profile at a given power rating. Thermal conductivity data along with heat capacity (C_V) density (ρ) of a material helps deriving its thermal diffusivity. The diffusivity data is useful in calculating the relaxation time of thermal transients in power ramp when the radial as well as axial temperature profile across fuel pin shoots up for a while resulting in augmented thermal stress in the fuel pin promoting thereby rapid redistribution and release of gas and volatile fission products.

The coefficient of thermal expansion (CTE) is required in understanding the reduction of fuel–clad gap under a given power rating. Thermal dilation of fuel reduces the gap facilitating heat conduction. However, it is necessary to critically maintain the diameter of fuel so that thermal dilation does not lead to mechanical stress on clad wall by making intimate contact. Mechanical stress on clad can lead to containment problem and uncontrolled stress leads to clad failure. In safety analysis, the values of thermal expansion data are essential in evaluating the dilation behavior of the fuel. The expansion coefficient (α) together with modulus of elasticity data decides thermal stress across a temperature gradient. The stress developed over temperature differential ΔT is expressed by $E\alpha\Delta T$, where E is modulus of elasticity. Added to the thermal stress is the mechanical one arising from fuel swelling due to accumulation of fission gases and reactive fps resulting in physical contact with clad wall. Irradiation-induced swelling further enhances the fuel–clad mechanical interaction (FCMI). Fuel pellet–clad interaction can severely damage clad wall leading to failure of fuel containment and release of radioactive fission products into the reactor coolant. The severity of FCI/PCI is more under power ramping.

For evaluation of mechanical stability of fuel pin, data of bulk modulus and yield strength of the fuel matrix are considered. The heat capacity values are referred for estimating the stored energy in fuel pin for a given temperature distribution and also for understanding the thermal relaxation property of fuel as indicated already.

The clad undergoes corrosion by the reactive volatile fission products. At a steep thermal gradient radially across the fuel rod, the fission gases, xenon and krypton, and the corrosive volatile products, iodine, tellurium, cesium, etc. undergo transport toward the fuel–clad interface and get released at the clad's inner surface. The major fraction of the gases remains dispersed inside the fuel matrix as microbubbles. However, for metallic fuels, the gas release can be as high as 70%. The volatile fission products like I and Te, though produced in small amounts, can cause corrosion and fatigue to the clad [8–10] resulting in degradation of its mechanical property leading to brittle fracturing and stress corrosion cracking.

An understanding of the vaporization behavior of fuels and fission products is essential for estimating the redistribution of various elements in the steep temperature gradients existing across the fuel pellets. The vaporization data are also important in the analysis of release of volatile components in off-normal situations of reactor. In general, the chemistry of cladded fuel with accumulating fission products must be known in order to analyze issues related to fuel containment over the entire burnup period. It is necessary to mention here that each nuclear fuel, whether it is oxide, carbide, nitride or metallic ones, when irradiated under a given power rating over a long time period, reveals its characteristic chemistry in the interactions of fuel-fps-clad constituents.

5.9 Basic Features of Thermophysical Properties

5.9.1 Thermal Conductivity

Thermal conductivity, K, of a solid material is a measure of its ability to transfer heat by conduction. Conductivity can be measured by steady heat flow, or alternatively, by transient heat flow methods. Steady radial heat flow method because of its simplicity is usually used to measure thermal conductivity of materials [11, 12]. The use of thermocouples to monitor the steady temperature gradient under a heat flux limits the applicability of the static method below 2000 K. The static method of thermal conductivity measurement has however the disadvantages of large waiting time in attaining steady state in ceramic materials that have poor conductivities. Transient heat flow methods have the advantage of quick performance of experimental measurements, since there is no need for waiting period in attaining steady state. For nuclear fuels, the laser flash method is generally used to measure the thermal diffusivity ($\kappa = K/\rho C_v$) of sample taken in the form of a thin disk. The method is based upon measurement of thermal signal due to abrupt temperature rise at the top face of the disk specimen when the bottom face is transiently heated by striking it with a short laser pulse. The diffusivity is expressible in terms of the signal characteristics [13]. This method requires thermal diffusivity measurement of a reference sample with known thermal conductivity, specific heat and density. Ideally, the reference material is to be selected with the thermal characteristics close to that of the sample. The transient heating method is suitable for different materials over a broad temperature range, ambient to 3000 K.

As for nuclear fuels, the thermal conductivity data helps analyzing the center-line temperature of fuel, when the surface temperature is constrained by the coolant flow. Thermal conductivity in the fuel pellets generally depend on temperature, burnup and porosity. Fuel porosity reduces thermal conductivity and gross swelling. Overall conductivity can be represented as $k_f = k_0 \times (f_D f_P f_R f_M)$, where the factors, f_D, f_P, f_R and f_M are respectively the effects of the dissolved fission products, precipitated fission products, radiation, and porosity [12]. Braced factors are all related to high burnup. Thermal conductivity deteriorates with burnup and results in increase in central line temperature of fuel operating at a given power rating.

Fig. 5.12 Thermal conductivities of typical ceramic and metallic fuels

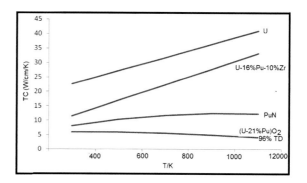

In crystalline solids, there are three major heat transport mechanisms: Transport at lower temperatures occurs mainly through lattice vibrations. At higher temperatures, radiation and electron contributions to the heat transport become important. As for UO_2 fuel, the temperature dependent transport is expressible [12] as $k_0(\text{Wm}^{-1}\text{K}^{-1}) = (0.0375+2.165 \times 10^{-4}T)^{-1} +4.715 \times 10^{9}T^{-2}Exp(-16361/T)$. For all oxide fuels, k_0 values are very low as compared to the metallic ones (Fig. 5.12).

Poor thermal conductivity of the oxides results in a large difference of center line and surface temperatures of fuel pins in order to remove the fission heat across large temperature gradient. The carbides and nitrides fuels have intermediate values of thermal conductivities, which change with the stoichiometries (C/M, N/M ratios); conductivity values of carbides are higher than those of nitrides. Besides dependencies on temperature, stoichiometry and porosity, their thermal conductivities depend on the presence of second phases (M, M2C3, MC2, M2N3, MN2, MO2, etc.) and residual O, N and C impurities. Thermal conductivity of MC reduces with M2C3 and oxygen contents and improves with higher pellet density. Conductivities of (U,Pu)C and (U,Pu)N fuels decrease with increase in plutonium contents. Because of high thermal conductivities, uranium and/or plutonium-based metallic alloys, carbides and nitrides with their high density fissile contents are used as fast reactor fuels.

5.9.2 Thermal Expansion

The dimensional change of materials with temperature is expressible to a first-order approximation as $\delta D = (\delta D/\delta T)_P\delta T$ where the quantity $(\delta D/\delta T)_P/D$ represents the isobaric expansion coefficient of the material in the chosen dimension, which can be its length, surface or volume. Evidently, for isotropic material, the volume (V), surface (S) and length (L) expansion coefficients are interrelated as $(\delta V/\delta T)_P/V = 3/2[(\delta S/\delta T)_P/S] = 3(\delta L/\delta T)_P/L = 3\alpha$. Measurement of the linear expansion property in isotropic materials helps evaluate their dimensional changes. To determine the expansion coefficient, two physical quantities, namely the linear displacement (ΔL) and temperature (T) are measured on a sample used in the form of solid cylinder

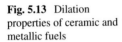

Fig. 5.13 Dilation properties of ceramic and metallic fuels

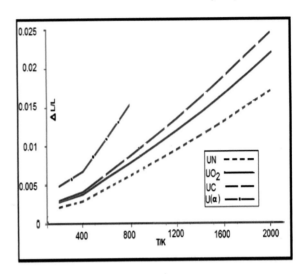

(rod/pellet) that is undergoing a well-defined thermal cycle. The linear expansions are usually measured by dilatometry, interferometry and thermomechanical analysis [14]. At extremely high temperatures, the optical techniques are used to monitor dilation property. It is necessary to mention here that the lattice parameters derived out of X-ray crystallography reflect the bulk expansion property but may not quantify the property for its use in practical applications such as in the analysis of fuel–clad compatibility in nuclear industry. Thermal expansion of fuel pellet is an important property which limits the lifetime of the fuels in reactors, because it can affect both the pellet and cladding mechanical interaction and the fuel–clad gap conductivity. For some of typical nuclear fuels, the thermal expansion properties are graphically presented in Fig. 5.13. The thermal expansion of nuclear fuel during irradiation depends on several other physicochemical factors besides temperature. The important factors are pellet cracking due to thermal stress, fuel stoichiometry, fission products accumulations and formation of high burnup structure in rim region of pellet. To address the fuel containment issue at high burnup, the simulation analysis involves all the factors besides the measured expansion property of virgin fuels. The thermal property study on high burnup SIMFUEL (simulated fuel composition) together with observational data of high burnup reactor fuels is considered in the fuel performance analysis.

5.9.3 Heat Capacity

The heat capacity of a substance is the heat energy required to increase its temperature by one-degree Kelvin. The temperature increasing operation can be carried at constant pressure (P), or alternatively, at constant volume (V); the energy requirement for a degree rise in temperature is higher at constant pressure because of the additional energy input in performing work associated with the consequential volume expansion under constant P. The heat capacities, C_P and C_V, are extensive properties unless they are mentioned specifically for unit mass or for one mole of the substance. For one mole, the properties are referred as molar heat capacities. C_P and C_V are always positive finite quantities when the substance is in single phase. Heat capacity increases with temperature. Temperature trend of heat capacity of solids was established from the consideration of thermal energy distributions over phonon states. The trend shows nonlinear increase of C_V with temperature rise from null value (at zero Kelvin) until attainment of Debye's temperature (T_D). Above T_D, C_V shows linearly increasing trend that ultimately flattens out at high temperature much above T_D value. The heat capacities are measured calorimetrically. Calorimetry under constant pressure is commonly used. If for practical reason the pressure is different from the standard value of 1 bar, then the measured value at a temperature is corrected to get the value at the standard pressure by using the pressure coefficient of the heat capacity.

Knowledge on fuel enthalpy and its temperature coefficient C_P are important quantities in determination of fuel behavior in normal reactor operation and also in presence of transients. Heat capacity data also help analyzing thermal conductivity from the measured thermal diffusivity values of materials. Heat capacity measurements of nuclear fuels are usually carried out using differential scanning calorimetry (DSC) and Calvet calorimetry [15]. The former is invariably used at first in order to get full knowledge of the presence of phase transitions particularly in the working range of the fuel. DSC also helps in generating heat capacity data of the material in quickest way fairly up to a temperature of 1200 K with special arrangement for higher-temperature measurement and control; the technique can be stretched up to 1800 K. The generated endothermic signal in DSC is analyzed for obtaining the heat capacity using suitable calibration procedure. With the DSC concept adopted in Calvet calorimetry [15], the heat measurement accuracy and sensitivity can be increased further. Calvet calorimetry is used to measure heat capacities of fuels fairly up to about 1900 K with good precision and accuracy. Enthalpy increment data of nuclear fuels can be directly generated by adopting sample drop mode in Calvet calorimetry. There, a known amount of the sample following its thermal equilibration in a temperature-controlled bath at T_0 (near room temperature) is abruptly dropped into the calorimeter vessel maintained at an elevated temperature T. The calorimetric measurement of heat flow until resumption of the temperature (T) is then carried out at one bar pressure, which is the ambient value. Measurements made this way at different set temperatures (T) yield the molar enthalpy increments function, $H(T) - H(T_0)$. C_P function is then derived from the temperature coefficient, $C_P = (\partial H / \partial T)_P$. Reported heat capacity values of some of the ceramic and metallic fuels as a function of temperature [16] are graphically shown in Fig. 5.14.

Fig. 5.14 Trend of specific heat capacity of different nuclear fuels

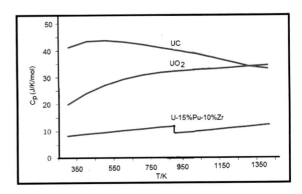

5.9.4 Vaporization Characteristics of Fuels and Fission Products Components

A fuel pin delivers thermal power by establishing a radial profile of temperature gradient. Fission products generated with burnup get redistributed across fuel pin. The redistribution depends on the temperature profile across pin and concentration of fission products. Fuel–clad interactions are essentially governed by the redistributions of reactive fps. In low burnup fuels, most of the fission products remain embedded in the fuel matrix. Only the reactive and volatile fission products tend to move due to temperature gradient. In oxide fuels, the more volatile components such as Cs/Rb-bearing oxides, halides and tellurides, and also the fps in their elemental states such as inert gases, Cd, Te, Sb, I, Br are distributed all along the pin radius. Through vapor transport process, they can come in contact with clad wall and initiate intergranular corrosion, which leads to reduction in mechanical strength of clad. Restructuring and redistribution in thermal reactor fuels is not as significant as in fast reactor fuels. In fast reactor pin because of very high central temperature, there is concern on the vaporization of both fuel and fps. The transport process leads to restructured fuel with drastic change in the composition, oxygen-to-metal ratio, grain reorientations and grain morphologies.

In metallic fuel, the fps mostly existing in solid solutions or in intermetallic states undergo transport in atomic forms. The vapor transport process occurs in helium bonded fuels. In sodium-bonded fuels, the migrating atoms dissolve in liquid sodium according to their solubilities and then chemically diffuse across intergrain gaps.

5.9.5 Vapor Pressure Measurement Techniques

The equilibrium vapor pressures of most of the chemical components of fission products, that generally produce condensable vapors on heating, are measured by dynamic methods. Static methods of monitoring the pressure using a manometric principle are

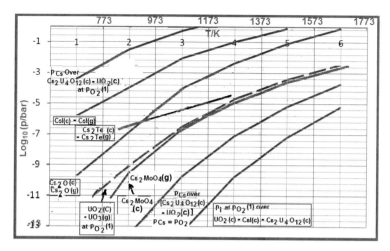

Fig. 5.15 Pressures of reactive volatile species in oxide fuels. ($p_{O2}(1)$ referred in the figure corresponds to the oxygen potential of Mo(c)/MoO$_2$(c))

difficult to implement with condensable vapors. Manometry involves pressure monitoring of a permanent gas relevant to the equilibrium property. In dynamic methods, the equilibrium state of the system is perturbed by adding a kinetic step of small but sustained escape route of vapor from the isothermally equilibrated system to such an extent that the intensive properties do not significantly deviate from their equilibrium values. Using well-established formalisms of vapor escaping kinetics, it is possible to obtain the steady-state pressure inside the enclosure. Knudsen effusion technique and vapor transpiration technique have been in use in the dynamic measurement of vapor pressures of materials as a function of temperature. The two dynamic methods described above are complementary techniques covering a large pressure range, from 1 bar down to about 10^{-12} bar. These methods have been employed for the characterization of most of the nuclear materials at high temperatures. Most of the pressure values plotted in Fig. 5.15 were derived by these characterization techniques.

5.9.6 Vapor Pressure Related Characteristics of Fuel Materials

In metallic fuels, in order to have observable effect of vapor transport across fuel pin under reactor irradiation, the concerned species should have sufficient vapor concentration initially at the grain surface. Considering the course of labyrinthine passage through interconnected pores/microcracks at the normal working temperatures of fuels, the transport over a long period of attaining high burnup will be significant if the species have vapor pressure above 10^{-10} bar. Through vapor transport the more volatile species such as Rb, Cs, Cd, Sr, Ba, I and Te concentrate at cooler periphery of the fuel. Moderately evaporating species such as Ag, Y and rare earth elements migrate out of the hottest central region. In helium-bonded fuel, the reactive components like rare earths having low melting points will migrate out of the central region through interconnected pores/cracks by capillary forces in their liquid state, but fps like Zr having high m.pt. can migrate out of the central region as rider on the moving fluid (in alloyed/nonalloyed forms). The alternative paths of slow solid-state diffusion apply to all the species. In sodium-bonded fuels, the reactive fps like Zr, Y, lanthanide elements, cesium, iodine, tellurium and cadmium, because of their high chemically driven diffusion in liquid sodium, migrate out of the central region and react with steel clad. In turn, the clad constituents migrate out into liquid sodium to react with the fuel forming actinide intermetallics at the peripheral region of the fuel slug. In oxide fuels, the reactive volatile components such as $Cs(g)$, $CsI(g)$, $Cs_2Te(g)$ and $Cs_2MoO_4(g)$ are known to have significant role in clad embrittlement. Failure of fuel containment at high burnup by stress corrosion cracking of clad in the presence of iodine, tellurium and Cs is established fact.

In fast reactor fuel, Pu and U can redistribute in the oxide phase, because of high steady-state temperature, particularly in the central zone of the fuel pin. Uranium concentration grows more toward fuel periphery, and it is transported through interconnected pores and cracks mainly in vapor phase as $UO_3(g)$ the pressure of which under the prevailing oxygen potential of high burnup fuel is significant. Besides UO_3, other vapor species such as UO_2, UO also contributes to uranium transport toward periphery. Additionally, Pu diffuses in the oxide matrix toward the central hot zone [17]. However, when O/M < 1.96, Pu will migrate down temperature gradient reducing the compositional disparity between center to periphery to some extent.

5.9.7 Thermochemical Aspects of UO_2, $(U,Pu)O_2$, $(U,Pu)C$, $(U,Pu)N$ and Metallic Fuels

Fissions of actinide isotopes in oxide, carbide and nitride fuels leave the electronegative element free in the respective cases. For example, in the case of oxide fuels two oxygen atoms are released in a fission event as: $^{235}UO_2 \rightarrow fps + neutrons + energy + 2O$. The fission released oxygen, carbon or nitrogen with their thermal energies diffuse through the fuel matrix in the respective cases and react, according to their thermodynamic potentials, among themselves, and with other fission products, fuel and clad. With increase of burnup, there is continuous chemical evolution and redistribution of the fission released elements and fps. Central melting, compositional variation, development of columnar and equiaxed grains, and lenticular voids are the general characteristics of restructured $(U,Pu)O_2$ fuels as noted in their high burnup states of irradiation in fast breeder reactors [18]. Thermal reactor fuels have relatively less restructuring; the peripheral part of $(U,Pu)O_2$ fuels form high burnup structure due to excessive local burnup by epithermal fast neutrons.

5.9.8 Oxide Fuels

The fission released oxygen (O) in the oxide fuels (MO_2) changes its thermodynamic potentials over the burnup period. The change is decided by taking into account the oxidations of reactive fission products, and also of fuel and clad. As O atoms diffuse through the matrix, they react with rare earths, alkaline earths, transition metals and alkali metals of the fission products. Oxygen can also react with the fuel matrix changing thereby its stoichiometry (O/M ratio). With the increase of O/M ratio, hyperstoichiometric fuel can share its oxygen with clads like zircaloy and stainless steel to result in formation of their oxide layers on inner surfaces of respective clads. Chromium component of stainless steel clad gets oxidized preferentially. It is important to mention here that the oxidation of the different elements of fps, fuel and clad occurs following the hierarchies of their chemical reactivities. With burnup, the chemical scenario in fuel changes in terms of varieties, proportion and distribution of the oxidized products. The O/M ratio of fuel is the indicator of O-potential and also of the prevailing chemical scenario inside fuel matrix.

Chemical states of oxidized fission products inside the widely used urania matrix are fairly well-established [19]. Table 5.4 presents the chemical states of fission products in a typically high burnup fuel. Fission process in ^{235}U-, ^{233}U- or ^{239}Pu-based fuels generates significant amounts of zero to trivalent fission products as compared to the tetravalent state of M in MO_2-based matrices. As a result, fission released two oxygen atoms per fissile element which is not fully consumed by the oxidations of reactive fps (see Fig. 5.16).

The leftover oxygen leads to hyperstoichiometry fuel, MO_{2+x}. A part of the oxygen undergoes chemical transport to the clad for its oxidation. As indicated in the beginning, oxygen diffusion across the helium filled fuel–clad gap involves chemical transport of the oxygen bearing species such as Cs_2O, $CsOH$, H_2O vapors that are

Table 5.4 Chemical states of fission products in typical oxide fuels [19]

Type of fission products	Chemical states
Nonvolatiles	
Rare earths (Nd, La, Ce, etc.) and transition metals (Y, Zr, Nb, etc.)	Dissolved state in fuel matrix (MO_2)
Alkaline earths (Sr, Ba) and Zr, U, Pu, Mo, Ce	Perovskites, M'MO$_3$ (gray) phase
Noble metals (Pd, Ru, Rh, Tc) and Mo	Alloy phases (white inclusions)
Gases and volatiles	
Inert gases (Xe, Kr)	As dispersed microbubbles in fuel Fractional release at fuel–clad gap and at plenum
Alkali metals (M = Cs, Rb)	MI, M_2Te at low-moderate burnup\ Conversion of MI, M_2Te to urinate/ molybdate at high burnup
Cd, Te, Sb	In alloy phases
Te, I, Br	As alkali metal compounds

Fig. 5.16 Tentative oxygen uptake by the fps, fuel and clad [19]

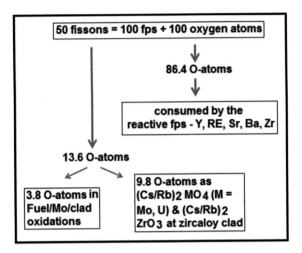

present in trace levels [19]. Reactive fission products such as Cs, Mo and to some extent Te form oxygen-rich compounds at high burnups.

The use of carbide, nitride or metallic alloy fuels is generally considered as advanced concepts to cater the strategic need of compact reactors. The fission released carbon and/or nitrogen from carbide, nitride and carbonitride fuels change the chemistry of fps and clad in a similar way as mentioned in the oxide case. Clad carburization and consequent deterioration of its mechanical property are required to be addressed in the simulation analysis of carbide fuels. The carbide fuel fabrication needs meticulous control on oxygen and moisture contents in the inert gas atmosphere maintained there. This is required because of its high pyrophoricity in oxygen atmosphere and of its susceptibility to hydrolysis. Presence of traces of oxygen also enhances carbon transport to clad via the vapor phase involving CO/CO_2 equilibrium; transported carbon carburizes clad surface. The spent carbide fuel reprocessing is equally problematic as it is difficult to dissolve it in nitric acid and the dissolution leaves behind organic complexes; this aspect has been detailed in separate chapters.

5.9.9 Metallic Fuels [16, 20, 21]

Metallic fuels have several attractive features such as high thermal conductivity, and high fissile and fertile atom densities which help accomplish higher breeding and the use of compact reactor core. Metallic fuels have favorable neutronic safety characteristics: High thermal expansion property of the fuels provides negative reactivity feedback during power ramp, low heat capacity reduces the stored energy of the core, and high thermal diffusivity facilitates the dissipation of the stored energy under loss-of-coolant event. U and Pu in their metallic states or even in alloyed states with some elements suffer from swelling under reactor irradiation due to anisotropic crystal structure and formation of low-melting alloys with clad materials. In order to surmount the stated problems, quite extensive studies were made in the past particularly in USA [20] to work out suitable alloys of U/Pu with several noble metal fission products (nick name fissium, Fs) and with transition metals like Mo and Zr. Some of studied alloys are U–5 wt% Fs, U–10 wt% Zr and U–19wt%Pu–10wt%Zr in their stable cubic phases. Zr addition helps increasing the solidus temperature of the base alloy (U–Pu) and improves upon containment. U–Zr was used as driver fuel in Experimental Breeder Reactor (EBR, USA), and the U–Pu–Zr alloy was thereafter tested for its performance as fuel in Integral Fast Reactor (IFR). The ternary alloy could perform satisfactorily up to a burnup of 20 atom% when the fuel density was to be compromised to some extent by introducing porosity. For commercialization of liquid sodium cooled metal reactors, it is desirable to have burnups of about 15atom%.

The extensive study on fuel–clad compatibility for safely attaining the burnup with a power rating of 450 W/cm prescribes that the alloy is to be fabricated with a smear density of 75%. The fuel slug of controlled porosity prepared by using injection casting technique is filled with sodium metal. Sodium bonding helps in thermal conduction of the porous fuel maintaining safe temperature profile within the fuel (~**1000 K** at center and **870** K at fuel–clad interface). Ferritic-martensitic-based advanced alloys are generally recommended as clad to withstand chemical corrosion from the metallic fps, lanthanides in particular, and also to withstand stress from fuel swelling mainly due to the fission gas entrapments. The gas entrapment increases internal pressure and reduces thermal conductivity perceptibly. The central temperature of virgin fuel rises from **943 K** to about 1080 K. The developed gas pressure at an early stage of burnup (>2atom%) becomes high enough to result in fuel slug swelling and mechanical stress development by making intimate contact with clad wall. Softened slug thereby undergoes distortion forming interconnected channels for the trapped gases to be released into plenum space.

At the normal operating condition, the U–Pu–Zr alloy fuel is seen to perform safely up to 20atom% burnup even under a peak clad temperature of 868 K. Zirconium from homogeneous alloy moves to cooler periphery to form a thin Zr-barrier layer on clad depleting thereby Zr content in alloy fuel. It is seen that under simulated power ramp when peripheral part of fuel attains temperature as low as 945 K, Zr depletion (depleted composition ~U–Pu–2atom%Zr) leads to fuel melting with failure of containment by breaching of clad. Lanthanide/actinide incorporations increase clad's brittleness. To avoid direct contact of fuel with clad wall, the Zr-sheathed fuel slug with good thermal contact with clad has been suggested. The sheathing minimizes Zr depletion in alloy fuel and reduces diffusion of Ln/Ac from slug. For recycling of actinides of spent fuel, the fuel-cycle program (integral fast reactor concept of EBR) adopts molten salt-based electro-refining process.

5.9.10   Carbide and Nitride Fuels [16, 21]

Mixed carbide (U,Pu)C and nitride (U,Pu)N drew attention since early years amidst extensive use of the oxide/mixed oxide fuels in the nuclear industries worldwide. Carbide and nitride fuels are equally attractive in terms of their properties, though a more extensive database exists for carbide fuels than for nitrides. The attractive features of these fuels are high fissile density with high breeding in liquid metal cooled fast reactors, superior thermal conductivity over that of oxide fuels. These fuels exist in cubic phases, and they are chemically compatible in liquid sodium, which can serve as coolant. Sodium bonded, or, alternatively helium-bonded fuels are fabricated with smear density less than 80%. As compared to helium-bonded fuels, the sodium-bonded state operates under low-temperature profile at a given power rating because of its higher thermal conductivity. Low operating temperature helps mitigating fuel–clad interaction due to less fuel swelling at high burnup. Fuel swelling makes good contact with steel clad and carburize it forming the brittle phase, $(Fe,Cr)_{23}C_6$. Swelled carbide matrix stresses on clad leading to FCI failure. Helium bonding, however, has the merit that the extent of carburization is less as compared to case of Na bonding. This is because of the only means of making carbon transport to clad is through vapor phase equilibrium of CO/CO_2. Liquid sodium, on the other hand, having finite solubility of carbon at high temperatures (~1.5 ppm at 800 K) transports carbon in dissolved and in dispersed states directly to clad besides the indirect means of the vapor phase transport.

Studies carried in several countries (USA, Russia, France, UK and Germany) concluded that the bonded UC, UN, (U,Pu)C, (U,Pu)N fuels with less than 20atom% Pu can be safely irradiated up to 10 atom% burnup. In some cases, it is claimed that peak burnup of 20 atom% has been achieved. FBTR in India has experienced with mixed carbide as driver fuel with the unique composition of $(U_{0.3}Pu_{0.7})C$. The cold worked steel cladded fuel has been running smoothly over three decades with the average power rating of 350W/cm (peak power 400W/cm); burnup attained thus far is about 18 atom%. Microstructural analysis of the irradiated fuels shows that clad carburization is quite low, and at the high burnup, the fuel porosity has diminished. The analysis leads to the conclusion that PuC component having low melting point the Pu-rich mixed carbide exhibits sufficient plasticity to accommodate swelling stress to clad. Carbon impurity in nitride fuel also carburizes the clad, but the deleterious effect to clad occurs at higher burnup as compared to that in carbide. Nitride fuels have the other attractive features of less fission gas release and also less swelling, provided the peak temperature of fuel does not exceed 1473 K. Above this temperature, fuel fragments extensively and causes FCMI failures. The nitride fuel unless fabricated by using ^{15}N isotope, the naturally occurring ^{14}N generates biologically hazardous ^{14}C isotopes that pose problem during refabrication of spent fuel. Nitride offers a higher fuel-particle margin to melting combined with the thermal and neutronic advantages of the fuel matrix. Nitride compounds are currently being investigated as fuels by researchers in Japan, Europe and Russia.

Exercises

1. List the different types of fuels used in the nuclear industry based on their chemical constitution and physical properties.
2. Explain the evolution of various phases in the system U–O, starting from UO2. Substantiate your answer with appropriate illustrations and figures.
3. A given metal carbon system has a carbide M_2C. Using a suitable expression, calculate the carbon potential of a mixture containing M and M_2C at 1073 K, if the Gibbs energy of formation of M_2C is -57 kJ mol^{-1} at 1073 K.
4. Draw the ternary phase diagram (schematic) of the system U–Pu–C at 1873 K. Explain the preferential segregation of Pu in the appropriate phases in the various two-phase fields with suitable thermodynamic basis.
5. In a mixture containing a metal M and its monocarbide MC that coexists with it in equilibrium at a given temperature, bring out the relation between the Gibbs energy of formation of the monocarbide and the carbon potential of this couple by writing the appropriate formation equation.
6. Explain (with the help of an appropriate expression) how the oxygen contamination of the monocarbide fuel would affect the carbon potential of the mixed carbide fuel that contains both MC and M_2C_3 phases.
7. Why is a small amount of zirconium added to the fuel meat while fabricating U–Al and Pu–Al fuels? What are the advantages and disadvantages of using the metallic fuel over the conventional oxide fuel?
8. Calculate the thermal efficiency of a PWR system at a power of 922 MW(e) if following are the data pertaining to this system with the following parameters. (A) Coolant flow = 60, 000 ton/h, (B) inlet temperature = 290 °C, (C) outlet temperature = 318 °C and (D) Cp of water = 5.63 kJ kg^{-1} K^{-1}.

References

1. D.E. Janney, C.A. Papesch, *FCRD Transmutation Fuels Handbook 2015* (INL National Library, INL/EXT-15-36520, September 2015), p. 13
2. D.E. Janney, C.A. Papesch, *FCRD Transmutation Fuels Handbook 2015* (INL National Library, INL/EXT-15-36520, September 2015), p. 8
3. D.R. Olander, *Fundamental Aspects of Nuclear Reactor Fuel Elements* (Published by Technical Information Center, Office of Public Affairs Energy Research and Development Administration, USA, 1976)
4. F. De Bruycker, High temperature phase transitions in nuclear fuels of the fourth generation. Ph.D. thesis, Universite d'Orleans, 2010, NNT:2010ORLE2060, p. 15
5. S. Rosen, M.V. Nevitt, A.W. Mitchell, J. Nucl. Mater. **9**, 137–142 (1963)
6. L. Olivares, World J. Nucl. Sci. Technol. **06**(01), 43–52 (2016)
7. Hj. Matzke, *Science of Advanced LMFBR Fuels* (North Holland, 1986)

8. Fuel rod internal chemistry and fission products behavior IAEA, Vienna, 1986 IWGFPT/25. Also: J. McFarlane, J.C. LeBlanc, D.G. Owen High-temperature chemistry of molybdenum, cesium, iodine, and UO_{2+x}, AECL-11708 (1996)

9. C. Guéneau, J.-P. Piron, J.-C. Dumas, V. Bouineau, F.C. Iglesias, B.J. Lewis, Fuel-cladding chemical interaction (Chapter 3), NEA/NSC/R5 (2015)

10. J. McFarlane, J.C. LeBlanc, D.G. Owen, High-temperature chemistry of molybdenum, cesium, iodine, and UO_{2+x} AECL-11708 (1996)

11. B.S. Fox, In-pile thermal conductivity measurement methods for nuclear fuels, A thesis submitted in partial fulfillment of the requirements for the degree of M.Sc in Mechanical Engineering, Utah State University, Utah (2010)

12. R. Othman, Steady state and transient analysis of heat conduction in nuclear fuel elements, TRITA-NA-E04051 (2004), p. 13

13. R.E. Taylor, J. Gembarovic, K.D. Maglic, Thermal diffusivity by the laser flash technique, in *Characterization of Materials*, vol. 1, ed. by E.N. Kaufmann (Wiley-Interscience, 2003). Also see the ppt presentation by Dr.-Ing. Wolfgang Hohenauer (wolfgang.hohenauer@phox.at)

14. J.D. James, J.A. Spittle, S.G.R. Brown, R.W. Evans, A review of measurement techniques for the thermal expansion coefficient of metals and alloys at elevated temperatures. Meas. Sci. Technol. **12**, R1–R15 (2001). Also see the ppt presentation by Dr.-Ing. Wolfgang Hohenauer (wolfgang.hohenauer@phox.at)

15. P.J. Haines, M. Reading, F.W. Wilbum, Thermal analysis and differential scanning calorimtery vol 1, chapter 5, principle and practice, in *Handbook of Thermal Analysis and Calorimetry*, M.E. Brown (Elsevier, Amsterdam, 2010). Also refer to (https://books.google.com › Science › Chemistry › Analytic, 2007) and to the ppt presentation by Dr.-Ing. Wolfgang Hohenauer (wolfgang.hohenauer@phox.at)

16. *Status and Trend of Nuclear Fuels Technology for Sodium Cooled Fast Reactors* (IAEA Nuclear Energy Series No. NF-T-4.1, Vienna, 2011)

17. M.H. Rand, T.L. Markin, in *Proceedings of UAEA Symposium Thermodynamics of Nuclear Materials*, Vienna (1967), http://fti.neep.wisc.edu/neep423/FALL99/lecture21.pdf, p. 637. Also refer to (fti.neep.wisc.edu › neep423 › FALL99 › lecture21)

18. H. Bailly, D. Menessier, C. Prunier, *The Nuclear Fuel of Pressurized Water Reactors and Fast Reactors-Design and Behavior* (Intercept Ltd., Paris, 1999)

19. D. Das, M. Basu, S. Kolay, A.N. Shirsat, Transport properties of gaseous and volatile fission products in thoria-based fuels, in *Thermophysical and Thermodynamic Properties, Fabrication, Reprocessing, and Waste Management*, ed. by D. Das, S.R. Bharadwaj. Springer, London. Also S.R. Bharadwaj, R. Mishra, M. Basu, D. Das (2013) *Thermochemistry of Thoria-Based Fuel and Fission Products Interactions. ibid.*

20. R.G. Pahl, D.L. Porter, C.E. Lahm, G.L. Hofman, Experimental studies of U–Pu–Zr fast reactor fuel pins in the experimental breeder reactor-II. Metall. Trans. **21A**, 1865–1870 (1990)

21. *Development Status of Metallic, Dispersion and Non-oxide Advanced and Alternative Fuels for Power and Research Reactors* (IAEA-TECDOC-1374, IAEA, Vienna, 2003)

Chapter 6
Post Irradiation Examination of Fuel

Prerna Mishra, V. Karthik, and Priti Kotak Shah

6.1 Introduction

Nuclear fuel undergoes continual physio-chemical changes during irradiation in a reactor. Post-irradiation examination (PIE) broadly refers to various inspections and examinations conducted on an irradiated material to understand its in-reactor behaviour and performance. PIE is a vital link in the nuclear fuel cycle as it provides feedback to the fuel designers, fabricators, reactor operators and fuel model developers on the performance of fuel and associated structural components, thereby helps in improving the quality, safety, reliability, and economics of nuclear fuel cycle. Various aspects of PIE are covered in this chapter.

6.2 Purpose of PIE

It is essential that the fuel elements maintain their integrity throughout their life cycle, beginning from their use in nuclear reactors, during storage after being discharged from the reactor till reprocessing or storage forever, in away from reactor (AFR) spent fuel storage pools. Such spent fuel elements are also required to withstand

P. Mishra (✉) · P. K. Shah
Post Irradiation Examination Division, Nuclear Fuels Group, Bhabha Atomic Research Centre, Mumbai 400085, India
e-mail: prernam@barc.gov.in

P. K. Shah
e-mail: pritik@barc.gov.in

V. Karthik
Metallurgy and Materials Group, Indira Gandhi Centre for Atomic Research, Kalpakkam 603102, India
e-mail: karthik@igcar.gov.in

stresses due to their handling, during storage, transportation and subsequent handling at reprocessing facilities.

The fuel elements in a nuclear reactor are subjected to a hostile environment due to a combination of radiation, temperature, chemical interactions and stress leading to degradation in their properties. Post-irradiation examination (PIE) of fuel and structural materials is aimed at generating data and its evaluation to provide insights into synergistic effects of reactor environment on performance of materials during irradiation and fitness for long-term storage and handling in spent fuel storage pools and while reprocessing. PIE also plays an important role in the following three aspects of reactor technology:

- Prediction and extension of life of nuclear fuel and structural components.
- Development of new types of fuels and structural components.
- Failure analysis of both in-core and out-of-core components.

6.2.1 Inputs and Outputs of PIE

PIE is expected to generate data on the physical and chemical changes occurring in the fuel, cladding and structural materials as a result of irradiation in the reactor core. Aggressive high-temperature exposure during reactor operations, corrosion/erosion effects on fuel cladding due to exposure to the flow of reactor coolants, pellet-clad interactions etc. play important role on fuel performance. Hence, details of various reactor-operating conditions, reactor physics data and fuel fabrication data like initial chemical composition and physical conditions are essential for arriving at a meaningful conclusion during PIE. At times, results of out-of-pile experimentation like fuel–coolant interaction at the temperature of operation, corrosion behaviour of structural materials due to coolant etc. are also useful in evaluating the PIE results. Various essential inputs for carrying out meaningful PIE and uses of PIE outputs are shown schematically in Fig. 6.1.

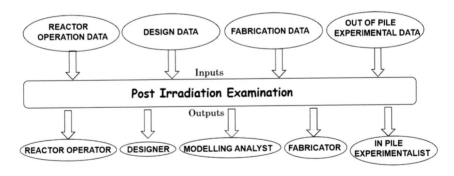

Fig. 6.1 Schematic depicting the inputs and outputs of PIE

PIE data is essential for a variety of agencies involved in the nuclear fuel cycle. The data provides essential feedback to the fuel designers, fabricators, reactor operators and fuel reprocessors. PIE can identify potential performance issues and help to increase the fuel burn-up, which is linked to fuel cycle costs. In the case of a commercially operating reactor system, where various fuel and structural material behaviour are well understood, PIE helps in optimizing the fuel performance by proper quality control during fuel/structural material fabrication and in adopting suitable reactor-operating conditions to avoid failures and undue outage of the reactor. PIE provides data for licencing new fuel designs as well as for material models/fuel performance codes that predict the fuel behaviour under normal and off-normal conditions.

6.3 Types of PIE

PIE can be broadly categorized as in-pile, pool-side inspection and hot cell-related examination based on the location at which examination is carried out.

6.3.1 In-Pile/On-Line Examination

In-pile/on-line examination gives first-hand information on the behaviour of fuel/structural materials, critical components and their interactions with each other under particular reactor-operating conditions. On-line measurements like temperature of the fuel, irradiation creep measurements, released fission gas pressure monitoring etc. are generally carried out. Owing to the highly hostile environment present in a reactor, it is challenging to access the particular point of interest for direct examination. A large number of non-destructive techniques (NDT) such as ultrasonic, acoustic emission and eddy current are used for in-pile/on-line monitoring of the reactor components.

6.3.2 Pool-Side Inspection

Inspection pool is generally a part of power reactors, where regular inspection of the spent fuel is carried out as a routine work. This is done to quickly assess the surface condition of irradiated fuel bundles and also to detect any gross abnormalities or failed fuel bundles. The irradiated material is taken out of the reactor and kept immersed in a pool of water. Basic leak testing of the fuel bundle can be carried out automatically in this immersion technique (maximum sensitivity up to 10^{-4}std cc/sec). Water acts as a good shield for radiation and also it provides both physical and dynamic containment for the irradiated fuel. Ease of material handling and viewing is an attraction of this technique. One of the important aspects of pool-side inspection is that the establishment and operating costs are very low when compared to other PIE facilities. With the use of remote-handling equipment, robotic devices with NDT equipment, remote viewing systems etc., the quality and accuracy of the pool-side inspection procedure have improved considerably. Pool-side inspection helps in screening a large number of fuel assemblies and in selecting the fuel elements for detailed examination in the hot cells.

6.3.3 PIE Using Hot Cells

PIE using hot cells is the most prominent and versatile method used for post-irradiation examination of highly radioactive materials, which include irradiated fuel elements and structural material removed from operating nuclear reactors as part of material surveillance and those irradiated for materials research and other components for nuclear reactors like, control rods, neutron flux monitors etc. Facilities for PIE make use of different categories of enclosures like concrete/lead-shielded cells, glove boxes, fume hoods, etc., for shielding from radiation as well as for containing the radioactivity, depending on the type of radiation and the toxicity of the material required to be examined [1]. Shielded enclosures designed with or without leak-tightness in which highly radioactive materials are handled in isolation from the operator's environment using remote-handling devices are known as 'hot cells'.

Hot cells are provided with physical and/or dynamic barriers, which establish the necessary containment required for the safe handling of radioactive materials. Physical containment is achieved by sealing the enclosure, whereas dynamic containment is maintained by suitable ventilation systems to ensure flow of air from outside to inside of the hot cell through suitably designed openings or unintentional leak paths like cable penetrations; gaps between sleeves and shielded plugs, door-frames and doors, transfer ports, sleeves provided for installation of remote handling equipment, etc. Normally, inside of the hot cells is maintained at some nominal negative pressure with respect to outside, by provision of dedicated exhausts. Table 6.1 gives the features of different containment systems for handling irradiated materials.

A series of hot cells is used in PIE, in which irradiated fuel is subjected to full-scale metallurgical investigation using remote-handling equipment. Hot cell walls are made of concrete or lead for shielding operator from radiation. Hot cells used for PIE are of two types, viz., β-γ and α-β-γ depending on the nature of radiation and the radio-toxicity of the material. Materials that emit predominantly β and γ radiations (e.g. irradiated natural/low enriched UO_2 fuel, structural material, etc.) are handled in β-γ hot cells, and materials that are radio-toxic and emit all three α, β and γ radiations (e.g. irradiated mixed oxide, e.g. UO_2-PuO_2 fuels) are handled in α-β-γ hot cells. The inside of the hot cells is lined with stainless steel plates for ease of decontamination. α-β-γ hot cells in addition to lining will have a secondary leak-tight enclosure within them for better containment of the radioactive material. Figure 6.2 shows the cross-sectional view of a typical hot cell. Hot cells are also used for reprocessing of spent fuel, production of radio-pharmaceuticals, in addition

Table 6.1 Features of fume hood/glove box/hot cell [1]

Enclosure	Material of construction	Radioactivity handled	Pressure (negative)	Ventilation rate
Fume hood	Steel, stainless steel, PVC, reinforced resin-bonded laminates	Low β,γ	20 mm of WC	0.5 m/s across a slash opening of 300 mm
Glove box	Stainless steel, Perspex, mild steel, plastic	α and low-level β,γ	10–20 mm of WC	15 to 20 air changes per hour
Hot cell	Stainless steel, steel, concrete, lead	Very high α,β,γ and neutron	15–25 mm of WC	20 to 40 air changes per hour

to PIE. Hot cells are generally constructed as a train connected in series or isolated depending on the intended use.

Operations inside the hot cells are carried out using mechanical and/or electrically operated remote-handling systems. Master–slave manipulators (MSM) are one such most commonly used mechanical remote-handling system, where the slave arm inside the hot cell mimics the movements of the master arm located in the operating area by use of tape-pulley combinations or mechanical linkages. To ensure confinement

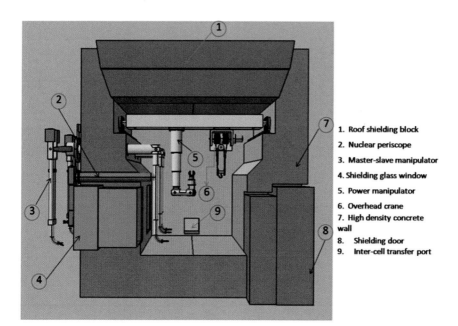

1. Roof shielding block
2. Nuclear periscope
3. Master-slave manipulator
4. Shielding glass window
5. Power manipulator
6. Overhead crane
7. High density concrete wall
8. Shielding door
9. Inter-cell transfer port

Fig. 6.2 Schematic showing various features of a hot cell

of radioactivity and to help in maintaining the negative pressure within the hot cells with respect to the operating area, the slave arm is enclosed in a tubular sheath (called booting) fixed to the through-tube of the master–slave manipulator. In some hot cells, remotely operated material-handling equipment like cranes and power manipulators are provided. For better leak-tightness in case of hot cells handling radio-toxic materials, sealed penetrations are provided in the cell wall for service lines, cables etc. Radiation shielding glass windows, periscopes, CCD cameras, etc. are provided to aid the operator for viewing inside the hot cells. Adequate lighting using sodium/mercury vapour/metal halide lamps is provided for comfortable viewing by the operator for carrying out the required tasks within the cells. Hot cells are also provided with transfer ports for docking the shielded fuel transport casks/storage casks for posting the irradiated material into them without compromising the radiation shielding and ventilation. Hot cells have provisions for transfer of materials in and out, personnel entry, introduction of tools and equipment etc.

The atmosphere within the hot cells depends on the chemical nature of the materials being handled; inert fuels like UO_2 and MOX are handled in air, while pyrophoric fuels like UC and PuC are handled in inert atmospheres. Elaborate ventilation systems are provided for hot cells and their surrounding areas with HEPA filter banks in the exhaust stream to limit the radioactivity of the exhaust to within permissible levels stipulated by the regulators.

6.4   Irradiation Damage in Solid Nuclear Fuels

Irradiation-induced changes in the properties of nuclear fuel elements are presented in this section. Nuclear fuels operate at high temperatures and radiation fields during their residence in reactors. The main source of radiation damage in nuclear fuels is highly ionized and energetic (165 MeV) fission fragments and fast neutrons with energy up to 10 MeV. Fission fragments and neutrons move rapidly through the lattice, exchanging their energy with the lattice atoms before they come to rest in the matrix.

The damage to the fuel by fission fragments is caused in the following three ways:

(i) Fission spikes: Over a distance of 6–10 microns in the fuel matrix, the highly ionised fission fragments exchange their energy with the lattice atoms till they come to rest. In this process, the fission fragments release all their energy in raising the fuel temperature to more than 3000 °C for a brief period of 10^{-11} s. This cylindrical region with radius of about 100 Å and fission fragment moving along its axis is called the fission spike. The end portion of the fission spike where the fission fragment comes to rest is characterized by severe disarrangement of lattice atoms and is often called displacement spike, while the rest is called thermal spike.

(ii) Lattice defects: Collision of fission fragments with lattice atoms of the fuel matrix produces primary knock-on atoms, which have sufficient energy to displace other atoms from the fuel lattice. This gives rise to creation of several types of point defects in the lattice which enhances all diffusion-controlled processes and creation of new transport processes that are active at relatively low temperatures. The lattice defects, namely the vacancy and the interstitial have a large effect on the fuel element behaviour.

(iii) Impurity atoms: Each fission event creates two atoms of fission products which are chemically different from the original material. Build-up of fission product atoms influences the properties of fuels.

6.5   Irradiation Effects in Fuels

The irradiation effects vary depending on the type of the fuel, viz., metallic and ceramic fuels and also vary with type of the reactor system (thermal or fast reactor) and the operating conditions. The chief irradiation effects in metallic fuels is the dimensional changes caused by fission product swelling, irradiation creep and irradiation growth, while in ceramic fuels, the major irradiation effects include (a) densification, (b) restructuring, (c) redistribution of fuel and fission products, (d) effects of fission gases: swelling and gas release and (e) irradiation creep [2].

6.5.1   Metallic Fuels

6.5.1.1   Fuel Swelling

Swelling of fuels is primarily due to replacement of uranium atom by its fission products. High-purity uranium swells significantly in the temperature range 350–600 °C with a breakaway swelling (swelling maxima) at about 450 °C. The swelling of uranium originates from the nucleation and growth of bubbles of fission gases, xenon and krypton which are insoluble in fuel matrix and tend to precipitate as bubbles on grain boundaries, dislocation and twins. As metal fuels operate at lower fractions of their melting points, fission gas diffusion is fairly slow. Due to higher thermal conductivity of fuel, the thermal gradient across the fuel is small, hence, the driving force for directed bubble migration is low.

6.5.1.2   Irradiation Growth

Irradiation growth refers to the change in fuel shape at constant volume without application of any external stress. The phenomenon of irradiation growth is caused by the anisotropic thermal expansion of alpha uranium. The thermal stresses induced in the fission spike due to the anisotropic expansion of uranium results in aggregation of vacancies and interstitials on different crystallographic planes. This causes elongation (due to interstitial clusters) in the <010> direction and contraction (due to vacancy cluster) in <100> direction. Irradiation-induced growth coupled with creep results in bowing of the uranium fuel element.

6.5.2 Ceramic Fuels

A typical oxide fuel operates with a high centre temperature and with a radial thermal gradient as high as 3000–5000 °C/cm. This leads to formation of radial cracks in the brittle ceramic fuel, which propagate inward from the fuel outer radius due to thermal stresses. In the beginning of life, due to differential thermal expansion between centre and rim, the pellet takes an hourglass shape. In high-temperature operation of the fuel, cracking and distortion of the fuel pellets lead to significant changes in the microstructure, composition, dimension and shape of the fuel [3].

6.5.2.1   In-Pile Densification

Densification is caused by fission spike-pore interaction which results in shrinkage and dissolution of pores in the fuel matrix. A fission fragment while traversing in the fuel grain re-solves the pores on its path into lattice vacancies. Vacancy migration to the grain boundary results in densification of fuel. Pores less than 2 μm size disappear from the fuel pellet during early life. It leads to gap formation in the pellet stack, increase in the pellet-clad gap and collapse of the cladding. It occurs in early life of ceramic fuels under irradiation. It occurs in the temperature range 400–1200 °C. Resintering at 1700 °C for 24 h gives same amount of densification. In-pile densification leads to increase in fuel clad gap, rise in fuel temperature, shortening of fuel stack and collapse of cladding.

6.5.2.2   Fission Gas Release Behaviour

A significant fraction (15%) of the fission products in nuclear fuel consists of noble gases, xenon and krypton, which have extremely low solubility in the fuel matrix. Swelling of the fuel is attributed to the volume of fuel occupied by the gas bubbles. Bubble formation depends on the mobility of gas, the minimum number of gas atoms required to form a stable nucleus and the rate at which lattice vacancies can be supplied to enhance the stability of a nucleated complex. A major part of the fuel swelling is caused by formation and growth of large intergranular bubbles in the process of gas release.

Gaseous fission products (xenon and krypton) are released by the ceramic fuel to the free spaces in the fuel pin. Release of fission gases is governed by the mobility of the gas as individual atoms or as bubbles and is strongly dependent on temperature. The initial process involves athermal release by ejection of atoms of gaseous fission

Table 6.2 Typical values of fission gas release measured in fuels [3–5]

Fuel	Reactor type	Burn-up (typical values)	Fission gas release (% of gases generated due to fission) (%)
UO_2	PHWR	8 GWd/t	7–9
UO_2/MOX	PWR	60 GWd/t	3–6
$(U, Pu)O_2$	Fast reactor	100 GWd/t	80–85

products at the free surfaces of the fuel, as a result of displacement cascades caused by fission product implantation. This process operates at fuel temperatures below 1000 °C. In the temperature range 1000–1600 °C, atomic diffusion of gas atoms from grain interior to the grain boundaries takes place and as bubbles grow in size, they interlink with each other forming a continuous channel for release of the gas to open surfaces. At fuel temperature above 1600 °C, almost all the fission gas generated in the fuel is released. The release occurs by movement of gas bubbles which are sufficiently mobile at such high temperatures. Apart from fuel temperature, other factors which influence the fission gas release are fuel grain size, burn-up, fission rate, fraction of open porosity, fuel density and stoichiometry. The typical fission gas release for various fuels is given in Table 6.2.

The fission gases retained in the fuel precipitate as gas bubbles and cause fuel swelling. The dimensional changes in fuel due to swelling lead to fuel clad mechanical interaction which could result in clad failure. On the other hand, as the gases get released from the fuel body, they increase the pressure inside the fuel element subjecting the cladding tubes to stress. In addition, these gases reduce the thermal conductivity of the fuel clad gap, thereby increasing the temperature of the fuel. Fission gas release is expressed as a percentage of the estimated total quantity of gases generated due to fission based on fuel burn-up. For a fuel with a given microstructure, density and burn-up, the quantity of released fission gases is a direct indicator of the operating temperature of the fuel.

6.5.2.3 Microstructural Changes

The microstructure of as-fabricated sintered ceramic fuel pellet consists of equiaxed grains of 5–15 micron diameter with uniformly distributed porosity. During reactor operation, the fuel is subjected to increase in power which leads to temperature increase, resulting in changes in fuel microstructure. Microstructure of the spent fuel evaluated during PIE can be used to estimate the maximum operating temperature of the fuel, which is an indicator of fuel performance and in-turn that of the reactor.

Irradiated fuel cross-sections when examined using a metallograph reveal concentric zones of altered microstructures depending on the operating temperature of the fuel. This change in microstructure, called restructuring, results in the creation of a central hole due to migration of pores from the mid radius region of the fuel towards the centre due to prevalent higher temperature. This central hole is surrounded by a concentric zone of elongated grains called columnar grains and a region of larger than as-fabricated equiaxed grains towards the periphery. The outermost zone at the periphery of the fuel consists of as-fabricated sintered grains. The restructuring involves movements of pores up the temperature gradient to the fuel centre by an evaporation–condensation mechanism. These restructured zones have fairly defined temperatures associated with their boundaries and is a direct indicator of the maximum operating temperature of the fuel element. Table 6.3 gives the various microstructural zones formed in a typical oxide and carbide fuel and the associated temperatures.

The columnar, equiaxed and as-sintered regions of restructured oxide fuel pellets having unique microstructures display varying thermo-physical behaviours. A schematic of the restructuring observed in irradiated oxide fuel is shown in Fig. 6.3.

Table 6.3 Microstructural zones observed in the cross-section of an irradiated ceramic fuel pellet [2]

Fuel type	Temperature range	Microstructural zones
$(U, Pu)O_2$	Below 1100 °C	As-sintered microstructure
	1100–1300 °C	Intragranular porosity
	1300–1700 °C	Equiaxed grain growth
	Above 1700 °C	Columnar grain growth and central void
$(U, Pu)C$	Below 1100 °C	As-fabricated microstructure and small bubbles
	1100–1400 °C	Grain growth, large gas bubbles, bubbles at grain boundaries
	1400–1700 °C	Radially elongated grains and pores
	Above 1700 °C	Irregular, big-rounded pores and central void

Fig. 6.3 Various zones of a restructured oxide fuel cross-section

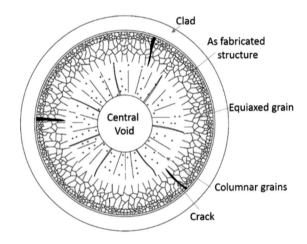

6.5.2.4   Redistribution of Fuel Constituents and Fission Products

Fission in nuclear fuel results in production of atoms of more than 30 elements and about 120 isotopes of these elements have appreciable half-lives. Due to steep temperature gradient in the fuel, some of the fission products move away from the location where they are produced. One of the important consequences of fission product redistribution is the possibility of chemical attack by fuel on the cladding material. Volatile fission products Cs, Rb, I, Te, Sb and Cd tend to migrate to the cooler regions in the fuel. Owing to high volatility, gross radial migration of Cesium occurs from high-temperature central regions to cooler periphery of the fuel, pellet-to-pellet interfaces and to the plenum in the fuel pin.

6.5.2.5   Plutonium and Oxygen Redistribution

In mixed oxide (MOX) fuel containing plutonium, oxygen redistributes under the influence of radial thermal gradient, which is governed by the initial stoichiometry of the fuel. Higher oxygen potential of irradiated MOX fuel and volatile fission products (Cs, Te, I, etc.) are the primary cause for initiation of fuel clad chemical interaction (FCCI).

   In plutonium-bearing ceramic fuels, the formation of a central void and the movement of pores lead to a redistribution of plutonium occurring during the early stages of irradiation, depending upon the initial stoichiometry of the fuel. In fuel having an initial O/M of 2.01, the plutonium segregates towards fuel centre, while in fuel having an initial O/M of 1.95, the plutonium segregation occurs away from the fuel centre. The main consequence of Pu enrichment at the edge of the central hole is a slight decrease in melting temperature and locally increased burn-up.

6.5.2.6   Irradiation Creep

The effect of irradiation on creep of ceramic fuels is twofold: (a) to enhance the thermal creep and (b) to develop creep under condition in which thermal creep is absent. After fuel clad gap closure, the cladding is stressed by the oxide fuel swelling, while irradiation creep of the fuel relieves the stresses from damaging the cladding [6].

6.6   Irradiation Effects on Fuel Cladding

Fuel cladding is the outer layer of the fuel rods, acting as barrier between the reactor coolant and the fuel pellets. Cladding constitutes one of the barriers in 'defence-in-depth' approach; therefore, its performance is one of key safety aspects. Cladding, end plugs and welds by design are expected to maintain the integrity throughout the life cycle of the fuel element. Zirconium-based alloys are used as fuel cladding in case of thermal reactors fuels, while stainless steels, both austenitic and ferritic-martensitic, are commonly used as cladding material for fast reactor fuels.

Neutron irradiation of cladding material leads to generation of defects in the crystal lattice. Claddings also pick up reaction products due to corrosion by coolant. Primarily, these bring about an increase in yield strength and ultimate tensile strength and decrease in ductility of the fuel cladding.

Some of the other, not so common, irradiation-induced changes in fuel cladding due to long-term irradiation like in fast reactors are (i) creep, growth and void swelling, all leading to dimensional changes in fuel cladding; (ii) segregation leading to changes in microchemistry; and (iii) dissolution of otherwise stable phases into the matrix and formation of new precipitates. These phenomena also affect the mechanical properties of fuel cladding resulting in increase of strength and reduction of ductility [7].

These changes are assessed during PIE and the feedback is provided to the designers, fabricators, reactor operator and to the agencies managing the storage pools and reprocessing.

In addition, chemical degradation caused by exposure to high-temperature water can cause phenomena such as general corrosion, hydriding, stress corrosion cracking crud deposition and localized forms of corrosion such as nodular corrosion and shadow corrosion and fuel clad interactions. These microstructural changes can severely impact fuel cladding properties such as strength, ductility and corrosion resistance [8].

6.7   Techniques for PIE of Fuels

Performance evaluation and failure analysis of irradiated nuclear fuels are carried out using various non-destructive and destructive techniques. The non-destructive techniques include visual examination, leak testing, profilometry, gamma scanning, eddy current testing, ultrasonic testing, X-ray radiography and neutron radiography. The destructive testing techniques include fission gas analysis (retained and released), metallography/ceramography, microhardness testing, β-γ autoradiography, α-autoradiography, scanning electron microscopy, electron probe microanalysis, radiochemical burn-up analysis and mechanical testing of cladding.

6.7.1   Non-destructive Testing Techniques

6.7.1.1   Visual Examination

Hot cell examination campaign on fuel elements starts with visual examination of the as-received fuel assembly. Visual examination is used to detect any gross distortions and to check integrity of the assembly. Individual fuel elements are subjected to further visual examination after dismantling of the fuel assembly. Inspection of the outer surface of the fuel elements is carried out to examine the condition of the appendages, end plug welds, spacers, corrosion and crud deposit on the cladding. In case of failed fuel elements, the type, size and location of the defect in the cladding are documented. The features identified from visual examination are useful to decide the region of interest for subsequent examinations.

Initial inspection is carried out directly through the hot cell viewing windows (Fig. 6.4). Detailed visual examination is carried out by using through-wall periscope and/or digital cameras with PTZ features.

Figure 6.5 shows the hydride blister perforation observed in the cladding of a PHWR fuel element during visual examination.

Fig. 6.4 Hot cell viewing windows and master–slave manipulators in the operating area of the hot cells [9]

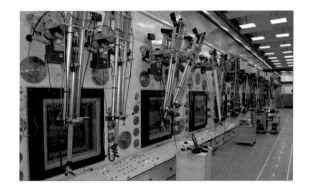

Fig. 6.5 Perforation in the cladding of a fuel element observed during visual examination

Fig. 6.6 Bubbles emanating from the failure location of a fuel pin during liquid nitrogen-alcohol leak testing inside the hot cell

6.7.1.2   Leak Testing of Fuel Elements

Leak testing of fuel elements confirms the presence of through-wall defect in the cladding. One of the methods for leak testing of short fuel elements is by liquid nitrogen-alcohol method. The fuel elements are kept immersed in liquid nitrogen bath for 10–15 minutes and then immersed in alcohol bath. In case of a leaky fuel, liquid nitrogen seeps and fills up the internal spaces. On transfer to the alcohol bath, the element warms up, converting liquid nitrogen into gas, which escapes through the leak as bubbles, thus enabling the identification of the location of the through-wall defect. Bubbles emanating from a failed fuel element immersed in alcohol bath are shown in Fig. 6.6.

6.7.1.3   Profilometry and Length Measurement

Profilometry is measurement of the diameter of the fuel element at closely spaced intervals along the length thereby enabling the generation of the outer profile. The instrument used is called profilometer. The output of the profilometer is used to evaluate the dimensional changes in the fuel element which is useful to determine the extent of fuel swelling, circumferential ridge height, ovality and bow. Both laser scanning profilometry (non-contact method) and LVDT transducer-based profilometry (contact method) are used in PIE. Figure 6.7 shows the diameter profile of a PHWR fuel element [10]. Increase in the length of the fuel element is measured using a digital vernier.

6.7.1.4   Gamma Spectrometry and Gamma Scanning

Gamma spectroscopy uses the uniqueness of the gamma energy spectrum emitted by radio-nuclides present in the fuel. These radio-nuclides produced due to nuclear fission are a measure of fuel burn-up. Gamma spectrometry helps to identify and estimate the quantity of gamma emitting radio-nuclides present in the irradiated fuel element. Axial gamma scanning of the fuel element provides information on the fuel column length, fuel pellet interface, power and burn-up distribution along the fuel

Fig. 6.7 Diameter profile of a PHWR fuel element after irradiation

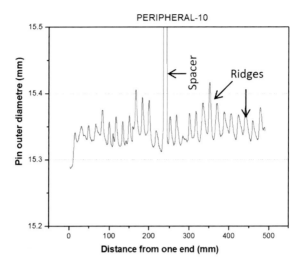

column, which can vary significantly in case of long fuel elements used in BWRs and PWRs.

The gamma scanning set-up consists of a scanning stage, collimator, detector and the recorder assembly. The fuel element is moved at a predefined speed using the scanning stage in front of a collimator fitted in the hot cell wall. The detector kept on the other side of the collimator records the gamma activity. Gross gamma scanning and gamma spectrometry is carried out using NaI(Tl) and high-purity germanium (HPGe) detector. Typical output of a gross gamma scan of an experimental CANDU fuel element is shown in Fig. 6.8 [11].

6.7.1.5   Eddy Current Testing

Eddy current testing with encircling coil is carried out on irradiated fuel elements to detect defects in the cladding, to study local dimensional variations in the fuel element, and to measure the oxide layer thickness on the outer surface of the cladding.

The eddy current test set-up consists of a differentially wound encircling coil type of probe which scans the circumference of the cladding as the fuel element is translated through the probe. The dimensions of the probe and test frequency are selected based on the diameter and material of cladding. Typical use of this technique in identifying defect in the cladding and confirmed by optical microscopy of a BWR-type fuel element is shown in Fig. 6.9 [12].

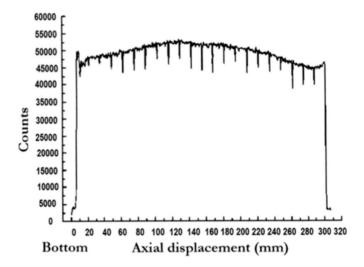

Fig. 6.8 Axial gross gamma scan of an experimental CANDU fuel element

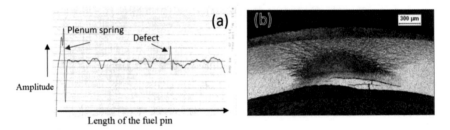

Fig. 6.9 (a) ECT signal and (b) micrograph at the corresponding defect location in a BWR fuel cladding

6.7.1.6 Ultrasonic Testing

Ultrasonic testing method is used to detect defects in a fuel element cladding and checking the integrity of the end plug to cladding-tube weld zone.

Immersion ultrasonic testing of end plug welds of the fuel elements is carried out using high-frequency-focused beam technique for volumetric scanning of complete circumferential weld. Ultrasonic probe inclined at an angle in axial direction is used to generate refracted shear waves to intercept the defects in the weld region. The ultrasonic beam passes through the sound welds, while it gets reflected from the defective end plug weld. Figure 6.10 (a) and (b) show the ultrasonic signals obtained from a sound weld and from a defective weld, respectively [13].

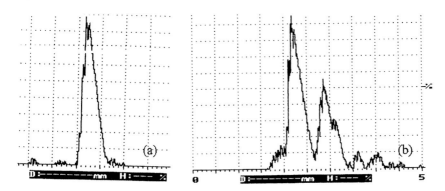

Fig. 6.10   Ultrasonic signals from (**a**) sound weld and (**b**) defective weld

6.7.1.7   Radiography

Radiography techniques are based on the general principle that radiation is attenuated on passing through matter. In radiography, the object under examination is placed in the incident radiation beam of X-ray, gamma ray or neutron. After passing through, the beam enters a radiographic film/detector that registers the fraction of the initial radiation intensity due to attenuation by each point in the object. Any inhomogeneity or an internal defect (e.g. void, crack, porosity, or inclusion) in the object will show up as a change in radiation intensity reaching the film/detector. Radiography is a versatile non-destructive technique for imaging the internal features of the fuel pin and for determining critical gaps and clearances available in the irradiated pin [14].

(a)   *X-ray radiography*

X-ray attenuation increases with increase in atomic number because of increased electron cloud density. Radiography using X-rays makes use of differences in density and thickness of various components of fuel pin internals to image the features. One of the major issues during X-radiography of irradiated fuel is the fogging of the radiographic film due to high gamma radiation leading to loss of image contrast. This is overcome by high-speed positioning and retrieval of the films and by optimizing the voltage and exposure (kV, mA, time) parameters. The film after processing is digitized and subjected to image processing for estimation of internal dimensions such as increase in fuel stack length and variations in pellet-to-clad gap. The increase in fuel stack length is a measure of the axial fuel swelling. X-radiography also reveals abnormalities like chipping and cracking of pellets. Figure 6.11 shows typical radiographic images of mixed carbide fuel pins irradiated in FBTR.

(b)   *Neutron radiography (NR)*

Penetrating nature of neutrons and their differential absorption by the material are used to obtain the details of the internal structure. Neutrons on passing through the object are differentially absorbed depending on the atomic number, thickness

Fig. 6.11 X-radiographs of
mixed carbide fuel pins
showing the different regions

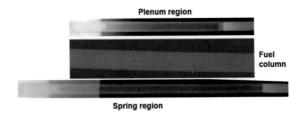

of the materials, homogeneity and composition. Neutron radiography (NR) enables detection of lighter elements in a dense matrix. Thus, the information provided by neutron radiography is complimentary to X-radiography and is ideally suited for irradiated components as imaging medium is insensitive to gamma radiation and therefore devoid of gamma fogging issues.

For neutron imaging of irradiated objects, the indirect or transfer technique is most commonly employed. In this method, a converter screen such as Dysprosium (Dy^{164}) or Indium (In^{115}) is positioned behind the object for neutron exposure. The screen which becomes activated is transferred to the dark room and placed in close contact with X-ray film. The radioactive emission from the screen [Dy^{164} (n, γ) Dy^{165}] subsequently produces an image on the film which is processed for analysis.

Neutron sources for radiography include radioisotopes (e.g. ^{241}Am/Be, ^{252}Cf), particle accelerators (DT source) and nuclear reactors like Kamini (at IGCAR, Kalpakkam), which is located beneath the shielded hot cell [15].

In PIE, the data from neutron radiography supplements the data from non-destructive tests and aids in planning for destructive examinations. Information that can be obtained from NR includes cracking of fuel pellets, fuel densification/swelling, fuel fragments in plenum, mix up of enrichment in fuel column, central hole/channel in annular pellets, pellet-clad gap, insulation pellets, internal hardware like end appendages, spring, etc. Hydriding in case of Zircaloy cladding can be clearly identified [16]. Typical neutron radiographs of mixed oxide fuel irradiated in fast reactor revealing the cracks and central channel of annular pellets are shown in Fig. 6.12a. A neutron radiograph revealing hydride blister in clad of a PHWR fuel pin is shown in Fig. 6.12b.

Recent developments in NR of nuclear materials include (i) real-time NR systems employing CCD-based cameras with neutron-sensitive scintillator screens and (ii) computed radiography making use of photostimulable phosphor image plate and a

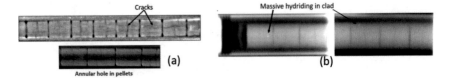

Fig. 6.12 Neutron radiographs showing the (**a**) internals of irradiated MOX fuel pin and (**b**) hydride blister in Zircaloy clad

scanner. The latter has been successfully employed for neutron imaging of irradiated objects.

6.7.2 Destructive Techniques

Once all the steps of non-destructive test are over, the fuel pins are subjected to destructive tests such as fission gas analysis, metallography/ceramography for microscopic examination, β-γ autoradiography, α-autoradiography, burn-up analysis and mechanical testing of cladding.

6.7.2.1 Fission Gas Release Measurement and Analysis

The first destructive test on the irradiated fuel pin is generally fission gas release (FGR) measurement. The fuel pins are punctured under vacuum, gas is collected and analysed [17]. The increase in pressure after puncturing is used for estimation of released volume of fission gases. The same set-up allows estimation of void volume in the fuel pin thus enabling the estimation of internal pressure of the fuel pin. Chemical composition of the gases is determined using a dual-column gas chromatograph. A quadrupole mass spectrometer is used for measuring the isotopic ratios of Xe and Kr. The percentage of fission gas released is calculated using the following formula [18]:

$$\%FGR = (V_R/V_G)^*100 \tag{6.1}$$

where V_R is the volume of the released fission gas released as measured by the experiment and V_G is estimated volume of fission gas generated in the fuel based on fuel burn-up.

6.7.2.2 Microscopic Examination

Metallographic examination of irradiated fuel is carried out using a remotized and shielded optical microscope to observe the changes in fuel microstructure due to irradiation, extent of cladding oxidation and hydriding and likely cause of fuel failure. Different steps involved in metallographic study are selection of specimen location (based on the information from preliminary NDT information), impregnation of fuel pin by cold-setting resin (to avoid falling of cracked fuel pieces during subsequent specimen preparation [19]), specimen cutting, mounting, surface preparation by mechanical grinding, polishing and etching. The metallographic examination consists of two stages:

In the first stage, examination of as-polished specimen is carried out to analyse the porosity distribution in fuel, fuel-cracking behaviour and dimensional changes.

Fig. 6.13 Micrograph showing restructuring in oxide fuel [20]

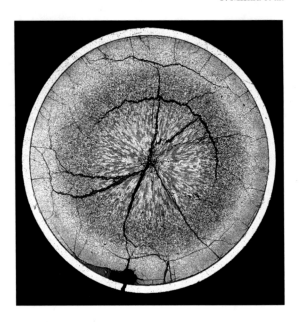

The cladding is examined to study the extent of oxide layer thickness, fuel clad interaction, clad deformation, cracking, thinning, etc.

In the second stage, microstructural changes in fuel are studied after etching of the fuel specimen. A micrograph (after chemical etching of fuel) showing restructuring in oxide fuel is shown in Fig. 6.13 [20].

The microstructure of unirradiated sintered ceramic pellet consists of equiaxed grains of 5–15 μm with uniformly distributed porosity. Parabolic temperature distribution in the fuel pellet leads to fuel restructuring. Restructuring of fuel refers to the alteration in the microstructure across the pellet cross-section due to the steep temperature gradient from centre to the periphery of the fuel. Cross-section of the irradiated fuel reveals concentric zones of different microstructures which are indicative of the maximum temperature experienced by the fuel during irradiation.

Restructuring observed in the fuel during PIE contains information about:

- Fuel centre temperature
- Radial temperature profile
- Redistribution of fuel and fission products
- Mechanism of release of fission gases.

6.7.2.3 Autoradiography

Autoradiography of fuel is based upon the effect of alpha and beta-gamma radiations of fuel and the fission products on the detector film/ plate. Autoradiographs are recorded using prepared metallographic fuel specimens. Alpha autoradiography

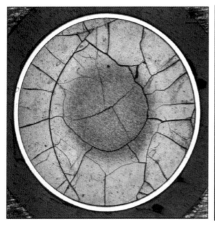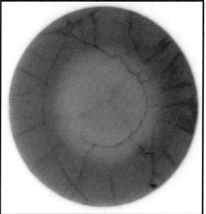

Fig. 6.14 Photomacrograph and the corresponding β-γ autoradiograph of fuel section [21]

is carried out to study the extent of Pu redistribution in the fuel cross-section. Beta-gamma autoradiography reveals the distribution of volatile fission products like Cesium across the fuel cross-section. A suitable detector film or plate is exposed to the fuel specimen for certain time to generate a latent image corresponding to the distribution of radioactivity within the sample. The latent image is developed revealing darker regions indicative of higher amount of fission products/Pu and lighter regions indicating a lower amount of fission products/Pu.

Typical photomacrograph of the cross-section from the outer fuel pin of a high burn-up PHWR fuel bundle along with the corresponding β-γ autoradiograph is shown in Fig. 6.14 [21]. The β-γ autoradiographs indicate that there is a considerable migration of fission products from the centre towards the periphery.

6.7.2.4   SEM Examination of Fuel

The behaviour of gases produced by fission is of great importance for nuclear fuel performance. Retained gas in the form of bubbles may lead to fuel swelling. Hence, it becomes important to study the fission gas bubbles evolution in the fuel. Characterization of the fission gas bubbles is carried out on fractured grains of irradiated UO_2 fuels under scanning electron microscope (SEM).

Figure 6.15 shows a grain of UO_2 taken from the central region of the fuel section of the outer fuel pin of a high burn-up PHWR fuel [10]. It reveals bubbles and channels on the grain faces. White precipitates observed in the micrograph are metallic fission products containing Mo, Ru, Tc, Rh and Pd.

Fig. 6.15 SEM micrograph
of a grain from the centre of
the fuel section of high
burn-up PHWR fuel showing
fission gas bubbles, channels
and metallic fission products

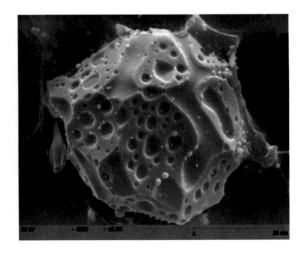

Fig. 6.15 SEM micrograph of a grain from the centre of the fuel section of high burn-up PHWR fuel showing fission gas bubbles, channels and metallic fission products

6.7.2.5   Cladding Examination

(a)   Cladding metallography

The metallographic samples of fuel cladding are examined in as-polished condition to measure the oxide layer thickness on both outer and inner surfaces. The samples are examined after etching to reveal the morphology and distribution of the hydride platelets. The defects in cladding, like hydride blisters (sunburst hydriding [22]) if present, are examined and their depth of penetration in the cladding is measured.

(b)   Mechanical testing of clads

One of the life-limiting factors of fuel clad is the decrease in ductility due to irradiation. In the case of thermal reactor fuel cladding, hydrogen picked up due to clad corrosion leads to formation of zirconium hydrides. These hydrides if precipitated in radial direction reduce the ductility of cladding. The claddings are subjected to circumferential stress due to pellet-clad interaction and therefore, the data on circumferential strength and ductility of nuclear fuel claddings is important. Standard tension test in circumferential direction is possible only by flattening the clad tube section which can lead to cracking in irradiated condition. Burst test by internal pressurization is another method for evaluating circumferential ductility. However, it requires a minimum tube length of 200 mm and consumes considerable volume of irradiated clad sample to provide a single data of ductility. Ring tension test is very commonly employed for irradiated cladding as it provides a measure of transverse ductility using a ring sample of less than 5 mm width taken from the cladding. Small volume of specimen for ring tensile permits extracting multiple specimens of cladding at close intervals for testing.

The ring specimens from the fuel cladding are tested using universal testing machine using special grips. The load–displacement data obtained from the test

Fig. 6.16 Typical engineering stress–strain plots for unirradiated and irradiated PHWR fuel clads tested at ambient temperature

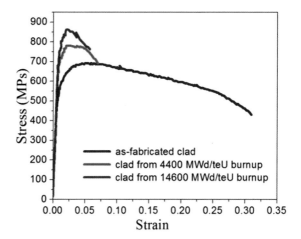

is analysed to estimate the transverse tensile properties of the clad. Typical engineering stress–strain diagrams obtained in ring tension testing of the unirradiated and irradiated specimens are shown in Fig. 6.16.

In case of fast reactor fuel claddings, the tensile properties are generally determined using longitudinal sections of clad tubes of typically 60 mm in length extracted from different locations of the fuel pin. After defueling by mechanical or chemical methods, mandrels are inserted from either ends of the tube and loaded on the UTM using special gripping tools for testing. The tests are conducted as per the ASTM E8 standards. The tensile test set-up in hot cell and typical stress–strain curves of irradiated SS316 claddings are shown in Fig. 6.17. In addition to longitudinal tensile testing of cladding, ring tensile method has also been employed for fast reactor cladding tubes, though to a lesser extent.

(iii) Hydrogen analysis of clad

Small pieces of clad are cut from the fuel element and are used for hydrogen analysis by any of the suitable methods like inert gas fusion (IGF), differential scanning calorimetry (DSC), or hot vacuum extraction quadruple mass spectrometry (HVE-QMS) method. The HVE-QMS method has the advantage of estimating the isotopic composition of the evolved hydrogen. This data comes handy during failure analysis in case of PHWR fuels.

6.7.2.6 Burn-Up Estimation

The burn-up of a nuclear fuel is defined as the energy produced per unit mass of fuel and, hence, is related to the inventory of fission products present in the fuel. It affects both physical and material properties. Burn-up is also an important parameter for design and operation of power reactors from safety point of view. Different units of burn-up are MWd/THM, atom % and fission/cc.

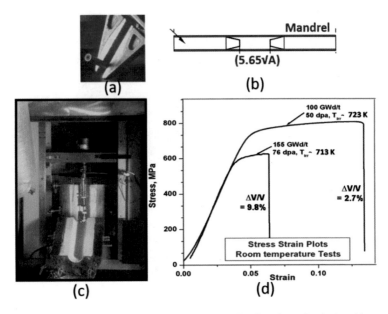

Fig. 6.17 Tensile testing of irradiated cladding: (**a**) clad tube, (**b**) schematic of tube with mandrels inserted, (**c**) UTM fitted with furnace in hot cell and (**d**) typical stress–strain curve of irradiated SS316 cladding

A wide variety of methods have been developed to determine burn-up of irradiated fuel. Principal procedures used so far have been based on measurements of [23]:

(1) Changes in isotopic composition of fissioning elements in a nuclear fuel brought about during irradiation. This is done by mass spectrometric method.
(2) Amount of fission products formed in a nuclear fuel during irradiation. This is done by radio chemical and non-destructive methods.

Determination of fuel burn-up by measuring the content of an isotope that accumulates through the fission process is explained here to give an idea how this burn-up is measured.

[148]Nd is usually selected for the burn-up determination for the following reasons [24]: It is a stable fission product (no decay corrections are necessary). It has a well-known fission yield. It is not volatile and has no volatile precursors. It has almost the same yield for [235]U and [239]Pu and the yield is independent of neutron energy [25]. The standard procedure normally comprises fuel dissolution, separation of U, Pu, Nd followed by determination of isotopic composition [26].

For the burn-up calculation, the following definition has been widely adopted. The definition is based on the number of fissioned atoms compared to the pre-irradiation number of heavy metal (HM) atoms in a given fuel [26]:

$$BU(atom\%) = \frac{number\ of\ fissioned\ atoms}{pre-irradiation\ number\ of\ HM\ atoms} \times 100 \qquad (6.2)$$

where

$$number\ of\ fissioned\ atoms = \frac{number\ of\ burn\ up\ monitor\ atoms}{yield\ of\ burn\ up\ monitor} \qquad (6.3)$$

6.8   Safety—Shielding, Radiological Safety

The radioactive material for PIE is handled in different containment systems to prevent radiation exposure and to minimize the spread of radioactivity to the working environment. The containment may be partial or total, depending on the amount of radioactivity being handled and the toxicity of the material. Of the various containment systems, hot cells are used for PIE of fuels.

6.8.1   Built-In Safety Features of Hot Cells

Hematite concrete (density 3.4 g/cc) is generally used for the construction of cell walls. The cells have lead glass viewing windows whose shielding thickness is specified to match the thickness of the concrete wall in which they are located. A pair of master–slave manipulators is fitted at each window. Piped services and electrical connections are brought into the cells through service ports and are operated from the front face. To compensate for the loss of shielding due to pipe embedment, additional lead shielding is provided in the concrete wall. All cells are lined on the inside with stainless steel sheet to facilitate decontamination. Fuel transfer ports are provided at the side walls to enable shielded transfer of radioactive fuels into the cells. The rear walls are provided with external transfer drawers for transferring material into the hot cells without compromising shielding. The cells are provided with steel doors for personnel entry. Roof plugs made of concrete or steel are provided for introduction of heavy machinery into the cells in case of need.

6.9   Fuel Modelling

Ensuring safe and economic operation of fuel rods requires prediction of fuel behaviour and life-time. However, many processes take place simultaneously and interact with each other in a complex manner and affect the behaviour of nuclear fuel during irradiation. Figure 6.18 [27] shows the complex inter-relation between various physical, chemical, mechanical and metallurgical phenomena which occur inside a nuclear reactor fuel element. A fuel modelling, involving the interactive processes

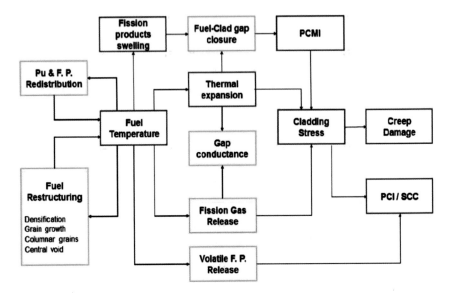

Fig. 6.18 Inter-relation between various physical, chemical, mechanical and metallurgical phenomena which occur inside a water reactor fuel element [27]

shown in the figure, predicts the irradiation behaviour of the fuel element and estimates the evolution of changes in the fuel element. A computer model can incorporate all the data of the material properties and the understanding of the phenomena and their mutual interactions. It is very useful tool for design of fuel elements, safety analysis and for the interpretation of the results of irradiation experiments. Modelling helps in reducing the experimental matrix and at the same time extending the predictive capabilities using the valuable PIE data generated. Modelling in combination with irradiation experiment studies helps to gain a better understanding of the fuel behaviour.

Exercises

1. What is the purpose of post-irradiation examination?
2. What are the types of PIE?
3. What are the features of a hot cell?
4. Mention the main irradiation effects in ceramic fuels.
5. What is restructuring of oxide fuels? Explain.
6. Name the techniques used in PIE of fuels.
7. Why is mechanical testing of clads important during fuel PIE?
8. What is meant by burn-up of nuclear fuel?
9. What are various factors that affect fission gas release?

10. What kind of information can be obtained by observing the restructuring in irradiated fuel?
11. With the help of box diagram, show the inter-relation between various physical, chemical, mechanical and metallurgical phenomena which occur inside a water reactor fuel element.
12. What are the effects of retained fission gas and fission gas release?

References

1. K.V. Kasiviswanathan, Hot cells, glove boxes and shielded facilities. Encycl. Mater.: Sci. Technol. 3830–3834, ISBN: 0-08-0431526
2. P.R. Roy, D.N. Sah, Irradiation behaviour of nuclear fuels. Pramana **24**, Nos 1 & 2, 397–421 (1985)
3. Nuclear Fuels, A nuclear energy division monograph. Commissariat à l'énergieatomique, ISBN 978-2-281-11345-7
4. U.K. Viswanathan, P.M. Satheesh, S. Anantharaman, Measurement of fission gas release from PHWR fuel pins. BARC Newsl. (306) (2009)
5. C.N. Venkiteswaran et al., Irradiation performance of PFBR MOX fuel after 112 GWd/t burn-up. J. Nucl. Mater. **449**, 31–38
6. W. Dienst, Irradiation induced creep of ceramic nuclear fuels. J. Nucl. Mater. **65**, 1–8 (1977)
7. P. Rodriguez, R. Krishnan, C.V. Sundaram, Radiation effects in nuclear reactor materials—correlation with structure. Bull. Mater. Sci. **6**, 339–367 (1984). https://doi.org/10.1007/BF0 2743907
8. H. Bailly, D. Menessier, C. Prunier, The nuclear fuel of pressurized water reactors and fast reactors design and behaviour. France: Commissariat al'Energie Atomique (1999)
9. A. Bhandekar, K.M. Pandit, M.P. Dhotre, P. Nagaraju, B.N. Rath, P. Mishra, S. Kumar, J.S. Dubey, G.K. Mallik, J.L. Singh, New hot cell facility for post-irradiation examination. BARC Newsl. 19 (2015)
10. P. Mishra, B.N. Rath, A. Kumar, V.P. Jathar, H.N. Singh, P.K. Shah, R.S. Sriwastaw, J.S. Dubey, G.K. Mallik, Performance of extended burnup PHWR fuel. IAEA-TECDOC-1865, p. 65
11. Silviu IONESCU, Octavian UTA, Marin MINCU, Gabriel GHITA, Ilie PRISECARU, Ring tests on CANDU fuel elements sheath samples, U.P.B. Sci. Bull., Series C, Vol. 79, Iss. 3, 2017
12. P. Mishra, J.L. Singh, J.S. Dubey, K.M. Pandit, V.P. Jathar, B.N. Rath, P.M. Satheesh, R.S. Shriwastaw, P. Shah, A. Bhandekar, S. Kumar, P.B. Kondejkar, H.N. Singh, S. Anantharaman, *Post-irradiation Examination of Thoria-Plutonia MOX fuel from AC-6 cluster*, BARC Report, BARC/2014/E/003 (2014)
13. P. Mishra, V.P. Jathar, J.L. Singh, D.N. Sah, P.K. Shah, S. Anantharaman, In-reactor degradation of fuel and cladding in fuel pins operated with weld defects. J. Nucl. Mater. **439**(1–3), 217–223 (2013)
14. B. Venkatraman et al., Radiographic techniques for post-irradiation characterisation of FBTR fuel pins, in *Characterization and Quality Control of Nuclear Fuels (CQCNF 2002)*, ed. by C. Ganguly, R.N. Jayaraj (Allied Publishers Ltd, 2004), pp. 650–658
15. N. Raghu, V. Anandaraj, K.V. Kasiviswanathan, P. Kalyanasundaram, Neutron Radiographic Inspection of Industrial Components using Kamini Neutron Source Facility. AIP Conf. Proc. **989**, 202 (2008). https://doi.org/10.1063/1.2906066

16. J.L. Singh, N. Mondal, M. Dhotre, K. Pandit, et al., Non-destructive evaluation of irradiated nuclear fuel pins at Cirus research reactor by neutron radiography, in *National Seminar & Exhibition on Non Destructive Evaluation, NDE* (2011), pp. 11–15

17. R.S. Shriwastaw, P. Mishra, P.K. Shah, J.S. Dubey, V.P. Jathar, H.N. Singh, B.N. Rath, K.M. Pandit, Anil Bhandekar, Ashwini Kumar, J.L. Singh, G.K. Mallik, Sunil Kumar, Arun Kumar, Post-irradiation examination of high burnup PHWR fuels, HOTLAB-2015

18. K. Viswanathan, D.N. Sah, B.N. Rath, S. Anantharaman, Measurement of fission gas release, internal pressure and cladding creep rate in the fuel pins of PHWR bundle of normal discharge burnup. J. Nucl. Mater. **392**, 545–551 (2009)

19. N.G. Muralidharan, C.N. Venkiteswaran, V. Karthik, P.A. Manojkumar, S. Sosamma, C. Babu Rao, V. Venugopal, K.V. Kasiviswanathan, Remote metallographic examination of mixed carbide fuel of fast breeder test reactor in radiometallurgy laboratory. Int. J. Nucl. Energy Sci. Technol. **1**(2/3), 191–196 (2005)

20. Post-irradiation examination of thermal reactor fuels. J. Nucl. Mater. **383**(1–2), 45–53 (2008)

21. S. Anantharaman, U.K. Viswanathan, S. Chatterjee, E. Ramadasan, K. Unnikrishnan, J.L. Singh, K.S. Balakrishnan, P.M. Ouseph, H.N. Singh, R.S. Sriwastaw, P. Mishra, D.N. Sah, Post-irradiation examination of water reactor fuel assemblies in India, IAEA TECDOC (1635)

22. P. Mishra, B.N. Rath, A. Kumar, V.P. Jathar, N. Kumawat, A. Bhandekar, K.M. Pandit, M.P. Dhotre, U. Kumar, V.D. Alur, H.N.Singh, R.S. Shriwastaw, P.K. Shah, P.S. Ramanjaneyulu, J.S. Dubey, G.K. Mallik, J.L. Singh, Post irradiation examination of a failed fuel bundle from TAPS-3 PHWR, BARC report, BARC/2017/E/003 (2017)

23. K.I Noue, K. Taniguchi, T. Murata, H. Mitsui, A. Doi, Burnup determination of nuclear fuel. Mass Spectrosc. **17**(4) (1969)

24. M. Betti, Use of ion chromatography for the determination of fission products and actinides in nuclear applications. J. Chromatogr. A **789**, 369–379 (1997)

25. P.J. Richardson, Methods for determining burn up in enriched Uranium-235 fuel irradiated by fast reactor neutrons, NASA TM X-2153

26. C. Devida, M. Bettt, P. Peerani, E.H.Toscano, W. Goll, Quantitative burnup determination: a comparison of different experimental methods, in HOTLAB Plenary Meeting 2004, September 6–8th, Halden, Norway

27. L. Luzzi, Fuel performance codes, in INSPYRE First Summer School, Delft, May 13–17 (2019). https://www.eera-jpnm.eu/inspyre/filesharer/documents/06%20Education%20and%20training%20activities%20(WP8)/4%20Summer%20schools/2019_05_13-17_First_INSPYRE_School_Delft/Public_documents/Presentations/13%20L%20Luzzi%20Fuel%20performance%20codes.pdf

Chapter 7
Nuclear Fuel Reprocessing

C. V. S. Brahmananda Rao, P. V. Achutan, and N. Sivaraman

7.1 What is Nuclear Fuel Reprocessing?

The fissile isotopes ^{235}U and ^{239}Pu undergo nuclear fission on bombardment with neutrons and thereby produce heat, which is harnessed in the turbines and converted to electricity. Fission involves the breaking of the heavy fissile nuclei (uranium (U) or plutonium (Pu)) into various elements of lower mass/atomic number known as fission products. The fission of 100 fissile nuclide yields 200 new nuclides (fission products). In addition, Pu and its isotopes, and the elements Neptunium (Np), Americium (Am), and Curium (Cm) are also produced in the reactor. The nuclear fuel resides for an optimum period in a reactor towards extraction of maximum possible energy. The fuel is replaced either periodically or on a continuous basis, based on its burn-up history in the reactor. The reasons for this are many fold [1]. The fuel undergoes several changes during its use in reactor reaching to high burn-up values which affect the reactor efficiency. The buildup of fission products, some of which are neutron absorbers and the formation of plutonium, transuranics and their higher isotopes due to neutron irradiation at higher burn-up, and chemical and mechanical interaction of the fuel with the clad material due to heat and irradiation are some of the factors that warrant the periodic replacement of the fuel. The spent fuel (removed after burning in the reactor) has typically about 50 elements and 400 isotopes. The composition of the spent fuel depends on the reactor fuel, type of the reactor (thermal or fast), and residence time of fuel in the reactor. The half-life of these radioisotopes

C. V. S. Brahmananda Rao · N. Sivaraman (✉)
Materials Chemistry and Metal Fuel Cycle Group, Indira Gandhi Centre for Atomic Research, Kalpakkam, India
e-mail: sivaram@igcar.gov.in

C. V. S. Brahmananda Rao
e-mail: brahma@igcar.gov.in

P. V. Achutan
Formerly Fuel Reprocessing Division, Bhabha Atomic Research Centre, Mumbai, India

© The Author(s), under exclusive license to Springer Nature Singapore Pte Ltd. 2023   213
B. S. Tomar et al. (eds.), *Nuclear Fuel Cycle*,
https://doi.org/10.1007/978-981-99-0949-0_7

varies from milliseconds to millions of years. The mass number (number of protons and neutrons) varies from 74 (Ge) to 161 (Dy). The yields of fission product isotopes vary from 10^{-6} to about 7.4%. The fission products include alkali, alkaline earth, transition metals, and lanthanides. The discharged fuel also contains unused uranium, plutonium generated due to neutron capture by ^{238}U followed by beta decay, and minor actinides (Np, Am and Cm). Among options available for dealing with the spent fuel and its disposal, the countries with abundant energy resources can afford to use once through cycle (at least as a short-term policy) wherein the spent fuel is not processed for the recovery of leftover fissile and fertile elements. For countries opting to conserve and augment their resources, a closed fuel cycle with reprocessing of the fuel to recover the left over fertile and fissile materials for recycle is more attractive and advantageous.

Nuclear fuel reprocessing is a vital link in the closure of fuel cycle. The success of the closed fuel cycle in enhancing nuclear power production, in fact, depends on the effective utilization of plutonium for power generation as it can increase the amount of energy that can be derived from a given amount of uranium.

The chemical reprocessing step in the nuclear fuel cycle involves removal of fission products and minor actinides from uranium and plutonium. The process also involves the separation of uranium from plutonium.

7.2 Reprocessing and Its Role in Closing the Fuel Cycle

Over the past seven decades, nuclear energy has been deployed throughout the world and developed into a matured technology. There are 31 countries which have operating nuclear reactors for power production. Currently, it is a viable alternative to conventional modes (coal and gas) and is a potential candidate to replace them as a greener technology. The acceptability of nuclear energy by the society will depend largely on the nuclear waste management. The availability of low-cost uranium in the international market has prompted some countries to follow a once through approach. The energy needs of developing countries are steeply rising with per capita energy consumption to improve the quality of human life and because of industrial and technological advancement. Depending on the resource availability, various countries have opted for different strategies for energy production. For example, India is following a three-stage strategy for nuclear power development based on the limited uranium resources and the abundant thorium deposits available. This strategy places utmost importance on recycling of resources; hence reprocessing forms a vital link between the three stages of nuclear fuel cycle.

7.3 Aqueous Chemistry of Actinides

Reprocessing deals with recovery and separation of actinides that are relevant to the fuel cycle, through chemical processes. The understanding of the chemistry of actinides is therefore important for an appreciation of the desired separations. The salient features of actinide chemistry [2] are discussed here.

7.3.1 Actinide Chemistry

Analogous to lanthanide series where the 4f-electrons are added beginning with cerium and ending with lutetium, a series of elements where 5f-electrons were envisaged by Glenn T Seaborg in 1944. The idea, combined with increasing stability of +3 oxidation states for the transuranium elements as the atomic number increases from $Z = 93-96$, led Seaborg to the conclusion that these new elements constituted a second rare earth series from actinium. The actinide elements begin with actinium and end with lawrencium. The lighter actinide elements exhibit a complex chemistry and have similarity with transition metals. They exhibit a number of oxidation states unlike the heavier actinides, which show more lanthanide-type behaviour. Among the actinide elements, thorium and uranium are naturally occurring in significant quantities. The transuranium elements are all artificially made. They have a large number of isotopes, and some of them can be isolated in isotopically pure form. They are produced by neutron or charge particle-induced nuclear reactions. Thorium and uranium were known prior to the discovery of nuclear fission and were in use for their chemical properties. Actinium and protactinium were discovered during the early studies with naturally occurring radioactive substances. Neptunium was discovered during an investigation of nuclear fission, and it led to the discovery of next element plutonium. The fissile nature of ^{239}Pu and its use as a nuclear weapon led to further thorough investigation of this series and resulted in the production of higher elements. In addition to chemical behaviour, other properties such as paramagnetic susceptibility, paramagnetic resonance, light absorption and reflectance, X-ray absorption, NMR, and crystal structure have provided a great deal of information on the electronic structure of the aqueous actinide ions and provided an understanding of their aqueous chemistry.

7.3.2 Actinide Oxidation States

The data on electronic structure of actinides and oxidation states exhibited by them are tabulated below (Table 7.1).

The main feature of the variation of electronic structure from ^{89}Ac to ^{103}Lr is the successive filling of the 5f-subshell. The electrons in the 6d and 7s subshells are more

Table 7.1 Electronic configuration of actinides

At. No	Symbol	Electronic configuration	Oxidation states[a]
89	Ac	$5f^0 6d^1 7s^2$	**3**
90	Th	$5f^0 6d^2 7s^2$	(3),**4**
91	Pa	$5f^1 6d^1 7s^2$	(3),4,**5**
92	U	$5f^3 6d^1 7s^2$	3,4,5,**6**
93	Np	$5f^4 6d^1 7s^2$	3,4,**5**,6,(7)
94	Pu	$5f^6 7s^2$	3,**4**,5,6,(7)
95	Am	$5f^7 7s^2$	**3**,4,5,6
96	Cm	$5f^7 6d^1 7s^2$	**3**,4
97	Bk	$5f^9 7s^2$	**3**,4
98	Cf	$5f^{10} 7s^2$	(2),**3**
99	Es	$5f^{11} 7s^2$	(2),**3**
100	Fm	$5f^{12} 7s^2$	(2),**3**
101	Md	$(5f^{13} 7s^2)$	2,**3**
102	No	$(5f^{14} 7s^2)$	**2**,3
103	Lr	$(5f^{14} 6d^1 7s^2)$	**3**

[a] The oxidation states shown in bold are the most stable

loosely bound than the electrons in the filled subshells and, in general, also than the 5f-electrons. In these outer shells, the binding energies are in the range of a few eV, that is, the same order of magnitude as is common in chemical bonding. Thus, Ac easily loses its $6d^1 7s^2$ electrons to form Ac^{3+}, and similarly Th loses its $6d^2 7s^2$ electrons to form Th^{4+}. For the subsequent elements, from Pa to Am, the situation is more complicated. The spatial characteristics of the f-subshell orbitals change abruptly at certain atomic numbers; that is, the f-shell electrons may be shielded more strongly in some elements than in others where the f-orbitals extend close to the surface of the electronic cloud (where chemical interaction occurs) and where the 5f-electrons are in closer contact with the d- and s-shell electrons. Thus, in the actinide series, the oxidation state +3 becomes increasingly more stable as atomic number increases. In the first half of actinide series (early actinides), the energy required for conversion of 5f to 6d is somewhat lower than that for the elements in the second half of the actinide series. Hence, early actinides exhibit higher oxidation state such as +4, +5, +6, and +7. For Th and Pa, the +4 and +5 are the stable oxidation states, respectively. For the actinides, U, Np, Pu, and Am, the higher oxidation states, that is, beyond +3 state, are stable. However, in the higher actinides, oxidation states beyond +3 are not stable in aqueous solutions.

7.3.3 Actinide Spectra

The electronic spectra of actinides can arise from electronic transitions from f–f, f–d, or charge transfer. The absorption spectra of ions in solution and crystalline form exhibit very narrow f–f bands in the near ultra violet to near infrared regions of the electromagnetic spectrum. The transition in the visible region gives rise to the observed colours of the ions. Each of the plutonium oxidation states has a characteristic colour in solution. The colours are specific and depend on the type and number of ligands.

The f absorption spectra of actinide ions provide distinctive fingerprint of the element and oxidation state. This has been extensively used for qualitative and quantitative analysis. The f–d and charge transfer transitions of actinide ions give rise to broad and intense absorption bands. These are responsible for the deep-coloured solutions in highly polarizing ligands.

7.3.4 Disproportionation

The chemical properties of the actinide elements have been intensely studied for the elements $Z = 89$–99, but much less for the heaviest members of the family ($Z = 100$–103). The pentavalent state of the actinides (except for Pa and Np) is less stable than the other states and normally undergoes disproportion in acid solutions. Plutonium is particularly interesting in the variety of oxidation states that can coexist in aqueous solutions.

The reactions are

$$2PuO_2^+ + 4H^+ \rightarrow Pu^{4+} + PuO_2^{2+} + 2H_2O \tag{7.1}$$

and

$$PuO_2^+ + Pu^{4+} \rightarrow PuO_2^{2+} + Pu^{3+} \tag{7.2}$$

7.3.5 Actinide Complexes

Since the differences in energy of the electronic levels are similar to chemical bond energies, the most stable oxidation states of actinides change from one chemical compound to another and the solution chemistry is sensitive to the ligands present in the complex. Thus, complex formation becomes an important feature of the actinide.

The chemical properties are different for the various oxidation states, while in the same oxidation state, the actinides closely resemble each other. These properties have been extensively exploited for the separation and isolation of the individual elements in pure form. The compounds formed are normally quite ionic. The ionic radii of actinide elements of different valence states decrease with increasing atomic number. Consequently, charge density of actinide ions increases with increasing atomic number, and therefore, the probability of formation of complexes and of hydrolysis increases with atomic number. This property is made use of in the inter-actinide element separation, where heavier actinides are eluted before the lighter ones. The pattern of stabilities of complexes in the tetravalent states follows the order of decreasing ionic radius

$$Th^{4+} < U^{4+} < Np^{4+} < Pu^{4+}$$

In the case of tributyl phosphate (TBP), dissolved in a diluent, e.g. kerosene, the extracted M(VI), M(IV), and M(III) species are $MO_2(NO_3)_2(TBP)_2$, $M(NO_3)_4(TBP)_2$, and $M(NO_3)_3(TBP)_3$, respectively. For the same element, the stability of the complexes varies with the oxidation state in the series

$$M^{4+} \geq MO_2^{2+} > M^{3+} \geq MO_2^{+}$$

The reversal between M^{3+} and MO_2^{2+} shows that the hexavalent metal atom in the linear $[OMO)]^{2+}$ is only partially shielded by the two oxygen atoms; thus the metal ion MO_2^{2+} has a higher charge density than $M^{3+}(\sim 3.2 \pm 0.1)$. Similarly in MO_2^{+}, the effective charge is $\sim 2.2 \pm 0.1$. This gives a reasonable explanation of the extraction pattern, though other factors (e.g. molar volume of the complex) also contribute to the phenomenon.

The extraction of trivalent actinides in general follows the sequence as the ionic radii decrease in that order.

$$Ac^{3+} < Am^{3+} < Cm^{3+}$$

The chemistry involved in the isolation and purification of the actinide elements from irradiated reactor fuel elements is discussed in later sections.

7.4 Reprocessing of Spent Fuel

7.4.1 History of Reprocessing of U–Pu-Based Fuels

Separation of uranium and plutonium from fission products is based on the complexing ability and the redox properties of actinides. These two properties form the basis for the separation of actinides. Separation methods based on precipitation, solvent extraction, and ion exchange have been used in fuel reprocessing industry [3].

7.4.1.1 Precipitation

The earliest method of separation adapted was precipitation. Glenn T Seaborg and his team were the first to separate plutonium from the irradiated samples by using a technique called carrier precipitation wherein one produces an insoluble salt which carries with it the material of interest. They employed bismuth phosphate ($BiPO_4$) which precipitated plutonium in the tetravalent state (Pu(IV)) from aqueous solution, while plutonium in the hexavalent state (Pu(VI)) remained soluble. The Manhattan Project was a research and development undertaken by the USA during World War II to produce the first nuclear weapon. The task was to produce fissile plutonium which can be used as nuclear weapon. The first reactor to produce ^{239}Pu was the X-10 Graphite Reactor, also known as the Clinton Pile, in 1943 at Oak Ridge, Tennessee, USA. In October 1943, the construction of nuclear reactor for plutonium production commenced at Hanford, Washington. The B Reactor, as it was known at that time, was completed in March 1945 and began producing plutonium for the implosion type atomic bomb which was used in "fat man" that was dropped on Nagasaki, Japan, on August 9, 1945. The problem encountered by the chemists of the Manhattan Project in the 1940s was the selection of a separation technique for preferentially removing plutonium from uranium.

The most stable oxidation state of uranium is U(VI), whereas as plutonium can exist in variable oxidation states of III, IV, and VI depending on the nature of solution. The oxidation states of fission products can vary from I to VII which make the separation of one element from the rest a daunting task. In the initial days of Manhattan Project, bismuth phosphate precipitation was used as a key route to separate plutonium from irradiated uranium. This process makes use of the fact that plutonium in Pu(IV) and Pu(III) states can be co-precipitated with bismuth phosphate, while Pu(VI) is not carried by the precipitate. Thus, the process involved reduction of plutonium to Pu(IV)/Pu(III) with a suitable reducing agent followed by co-precipitation with bismuth phosphate. The precipitate was dissolved in nitric acid and oxidized to VI state followed by reduction back to IV state. This process was repeated until the desired purity of the plutonium was achieved. This was the first method used to separate plutonium from fission products on a large scale at Hanford from 1945 to 1951. The disadvantages of this process are that it works in a batch mode, the plutonium has impurities, uranium is not recovered, resulting in large amount of waste, and the decontamination factor (ratio of beta and gamma activity per gram of Pu or U in feed to that in product) is relatively lower.

7.4.1.2 Solvent Extraction

Solvent extraction or liquid–liquid extraction [4] involves the partitioning of solute between two immiscible phases. In most cases, one of the phases is aqueous, and the other phase is an organic solvent consisting of an extractant in its pure form or dissolved in a suitable diluent/solvent. The phases are immiscible and form two layers, with the heavier phase at the bottom. The solute is initially present in one phase, but after extraction it is distributed in both phases. The efficiency of a liquid–liquid extraction is determined by the equilibrium constant for the solute's partitioning between the two phases.

The process of solvent extraction is an equilibrium process which can be described as

$$M + E \rightleftharpoons ME \tag{7.3}$$

In the first step (extraction stage), the metal, M, is transferred from an aqueous phase to an organic phase E as complex ME, in which case the process requires that the equilibrium position in this equation be shifted to the right. The second step is the reverse of the first, that is the metal is transferred from the organic phase back to an aqueous phase (stripping stage), in which case the process requires that the equilibrium position be shifted to the left.

The aqueous phase containing the metal is equilibrated with an organic phase into which the metal is extracted. The metal is distributed between the two phases. The distribution may be chemical or physical in nature depending on the system. Physical processes are those involving the extraction of simple, uncharged covalent molecules such as halides of Sb(III), Ge(III), and Hg(II) into organic solvents such as carbon tetrachloride. In such cases, the Nernst distribution law is usually valid and the distribution coefficient, K_d, is independent of both the total solute (metal) concentration and the phase ratio (ratio of the volumes of the aqueous and organic phases). The K_d is a ratio of the solute concentration in the two phases:

$$K_d = \frac{[M]_{org}}{[M]_{aq}} \tag{7.4}$$

The Nernst law depends only on the solubility of the metal species in the solvent phase. There is no chemical reaction between the metal species and the organic phase, i.e. the solute is in identical chemical form in both the phases. There are a few systems which behave without changing their form in the organic phase.

However, in majority of cases, there is a chemical reaction between the metal species present in the aqueous phase and one or more components of the organic or solvent phase. It is more meaningful to describe distribution ratio, D, which is the ratio of the concentrations of all the species of the solute in each phase.

$$D = \frac{[M]_{org}}{[M]_{aq}} \text{ in all forms} \tag{7.5}$$

Equipment for Solvent Extraction

Solvent extraction (SE) or liquid–liquid extraction (LLE) is the heart of nuclear fuel reprocessing. The process involves contact between two immiscible liquids, so that the solute of interest is transferred from one phase to another. The rate of transfer (mass) of solute to the solvent depends on the difference in concentration and the interfacial area of contact. Continuous operation of process requires the use of liquid–liquid contactors as solvent extraction equipment.

The purpose of liquid–liquid contactors is to create optimal conditions for mass transfer between the two phases. This includes bringing the two phases in contact, creating the droplets of dispersed phase to provide interfacial area for mass transfer and separation of liquid phase after the completion of the extraction. A variety of contactors are available for solvent extraction applications. In these contactors, the two immiscible liquids are brought into intimate contact to transfer the solute of interest from one phase to another. Most of these are operated in counter-current mode, where the two phases travel in opposite direction.

The equipment used in nuclear industry for LLE is classified as mixer–settlers, liquid-pulsed columns, and centrifugal extractors [5].

The process of solvent extraction requires two sub-processes—mixing and settling. During the process of mixing, immiscible pair of heavy (typically aqueous) and light (typically organic) is contacted vigorously so that phase with lesser throughput is distributed (dispersed) in the phase with higher throughput (continuous) in form of drops. Since interfacial area is significantly enhanced by formation of small drops, mass transfer takes place rapidly. The mixer unit output, in form of a mixed phase having dispersed phase along with continuous phase, is taken to a volume with practically nil energy input. As there is no further mixing energy input resulting in rapid coalescence of drops, the mixed phase breaks up into two immiscible phases. This process is called settling. Therefore, any solvent extractor will have essentially these two subunits; mixer and settler.

To get a better recovery of desired solute, several stages of extractors may be needed. In a discrete staged system, these stages may be arranged in a counter-current fashion. The extractor units, viz. mixer–settlers, pulse column, and centrifugal extractors, are briefly described below.

Mixer–Settlers

The mixer–settler is an extractor having discreet stages. It is an actual combination of several mixers and an equal number of settlers with organic and aqueous flows in a counter-current fashion, as shown in Fig. 7.1. The advantages of mixer–settlers include easy and simple construction and robust operation. The major shortcomings are large requirement of floorspace, large residence time, separation in a single gravity system, and often entrainments/other phase carryovers requiring separation pots or other techniques for recovery of entrained phase.

Pulse Column

A pulse column is a device with an elongated mixing zone and two settlers on each end. The top settler will be engaged in settling process if lighter phase is dispersed in the form of small drops. On the contrary, the bottom settler will be engaged in settling process, if the heavier phase is dispersed in the form of small drops. The heavy phase, fed from top, flows towards lower settler. The light phase, fed from bottom, flows towards top settler. The phase with lower flow rate will be dispersed phase. The sieve-plate internals with around 20% free area in form of approximately

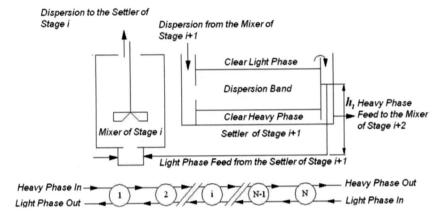

Fig. 7.1 Schematic of a mixer–settler equipment and counter-current arrangements of the mixer–settler in a cascade

3 mm diameter holes do not allow free flow due to interfacial forces, and pulsing energy is to be provided for intense mixing and counter-current flow of heavy and lighter phases in the respective right directions. The right amount of pulsing energy input does not allow settling of drops inside the mixing zone and drops of dispersed phase coalesce only in the settler where effect of pulsing is nullified by increasing settler diameter several times over the mixing zone diameter. This operation is termed as quasi-emulsion or stable operation. A lower than optimum pulse energy input may throttle the mutual flow, and wrong phase may exit from the wrong port. This condition is called flooding due to insufficient input. Energy excess to the optimally required creates very small drops which cannot be coalesced in the given experimental geometric and operational conditions. This condition is detrimental to recovery and would result in huge loss of dispersed phase. Figure 7.2 shows a schematic of a pulse column equipment.

Centrifugal Extractor

Centrifugal extractors are contactors which are particularly important in fast reactor fuel reprocessing. The centrifugal extractor utilizes centrifugal force for settling the minutes drops of the dispersed phase. The prevalent centrifugal force may be 200–400 times the natural gravity-based settling force used in the other contactors like mixer–settlers and pulse columns. This high-g operation results in a very clean separation and better decontamination factors.

The annular centrifugal extractor also consists of two essential parts—a mixer and a settler like a conventional extractor. In the annular centrifugal extractor, shown in Fig. 7.3, the mixer portion is an annular mixing zone between the inner rotary cylinder and the outer stationary cylinder. The mixed phase enters at the centre of the bottom of rotary settler. The mixed phase is subjected to a rapid coalescence due to high-g settling field. The slippage between the flow and the rotary cylinder is prevented by four axial baffles. The lighter phase (shown as organic) is collected

Fig. 7.2 Schematic of a pulsed column

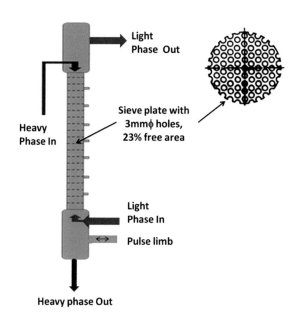

from the central organic weir located just above the settling zone. The heavier phase (shown as aqueous) traverses through an underpass to the upper chamber and is collected from the aqueous weir. A schematic of a typical centrifugal extractor is shown in Fig. 7.3.

Desirable Characteristics of an Extractant

The extractant to be employed must possess the following desirable properties for its usage in the solvent extraction mode [6].

1. The extractant should be selective for the metal ion of interest.
2. The extractant should have good capacity for extraction; i.e. distribution ratios in the extracting stage should be higher than 1.
3. The metal loaded into the organic phase should be readily stripped; i.e. distribution ratios in the stripping stage should be as low as possible.
4. It should be relatively immiscible with water, in order to reduce solubility losses.
5. Density of the extractant should be appreciably different from that of water and should have low viscosity and fairly high interfacial tension. These physical properties are important in promoting separation of phases following contact.
6. It should be relatively non-volatile, non-flammable, and non-toxic.
7. It should be readily purified, preferably by fractional distillation.
8. It should be stable in the presence of chemical agents used in the process, such as nitric acid.

Fig. 7.3 Schematic of
annular centrifugal extractor
equipment

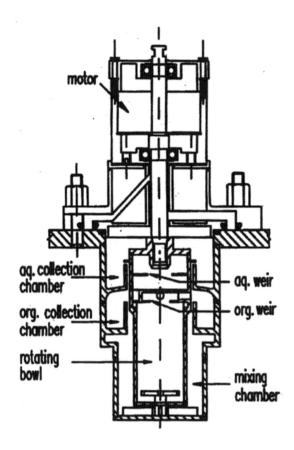

9. Solvents used for radioactive materials should also have good radiation stability.
10. It should be less expensive.

The process of separation of plutonium by precipitation from the fission products is a batch process, so the Hanford plant switched to solvent extraction in 1951. The first extractant used was Methyl Isobutyl Ketone (MIBK) or Hexone which extracts U(VI) and Pu(VI). In this process, the organic phase is neat or pure MIBK and the aqueous feed solution used in this process is 0.3 M nitric acid and 0.1 M sodium dichromate with 1.3 M aluminium nitrate (salting out agent). The loaded organic phase is partitioned by ferrous ammonium sulphamate wherein the plutonium is reduced and stripped back into the aqueous phase. This process was used at Hanford until 1966. Although the process was a continuous process, MIBK had the disadvantages of being unstable to nitric acid at higher concentrations (hence $Al(NO_3)_3$ is used as salting agent), low flash point, and high solubility in water (19.1 g/L). The other processes used in the early days of reprocessing industry were dibutyl carbitol (Bis(2-butoxyethyl) ether) or Butex process by the UK and ether-based "Trigly" process by Canadians.

7.4.2   Aqueous Reprocessing

The nuclear reactor fuel can be processed either through an aqueous reprocessing route by dissolving the spent fuel in nitric acid medium followed by precipitation, solvent extraction, etc., or by non-aqueous route using molten salt electrorefining process. In the early days, the aqueous processing routes were followed to separate pure plutonium for defence applications and the development continued through the cold war era, and today this is the only established route for reprocessing of spent fuel.

7.4.3   PUREX Process

The main objective of fuel reprocessing [7] is to recover the valuable and strategic fissile and fertile elements still present in spent fuel, and it aims to recover these elements, viz. mainly uranium and plutonium, selectively with minimum loss. It entails their separation together from bulk of fission products, followed by their individual separation and purification. The remaining fission products and waste arising need to be safely stored and disposed in an environmentally acceptable manner, keeping in view their long-term radiological hazards. Towards meeting this objective, the waste volumes should be as small as possible.

The reprocessing technology has evolved over the years, and its requirements are more demanding than the usual chemical processes. The chemical (PUREX) process employs a solvent extraction route with TBP as extractant and is basically a simple and rugged process. But the special requirements arising out of the highly radioactive and toxic nature of the elements under processing and the possibility of the "uncontrolled" chain fission reactions (criticality) occurring with the fissile materials ($^{235}U/^{239}Pu$), while being handled in significant amounts and their safe handling for its prevention, make fuel reprocessing a very complicated and challenging task. This calls for very sophisticated fail safe, maintenance free engineering designs and controls. The facility in turn needs appropriate infrastructure such as lead-shielded or concrete cell housings, corrosion and maintenance free stainless steel equipment, containment vessels and transfer modes with high integrity, designed to remain ever safe with respect to criticality, and wherever required with in-built redundancy.

In addition, remotely operated, radiation, and fire-resistant electronic instrumentation controls, alarms for monitoring the process operations and the fissile material containment, tracking, and surveillance are vital for the smooth and safe operation of the process.

To ensure safe and efficient material recovery, the chemical flow sheet design and process monitoring systems should take into account the special chemical characteristics and behaviour of elements like plutonium in solution and maintain their concentrations and quantities below the prescribed ever safe limits (with respect to criticality) in any given process vessel at all times, to avoid any inadvertent accumulation leading to accidents. Any such unintended accumulation should be promptly detected and corrected. This is achieved by following very stringent monitoring and analytical control protocols for nuclear material accounting and material balance assessment in specified accounting areas of the plant during the operation of the process.

In general, the safety from criticality hazards in the facility is achieved by control of fissile mass, its concentration level, employing vessels with ever safe geometry (like the use of ever safe diameter cylindrical columns, slab-type tanks with prescribed thickness, concentric annular storage tanks and pulse column contactors, bird cages for final oxide storage, etc.) with design provision of minimum distance between vessels for the escape of spontaneous fission neutrons. In select cases, use of neutron absorbers dissolved in process streams or incorporated in vessel design are also useful. Prior criticality safety clearance should be obtained for the quantities that can be handled in any task area. Strict administrative control protocols are to be followed during material transfer from one task area to another.

The major unit operations in the process include a mechanical chop leach task for dissolution of fuel, solvent extraction steps employing air-pulsed extraction column contactors, evaporation in pot and thermosiphon evaporators, and fluid transports using different modes (modes with no moving parts are preferred ones). The engineering and chemical flow sheet design should have provisions to guard against any mal operating conditions that may be encountered during these unit operations, which otherwise can lead to disastrous consequences because of the hazardous environment involved.

Further, the ventilation systems with well-designed air flow patterns and vigorous personnel health physics monitoring and radiation safety practices protocols should be in place to ensure the safety of the operating staff and the public at large. The radioactive releases in various forms to the ambient environment should not only be kept below the permissible levels but also aim at levels as low as reasonably achievable below these permissible limits.

Tri-n-butyl Phosphate (TBP) as Extractant

TBP belongs to organophosphorus family of compounds. It is an ester of phosphoric acid which can be prepared by the reaction of $POCl_3$ with 1-butanol in the presence of pyridine. The hydrochloric acid liberated after the esterification forms an adduct with pyridine. The product is vacuum distilled to get the pure product. The extraction of the metal ion (uranium, plutonium) is through the oxygen of the phosphoryl group present in the molecule. The coordinate covalent bond formation is through the donation of lone pair of electrons from oxygen atom of phosphoryl group to the metal. The TBP acts a Lewis base, and the metal is a Lewis acid.

The metal ions with +4 and +6 oxidation states are readily extracted by TBP, the and metal ions with +1, +2, +3, and +5 are poorly extracted by TBP. The common counter ion is nitrate.

The first application of TBP came to light in 1949, when James C Warf published the extraction studies of cerium, thorium, and uranium with TBP in nitric acid medium. The Oak Ridge National Laboratory (ORNL) used TBP for the recovery of uranium from reprocessing plant waste. They used TBP in hexane as the extractant, and the process was improved and was used at Hanford for recovery of uranium from waste solutions. Meanwhile, ORNL developed the process for recovery of uranium and plutonium from dissolver solution and named the process as PUREX process (plutonium, uranium reduction extraction). This process is used for complete recovery of uranium and plutonium from the fission products with a high decontamination factor (of the order of 10^8) and is suitable for continuous, large-scale, and remote operations [8].

TBP has been a workhorse in nuclear industry for the recovery and purification of actinides for the past seven decades. PUREX process uses TBP as an extractant in a long chain hydrocarbon (e.g. dodecane) as diluent. TBP has advantages over MIBK as nitric acid is used as salting out agent instead of aluminium nitrate, and the acid here can be recovered and reused thereby reducing the waste volumes. TBP is less volatile, less flammable, and more stable against attack by nitric acid than MIBK.

TBP is a universal choice for reprocessing industry. It is highly selective for uranium and plutonium and has very low extraction for the fission products. It is readily available, easy to purify, possesses high boiling point (289 °C), and has low aqueous solubility. It has good chemical, thermal, and radiation stability. The drawbacks are its high density (0.973 g/cm^3) and viscosity which is compensated by diluting with inert hydrocarbon diluents, e.g. n-dodecane. The diluent for TBP should be non-polar, non-reactive towards nitric acid and nitrous acid and stable to radiation. The straight chain aliphatic hydrocarbons containing 12–14 carbon atoms or n-paraffin mixtures are preferred. A solution of 30% (v/v) TBP in n-dodecane diluent is generally used. The plutonium purification cycles often used a lower percentage of TBP in dodecane medium (~20%).

The steps involved in the PUREX process are [9]

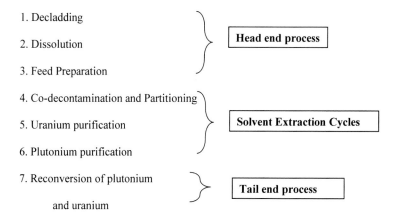

1. Decladding
2. Dissolution
3. Feed Preparation

Head end process

4. Co-decontamination and Partitioning
5. Uranium purification
6. Plutonium purification

Solvent Extraction Cycles

7. Reconversion of plutonium and uranium

Tail end process

The schematic of PUREX process is shown in Fig. 7.4.

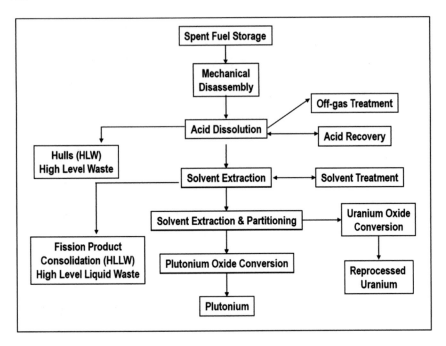

Fig. 7.4 Schematic of PUREX process

The schematic of the solvent extraction part of PUREX process in shown in Fig. 7.5.

7.4.3.1   Head-End Process Steps

The first three steps together are called head-end treatment, which brings the fuel into dissolved aqueous state in nitric acid medium suitable for PUREX process.

After discharging from reactor, the fuel assembly is stored in the reactor pond to allow decay of the short-lived radioactivity. Further this is transported in a cask to a storage place or a reprocessing plant as per the requirement. At the reprocessing plant, the fuel is taken to a head-end plant where the fuel assembly is taken into a shielded facility fitted with cranes and other remote handling equipment. Since the fuel is cladded, the fuel has to be exposed prior to processing. This is carried out by two methods, viz. chemical and mechanical, depending on the material of cladding.

Chemical Decladding

Aluminium cladded reactor fuel is decladded by chemical treatment. The decladding of the fuel elements is done by a mixture of sodium hydroxide (NaOH) and

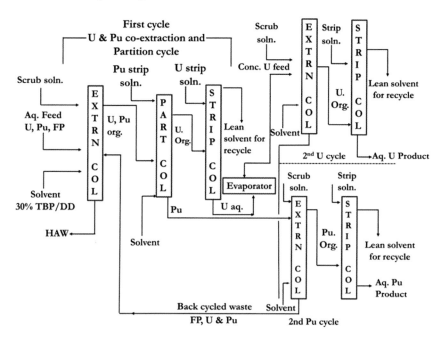

Fig. 7.5   Schematic of the solvent extraction part of PUREX process

sodium nitrate ($NaNO_3$). The NaOH dissolves aluminium to form sodium aluminate, $NaAlO_2$, in highly alkaline medium, and hydrogen evolved in the process reacts with $NaNO_3$ to yield ammonia as off-gas, which is safer than hydrogen. The net reaction is

$$20\,Al + 17\,NaOH + 21\,NaNO_3 \rightarrow 20\,NaAlO_2 + 18\,NaNO_2 + 4H_2O + 3NH_3 \tag{7.6}$$

The contents are kept agitated by air sparging. The dissolver is kept under negative pressure. The ammonia and air off-gas are passed through Down Draft condenser, NH_3 scrubber, and dehumidifier before being let off through stack after HEPA filtration. The progress of dissolution is monitored by analysing the aluminium and sodium hydroxide contents of the declad solution. The solution is drained, and the dissolver is washed with water or dil. acid before chemical processing of the exposed fuel. The declad waste contains very little uranium, plutonium, and fission product activity. The waste generated is high in salt content and not amenable for concentration and is difficult to decontaminate. Hence, this is not the preferred route for commercial operations.

Mechanical Decladding

During mechanical decladding, the fuel assemblies are chopped into small pieces and the fuel core exposed for further processing. The chopping is usually achieved with a shearing knife. This method is adopted for zircalloy clad fuels from power reactors and stainless steel clad fast reactor fuels.

Dissolution

The dissolution is the transfer of soluble nitrates of various elements present in the fuel into solution. The resulting solution is known as dissolver solution. This step is carried out in batch mode or continuous mode in concentrated nitric acid. The dissolution is exothermic, and the reaction is controlled by the concentration of acid and the temperature of reaction. The oxides of nitrogen (NO_x) liberated during the dissolution are deoxidized and returned for reuse. The volatile fission products with some traces of nitric acid are passed through an elaborate off-gas treatment system and finally filtered off by filter bank and let off through a tall stack.

UO_2 dissolves in nitric acid by the following reactions:

$$3UO_2 + 8H^+ + 8NO_3^- \rightarrow 3UO_2(NO_3)_2 + 2NO + 4H_2O \sim 8M \qquad (7.7)$$

$$UO_2 + 4H^+ + 4NO_3^- \rightarrow UO_2(NO_3)_2 + 2NO_2 + 2H_2O > 8-10M \qquad (7.8)$$

It is possible to dissolve uranium without any net evolution of gaseous products except the gaseous fission products by sparging oxygen as reactant, and this type of dissolution is called fumeless dissolution.

$$UO_2^{2+} + 2H^+ + 2NO_3^- + \frac{1}{2}O_2 \rightarrow UO_2(NO_3)_2 + H_2O \qquad (7.9)$$

Feed Preparation

The solution prepared by the dissolution of the fuel in nitric acid medium must be adjusted before the next step of solvent extraction. The dissolver solution is subjected to the following operations:

1. Solid removal or feed clarification to avoid choking, plugging, and emulsification during extraction.
2. Adjustment of feed acidity (2–3 M) for the extraction of uranium and plutonium.
3. Adjustment of salting strength by dilution or evaporation to the desired uranium concentration.
4. Adjustment of oxidation state of plutonium to Pu^{4+}.

After the dissolution of spent fuel in nitric acid, uranium exists as U(VI) while plutonium exists mainly in the Pu(IV) oxidation state. Plutonium has variable oxidation states ranging from III to VI. All the plutonium present in the dissolver solution is adjusted to IV state by the addition of sodium nitrite ($NaNO_2$), and the solution is maintained at 50 °C.

$$PuO_2^{2+} + NO_2^- + 2H^+ \rightarrow Pu^{4+} + NO_3^- + H_2O \text{ (Reduction)} \tag{7.10}$$

$$Pu^{3+} + NO_2^- + 2H^+ \rightarrow Pu^{4+} + NO + H_2O \text{ (oxidation)} \tag{7.11}$$

The nitrite has a dual role, wherein Pu(VI) is reduced to Pu(IV) and Pu(III) is oxidized to Pu(IV). NO_2 gas is an ideal candidate for the above process instead of sodium nitrite salt.

7.4.3.2 Co-extraction and Partitioning of Uranium/Plutonium

Solvent Extraction

The good extractability of nitrate complexes of $UO_2{}^{2+}$ and Pu^{4+} coupled with the non-extractability of fission products by TBP forms the basic principle of PUREX process. As Pu(III) is weakly extracted, further separation of uranium from plutonium is achieved by reducing Pu(IV) to Pu(III), in which state it gets stripped back from the organic phase.

If M is the metal atom, the following are the extracted species formed during the reaction with nitric acid in case of different oxidation states.

Trivalent state: $M(NO_3)_3$.3TBP
Tetravalent state: $M(NO_3)_4$.2TBP
Hexavalent state: $MO_2(NO_3)_2$.2TBP.

Nitric acid is extracted as $TBP.HNO_3$, and it is salted back from TBP during the loading of the metal ion into the organic phase. The extraction of U(VI) and Pu(IV) with TBP is given below:

$$UO_2^{2+} + 2NO_3^- + 2TBP \leftrightarrows UO_2(NO_3)_2 2TBP \tag{7.12}$$

$$Pu^{4+} + 4NO_3^- + 2TBP \leftrightarrows Pu(NO_3)_4 2TBP \tag{7.13}$$

The distribution ratio (D) of uranium and plutonium is directly proportional to the square of TBP (free) concentration and to the square (for uranium) and fourth power (for plutonium) of nitrate ion concentration in the aqueous phase. D values for extraction of nitrates of uranium and plutonium with TBP increase sharply with nitric acid, pass through a maximum, and decrease a little thereafter (Fig. 7.6). The initial rise is due to the increase in the concentration of nitrate ions which act a

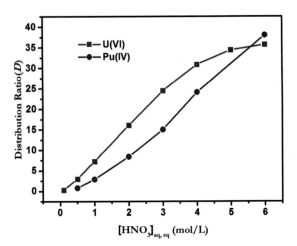

Fig. 7.6 Variation of distribution ratio for U(VI) and Pu(IV) with equilibrium aqueous phase nitric acid by 1.1 M TBP in n-dodecane at 303 K [From study carried out at IGCAR, Kalpakkam, India]

salting agent. But at higher acidities, the nitric acid itself competes with uranium and plutonium, resulting in the fall of D values. In the presence of large amounts of extracted species, the available free TBP decreases retarding further extraction.

Co-decontamination and Partitioning

The aim of the co-decontamination and partitioning step is to extract uranium and plutonium together to separate them from rest of the fission product impurities and then further separate uranium and plutonium from each other in the partitioning step. Uranium and plutonium are purified further individually, before conversion to their respective oxide form in the further subsequent steps. Since pulse column is the most prevalent contactor for reprocessing, the description of the solvent extraction process in following sections presumes the use of pulse columns.

Co-extraction

After conditioning of the dissolver solution (known as feed), it is fed to the centre of the contactor. The contactor has two sections comprising of extraction and scrub section. Solvent, 30% TBP/n-dodecane is fed at the bottom of the contactor, and 2–3 M nitric acid scrub is fed from the top of the contactor. Uranium and plutonium are extracted by the solvent, and the purpose of the scrub is to remove the traces of the fission products that get extracted into the solvent. The fission products left in the aqueous phase (high active waste, HAW) leave the contactor from the bottom and the loaded organic from the top. An additional scrub column is often provided to improve the decontamination from fission products. While processing long cooled fuels, ^{90}Sr, ^{90}Y, ^{95}Zr, ^{95}Nb, ^{106}Ru, ^{106}Rh, ^{137}Cs, ^{144}Ce, and ^{144}Pr are some of the fission products that affect the product purity. Among these, the behaviour of ^{106}Ru, ^{106}Rh, ^{95}Zr, and ^{95}Nb often determines the overall process performance. During operation, variations in acid profile are possible for better decontamination from Zr and Ru. A high loading of uranium in organic solvent phase reduces the extraction of fission products; however too high loading, on the other hand, will hinder the extraction of plutonium. Hence, an optimum saturation of 60–80% (of the theoretically possible 119 g U per litre) is generally preferred. Neptunium, Americium, Curium, and the other actinides produced in the reactor (known as minor actinides) are also present in spent fuel. In PUREX plants with appropriate redox conditions, Neptunium can be made to follow either plutonium or uranium stream for further separation. Am(III) and Cm(III), like the trivalent lanthanides, are poorly extracted by TBP and remain in the HAW. The HAW leaving the column is washed free of entrained TBP by diluent and concentrated by evaporation to form the high-level liquid waste (HLLW). The enhancement in acidity during evaporation is reduced by "killing" the acid using non-salt-forming chemicals such as formaldehyde, under controlled conditions.

Partitioning

The solvent loaded with uranium and plutonium from extraction column is taken to partitioning contactor for their mutual separation. The back extraction of plutonium from TBP is achieved by reducing it to trivalent state with an aqueous solution containing a suitable reductant like ferrous sulfamate or uranous nitrate stabilized with hydrazine.

Partitioning with ferrous sulfamate: Pu(IV) to Pu(III) conversion is carried out using ferrous sulfamate; the sulfamate is also a scavenger of nitrous acid, which is invariably present in nitric acid. Nitrous acid can convert the Pu(III) back to Pu(IV) making the stripping of the plutonium from organic to aqueous phase difficult.

$$NH_2SO_3H + HNO_2 \rightarrow N_2 + H_2SO_4 + H_2O \qquad (7.14)$$

Ferrous sulfamate was used extensively in PUREX process for partitioning of plutonium from uranium. The typical concentrations used for partitioning are 0.03 M Fe $(NH_2SO_3)_2$ + 0.1 M HNO$_3$, which is introduced at the end of the contactor from where organic phase exits.

Ferrous sulfamate has been discontinued as it has a disadvantage of introducing iron salt into the system, adding to solid waste burden, and sulphate (produced by the hydrolysis of sulfamate) into the solution which enhances the corrosion.

Partitioning with U(IV)-hydrazine: U(IV) is a stronger reducing agent than Fe(II) for the reduction of Pu(IV) to Pu(III), and it does not introduce any additional metal ion into the system. 0.1 M solution of hydrazine is added to the uranous solution to stabilize U(IV) oxidation state by destroying the nitrite ion present in the solution.

$$2Pu^{4+} + U^{4+} + 2H_2O \rightarrow 2Pu^{3+} + UO_2^{2+} + 4H^+ \qquad (7.15)$$

$$U^{4+} + NO_3^- + H_2O \rightarrow UO_2^{2+} + NO_2^- + 2H^+ \tag{7.16}$$

$$U^{4+} + 2NO_2^- \rightarrow UO_2^{2+} + 2NO \tag{7.17}$$

$$2NO + NO_3^- + H_2O \rightarrow 2H^+ + 3NO_2^- \tag{7.18}$$

Hydrazine destroys nitrite ion as follows:

$$N_2H_5^+ + NO_2^- + H_2O \rightarrow HN_3 + 2H_2O \tag{7.19}$$

$$HN_3 + H^+ + NO_2^- \rightarrow N_2O + N_2 + 2H_2O \tag{7.20}$$

When uranous nitrate stabilized with hydrazine is used as the reductant, the residual HNO_2 carried over by TBP from the extraction/scrub columns can oxidize the reductant U(IV) and the stripped Pu(III) in the partition column. The oxidation of Pu(III) is faster than that of U(IV). Also, the oxidation of U(IV) by nitrite is auto-catalytic as can be seen from the reactions, resulting in the generation of more nitrite ions than what is being consumed. Hence, the presence of hydrazine is essential as a stabilizer to rapidly destroy nitrite present/generated in the system. Acidity of the aqueous phase is maintained at 0.6–2 M; a higher acid profile might lead to reoxida-tion of Pu(III), increasing the possibility of plutonium reflux in the bottom sections of the column. Because of its extractability and poor stability in the TBP phase, U(IV) is used in excess of the amount stoichiometrically required for the plutonium present in the solution. For effective utilization of U(IV), it is introduced near the centre of the partition column just above or near the organic feed point, (and if possible, at additional entry points by splitting the total input of reductant solution) to reduce and strip the bulk of the plutonium to aqueous phase. From here, U(IV) being fairly extractable in TBP is carried in both directions by the two streams to all stages of the contactor. An aqueous scrub from top of the column, with U(IV), and hydrazine in dilute nitric acid removes the residual plutonium from the uranium-bearing organic phase that leaves the column from the top.

The stripped Pu (III) is further purified from the accompanying U(IV) and U(VI) in the bottom scrub section, by a fresh TBP scrub stream moving upward, which extracts and removes uranium. After the partitioning and scrub, the aqueous plutonium product leaving the bottom of the column is sent for final purification.

As alternative approaches for uranium/plutonium partitioning, hydroxylamine nitrate at low acidity, a kinetically slow reducing agent, and in situ electrolytic reduc-tion of plutonium in specially designed electropulse columns and mixer–settlers have also been tested extensively.

Stripping of Uranium

Uranium-bearing organic stream is stripped of uranium in the strip contactor. The uranium-bearing solvent is fed from the bottom of the contactor. The dilute strip stream flows from the top. An acid solution of pH = 2 is used to strip the uranium from the organic phase. The organic feed contains some nitric acid which is stripped from organic phase, thereby increasing the acidity of aqueous phase. The solvent after this cycle is sent for alkali washing to remove the degradation products of TBP, if any, before recycling the solvent.

Uranium and Plutonium Purification and Tail-End Processing

Uranium Purification

The aqueous uranium stream with trace level of plutonium is taken to evaporators. After evaporation, the concentration of solution is adjusted to 1 M in nitric acid and uranous nitrous stabilized with hydrazine is added to reduce Pu(IV) to Pu(III). This solution is fed into the contactor wherein uranium is extracted by TBP. Pu(III) and fission products remain in aqueous solution. The loaded solvent containing uranium is stripped with 0.01 M nitric acid, and the aqueous product containing uranium is concentrated by evaporation and passed through silica gel as tail-end polishing to decontaminate from Zr and Nb. The final uranium product is concentrated and precipitated as ammonium diuranate (ADU), filtered and calcined to get uranium oxide which is packed and stored.

Plutonium Purification

The plutonium after its separation from uranium is again passed through a solvent extraction cycle to remove the traces of uranium and fission products and to get the desired concentration required for precipitation. 20% TBP is a preferred extractant as the plutonium feed to be processed has low content (3–5 g/L). After stripping the plutonium from the organic phase to the aqueous phase, the plutonium is precipitated with oxalic acid, to get plutonium oxalate, which is filtered and calcined to give PuO_2, which is stored in bird cages (which has high surface area for neutron leakage).

For plutonium purification, anion exchange route was used for research reactor fuels which give the best purity of the product. The disadvantage of this process is the generation of large volumes of wash acids after each loading that is to be used or recovered. It is not suitable for large-scale continuous processing as several units of loading, washing, and elution cycle would be required.

Solvent Treatment

The solvent (TBP in n-dodecane) undergoes acid hydrolysis and radiation degradation, and the products of the degradation have an adverse effect on the performance of the process. The degradation products of TBP are dibutyl (DBP) and monobutyl (MBP) phosphoric acids which have the affinity to bind some of the metal ions. The presence of these products reduces the decontamination factor and loss of uranium and plutonium to the waste streams. These degradation products also hinder the process because of the tendency to form emulsion and interfacial crud. The diluent also undergoes degradation and reacts with acid to produce a variety of products such as alkyl nitrates, nitroalkanes, ketones, esters, alcohols, and carboxylic acids and through secondary reactions they generate surfactants, complexants, emulsifier, and crud formers, which interfere in the process. In the process plant, the most common method for solvent treatment is carried out by washing the solvent with 5% sodium carbonate solution. This step is effective in the removal of dibutyl and monobutyl phosphoric acids from the organic solvent. Hence after each run, the organic phase is analysed for the amount of DBP and MBP, and after treatment with sodium carbonate, the solvent is ready for the next cycle. After few cycles, this treatment is no longer effective because the degraded parts of the diluent retain activity which cannot be removed by alkali wash and the fresh solvent is used.

7.5   Fission Product Chemistry

A large number of fission products belonging to almost all the groups in the periodic table are produced and are present in the spent fuel. The fission products which co-extract along with uranium and plutonium into the organic phase are of some concern and are discussed below.

7.5.1   Ruthenium (Ru)

Ruthenium has the most complex chemistry of all the fission products, owing to the multitude of oxidation states and extractable complexes that are stable in nitric acid. It is the most troublesome fission product in the PUREX process. ^{106}Ru ($t_{1/2} = 371.5$ days) isotope is produced in appreciable quantity and is in secular equilibrium with the daughter ^{106}Rh ($t_{1/2} = 30$ s). Each of them is a highly energetic beta–gamma emitter and contributes a substantial fraction to the residual activity in uranium and plutonium and also in the recycled solvent.

7.5.2   Zirconium (Zr)

^{95}Zr ($t_{1/2} = 65$ d) is also a concern in the PUREX process. It is produced in high yield and has a high specific gamma activity. ^{95}Zr decays to ^{95}Nb ($t_{1/2} = 35$ d), another strong gamma emitter, and together they contribute to a large fraction of the total gamma activity in the aqueous feed in the PUREX process. When the PUREX flow sheet is optimized for Ru removal, uranium and plutonium decontamination, with respect to zirconium, is significantly degraded (and vice versa), so typical flow sheets represent a compromise for both elements. A simple solution to the problem is to reprocess only long cooled (>5 years) fuel, which allows the zirconium isotope, ^{95}Zr to decay. Stable zirconium isotopes of mass numbers 90, 91, 92, 94, and 96 are also present in PUREX process aqueous feed solutions. These stable isotopes originate from the fission process or from the zircalloy used as cladding material.

7.5.3   Technetium (Tc)

The chemistry of Tc extraction by TBP in the absence of multivalent cations is straightforward, and distribution coefficients are low under the conditions used in the PUREX process. However, there is a unique feature in the solvent extraction chemistry of Tc that sharply reduces the decontamination of uranium and pluto-nium with respect to the element. Pertechnetate ion (TcO_4^-) co-extracts with U(VI), Pu(IV), and other mutivalent cations. Thus, the high solvent loadings that are used in the PUREX process to suppress the extraction of other fission products actually enhance the extraction of Tc. ^{99}Tc ($t_{1/2} = 2.1 \times 10^5$ years) is produced in about 6% fission yield and is the only Tc isotope of importance in the fuel cycle.^{99}Tc is a low energy beta emitter that does not add appreciably to the activity of the various PUREX process streams. ^{99}Tc can cause substantial operating problems and release to the environment.

7.5.4   Niobium (Nb)

Niobium behaviour under conditions found in PUREX process streams indicates that the decontamination of uranium and plutonium with respect to niobium is limited by factors other than the solvent extraction chemistry of this fission product. Early gamma spectrometers could not easily resolve ^{95}Zr and ^{95}Nb; hence, the two were treated as one element. Even recent workers have found it difficult to obtain good analytical results for Nb due to its tendency to adsorb on the walls of the experimental apparatus and sample bottles. Nb can be a limiting factor in PUREX product stream decontamination, but this is due to the formation of colloidal particles and adsorption on solids (especially silicates) that are interfacially active and are easily transported through the process.

7.5.5   Yttrium and the Lanthanide Elements

Yttrium and lanthanides behave similarly with respect to solvent extraction by TBP. Trivalent lanthanides are extracted as a trisolvate

$$M^{3+} + 3NO_3^- + 3TBP \leftrightarrow M(NO_3)_3 3TBP \tag{7.21}$$

These elements are not normally a limiting factor in uranium or plutonium decontamination but are difficult to separate from the trivalent transplutonium elements. Extraction of these elements increases as the nitric acid or non-extractable nitrate salt concentration increases, but high solvent loading effectively prevents lanthanide extraction.

7.5.6   Iodine (I)

^{131}I ($t_{1/2} = 8.05$ d) and ^{129}I ($t_{1/2} = 1.6 \times 10^7$ years) are the major iodine isotopes produced in fission process. Short-lived ^{131}I can be allowed to decay by cooling the spent fuel. The major solid form of iodine in reactor fuel is CsI. These iodides are all readily dissolved and oxidized to elemental iodine in hot nitric acid-NO_x dissolver solutions. Most of the iodine is removed from these solutions by sparging air and can be removed from the off-gas by a variety of treatments. A small amount of iodate is formed during dissolution but is not extracted by TBP. Iodine reacts with hydrocarbon diluents to form organic iodides which cannot be stripped and is not removed by common solvent cleanup methods.

7.5.7   Molybdenum (Mo)

Molybdenum isotopes produced as fission products are mostly stable. Part of the molybdenum remains as insoluble dissolver residue, while the remainder is dissolved as Mo(VI), which is not appreciably extractable by TBP from nitric acid solution.

7.5.8   Noble Gases

Large amounts of krypton (Kr) and xenon (Xe) are produced in fission, but most of the isotopes are stable and short lived. After a month of cooling, the major activity remaining in the spent fuel is mainly ^{85}Kr, which is a low energy beta emitter that decays to stable rubidium. All the noble gas inventory is evolved during dissolution and can be separated by a variety of methods.

7.6 Challenges in Reprocessing of U–Pu-Based Fuels

At the head end of the PUREX process, volatile radioactive gaseous products get released during decladding and dissolution. Currently, an elaborate trapping system is in place to remove most of these toxic gases. But due to the difficulty in trapping some of these gases, a small fraction is released under stringent regulations. With increased nuclear energy production, these discharges may have to be brought down to as low as possible to preserve the safe environmental parameters. Radionuclides of concern are ^{3}H, ^{14}C, ^{85}Kr, and ^{129}I. Decades of studies have resulted in the development of new sorbents for iodine; investigation of iodine capture from very dilute streams; development of solid sorbents to capture and separate xenon and krypton at near room temperature.

Currently, PUREX process is well entrenched to meet the challenges of uranium and plutonium recovery with recoveries well exceeding 99.5% and almost approaching 99.9%. But some of the left out long-lived fission products and actinides are of major concern. The important among them are ^{241}Am, ^{237}Np, ^{243,244}Cm, ^{99}Tc, ^{107}Pd, and ^{135}Cs. Out of these, ^{237}Np and ^{99}Tc are having significant extraction in the TBP system under most of the current PUREX operating parameters. Under optimized process parameters, ^{237}Np can be either extracted completely along with uranium and plutonium or can be rejected to the high active waste. Similarly, significant amounts of ^{99}Tc get extracted into the TBP system under the current conditions. A special high acid strip for scrubbing out the Tc has to be incorporated into the standard PUREX flow sheet for almost quantitative transfer of Tc to the waste. The extracted ^{237}Np can be separated from the uranium product stream.

New approaches to make PUREX process proliferation resistant and economically cheaper are being experimented. Currently, depending on the cooling of the spent fuel, three or five cycles of operations are carried out to achieve the goal of separating pure uranium and plutonium with decontamination factors exceeding 10^7 with respect to fission products. Adopting a reprocessing to recycle strategy will allow longer cooling periods to the spent fuel with significantly lower fission product activity. This approach will help in reducing the number of cycles needed for purification. One cycle operation with a slightly reduced decontamination factor is the goal of the approach. This will necessitate the use of the state-of-the-art technology with respect to remotization and automation. One of the drawbacks of TBP is the higher affinity of uranium than plutonium under ambient conditions. This can be reversed by slightly raising the temperature. Hence, jacketed columns can be used to avoid the plutonium pileup during loading. The state-of-the-art multiprobe computer-controlled column monitoring with online parameter changes can go a long way in improving the process efficiency. After efficient scrubbing of the loaded organic phase, an acid-controlled stripping of plutonium along with required uranium can be adopted for obtaining a master blend for co-processing and subsequent fuel fabrication. This may reduce the DF, but will give enough proliferation resistance to the path as this product can be handled only by remotely operated automation systems. Finally, the excess uranium can be stripped as usual and can be converted to fuel material for recycle as blanket or for dilution of the MOX blend.

Waste volume reduction is another area under active development. Tail-end purification of MOX and U is carried out in this approach either by direct denitration or by microwave denitration depending on the maturity of the process for industrial application. Uranyl nitrate crystallization from concentrated product solution is also being pursued as an alternate step to uranium purification. Solvent cleanup techniques with salt-free reagents are another area under experimentation. Final cleanup of the solvent by vacuum distillation results in the minimization of the spent solvent.

7.7 Aqueous Processing of Thorium-Based Fuels

A striking parallel exists between natural uranium and thorium except for the fact that natural uranium has a 0.72% abundance of fissile ^{235}U, whereas natural thorium is mostly mono isotopic and is almost 100% ^{232}Th. The major advantages of thorium fuel cycle are [10, 11]

1. The generated isotope, ^{233}U, is the best fissile isotope in the thermal neutron spectrum.
2. The minor actinide production from thorium fuel cycle will be much lower than that from uranium cycle.
3. Thorium oxide is the best-known refractory, has better thermal characteristics, and can withstand higher burn-ups.
4. Thorium oxide, uranium oxide, and plutonium oxide have similar physical characteristics and can co-crystallize in the cubic centred form.

The major difference in the irradiation properties of thorium and uranium comes from the (n, 2n) reactions. In case of ^{238}U, this will lead to the formation of ^{238}Pu (~87.7 years), whereas ^{232}Th will produce ^{232}U (68.9 years). The (n, 2n) reaction of ^{233}U itself can also generate ^{232}U. Thus in the thorium fuel cycle, ^{232}U is an integral factor. The amount of ^{232}U in the product depends on the burn-up and the neutron spectrum encountered in the reactor. Unlike ^{238}Pu, the decay chain of ^{232}U has a noble gas and hard gamma emitters. This will call for closer control of the radiological aspects of this fuel cycle.

Several specific features of thorium fuel cycle have to be addressed especially during reprocessing and subsequent operations with the recovered fuel materials. The first and foremost is the minimum cooling period requirement of the irradiated fuel. Unlike the irradiated uranium-based fuels, where long-lived Pu isotopes are present, in irradiated fuel based on thorium, ^{233}Pa, formed by (n, γ) reaction on ^{232}Th, is present, and this isotope decays with a half-life of 27 days to fissile ^{233}U. This necessitates a longer cooling period for optimum recovery of fuel values. The refractory nature of thorium adds to the difficulty in dissolution of irradiated fuel prior to chemical processing. Further, because of (n, 2n) reactions, ^{232}U formed will always contaminate the ^{233}U product and the first decay product of ^{232}U, viz. ^{228}Th, will accompany the recovered thorium. Hence, both these products will have significant radiation dose and will have to be handled with appropriate systems from safety viewpoint. Yet another difference in chemical processing of thorium is the lack of variable valence states unlike that of plutonium. Plutonium separation from uranium is achieved by a change in valence state. Hence, selectivity difference has to be employed to separate thorium from uranium, and a clean separation of uranium from thorium is difficult to achieve using only solvent extraction. The limited solubility of thorium in organic solvent also is a constraint in chemical processing. If not properly controlled, this can lead to third phase formation impeding with remote operation.

7.8 Reprocessing of Irradiated Thorium/Thoria Fuels and Targets

Solvent extraction processes based on TBP diluted in an inert diluent are the natural choice due to the success of TBP-based PUREX process [10, 11]. The process, thorium uranium extraction (THOREX) process, has the goal to establish conditions needed to separate and decontaminate ^{232}Th, ^{233}U, and ^{233}Pa from irradiated thoria and thorium targets. The studies resulted in three different flow sheets, viz. Interim 23, THOREX-1, and THOREX-2 processes. Compared to PUREX process, the recovery and recycle of fuel values from irradiated thorium fuels and targets using TBP as extractant have several special features and requirements that need attention. However, major steps generally followed while processing irradiated thoria are similar.

7.8.1 History of Thorium Reprocessing

7.8.1.1 Interim 23 Process [10]

This process was developed to recover and decontaminate ^{233}U from thorium using 1.5% TBP from feed solutions of irradiated thoria. Silica gel was used to remove Pa, and Dowex 50 cation exchange resin was used for final product concentration. An overall loss of 0.5% was estimated with thorium separation factor >10^5. Solvent extraction provided fission product separation factor >10^5 and 10^7 after silica gel and concentration step.

7.8.1.2 THOREX-1 Process [10]

The aim of this process was to separate ^{233}Pa–^{233}U–^{232}Th and had the following steps. Diisobutyl carbitol was used for the separation of ^{233}Pa. ^{233}U extraction was carried out using 5% TBP. ^{232}Th extraction was done with 45% TBP. Although the products recovered had adequate decontamination, the process had engineering difficulties like handling different solvents, separate chemical treatments for cleanup of solvents, and handling of hot solutions in multiple stages of processing. Moreover, this process had difficulties like stripping of Pa, use of benzene to avoid third phase formation, etc.

7.8.1.3 THOREX-2 Process [10]

An alternate process was developed using single TBP solvent which involved thorium and ^{233}U separation using 41–55% TBP followed by separation of Pa by adsorption on silica gel column from HAW [10]. Thorium partition from ^{233}U was carried out using preferential stripping using dilute nitric acid. ^{233}U is stripped with very dilute nitric acid. Extraction condition for the first step had to maintain marginally acid-deficient condition so as to limit the Pa and FP extraction without effecting thorium precipitation as hydroxide, with an acid-deficient aluminium nitrate scrub step to keep the Pa in HAW. Proper flow rates were maintained to avoid third phase formation issues of Th without the use of benzene. Silica gel operation for Pa recovery did not require any feed adjustment. Th/U partitioning was done with proper acidity adjustment and flow rate control. ^{233}U was stripped with dilute nitric acid.

7.9 Fast Reactor Fuel Reprocessing

As compared to the fuel reprocessing of thermal reactors, international experience on reprocessing of fast reactors is very limited, since fast reactors have been operated in only a few countries. The fast reactor fuels need to have a high fissile content, to compensate for the small cross section for fission by fast neutrons. Accordingly, fast reactors using U–Pu mixed oxide fuels have up to 30% Pu content in the fuel, and fast reactors using enriched uranium oxide as the fuel have fissile enrichment typically of the order of 20%, and much higher enrichment is encountered in low power fast reactors such as test reactors (For example, the Fast Breeder Test Reactor at Kalpakkam, with a thermal power output of 40 MW, uses mixed uranium–plutonium carbide fuel with plutonium content as high as 70%).

The success of FBR programme depends on the reprocessing of irradiated spent fuels. Since the fissile content of the fuel before and after irradiation is high, fast reactors can be economical only if a closed fuel cycle with multiple recycling is adopted. Such multiple recyclings will also lead to enhanced utilization of uranium. The experience gained from thermal reactor fuel reprocessing (TRFR) will be useful for fast reactor fuel reprocessing (FRFR) programme. However, there are major differences in reprocessing spent fuels of fast reactors compared to thermal reactors due to the higher burn-up and higher plutonium content encountered with the fast reactor fuels [12]. Thermal power reactors use uranium oxide or U, Pu mixed oxide with small % of Pu as fuel. In the case of fast reactors, other ceramic fuels such as U, Pu mixed carbide or nitride, and metallic alloys such as U–Pu–Zr are also candidate fuels, and reprocessing of these "advanced" fuels is more complex as compared to oxide fuels. While aqueous reprocessing routes similar to those used for thermal reactor fuels can also be applied (with some additional steps or modifications) to the other ceramic forms of fast reactor fuels, pyrochemical reprocessing schemes are more appropriate for metallic alloy fuels and this aspect is dealt with later in this section.

7.9.1 General Issues in Fast Reactor Fuel Reprocessing

Fast reactor fuel reprocessing differs from thermal reactor reprocessing in many aspects. The differences arise mainly from the following characteristics of spent fast reactor fuels:

- High specific activity and high decay heat generation due to high burn-up (typically 10 atom% fission).
- Presence of sodium on subassemblies in the case of sodium cooled fast reactors resulting in additional steps to completely remove sodium.
- Enhanced formation of platinum group metals during the course of fission in a fast reactor; dissolution of fast reactor spent fuel in nitric acid medium generally results in insoluble residues, mainly from platinoids.
- Inert atmosphere to be maintained while dismantling and decladding fast reactor fuel pins, particularly in case of pyrophoric fuels, e.g. U/Pu carbide.
- The fissile material content in a fast reactor fuel can be as high as 20–30% compared to 0.7–4% in thermal reactors, thus making criticality an issue during reprocessing of spent fuels.
- Higher plutonium content in the spent fuel can lead to "third phase formation" during solvent extraction process; plutonium hydrolysis will be a concern during the stripping cycle of solvent extraction process. The higher plutonium content will also demand a greater emphasis on tracking fissile materials in the plant and avoiding criticality.
- The quantity of fission products (and radioactivity) will be much higher in the spent fast reactor fuels, leading to higher solvent degradation.

Hence, modifications in the plutonium uranium extraction (PUREX) process are required to meet the above challenges. Table 7.2 summarizes the major differences between thermal and fast reactor fuels.

7.9.1.1 Head-End Steps

The challenges posed for fast reactor fuel reprocessing are the higher burn-up encountered in the fast reactor (typically 10 atom per cent fission) and the proportionate

Table 7.2 Difference between TRFR and FRFR

Properties	LWR-UOX	FBR-MOX
Cooling time/years	4	7
Burn-up (GWd/t HM[a])	50	185
Total decay heat (W/t HM[a])	3.48	21.77
Dose by aqueous phase (Wh/L)	0.57	6.72
Dose received by solvent (Wh/L)	0.20	2.40
Pu/(U + Pu)	0.004	0.15–0.7
Specific β, γ activity	~200 Ci/L	~1000 Ci/L
Third-phase formation with plutonium	–	Expected under certain conditions

[a] heavy metal

increase in the amount of fission products. The high plutonium content of fast reactor fuel demands the design of equipment to be critically safe. The decay heat associated is generally high for fast reactor fuels due to short cooling times. Hence issues related to transportation can be minimized by locating reprocessing plant near the reactor itself. Cooling the fuel bundles is mandatory for removal of decay heat, and it can be done using either forced air or water. Plutonium concentration, extent of sintering, porosity, irradiation levels, and the dispersion of plutonium in the matrix of U + Pu determine the rate of dissolution of FR fuels (MOX or MC) in nitric acid. It has been reported that it may be extremely difficult to dissolve MOX fuels containing Pu > 40 wt % using boiling 10 M nitric acid without addition of hydrofluoric acid, and the insoluble residues contain higher % of plutonium. Alternately, oxidation using electrogenerated species such as Ag(II) and Ce(IV) can be used to enhance the rate of dissolution and also achieve complete dissolution [13].

The fission of ^{239}Pu results in the formation of higher proportion of noble metals such as Ru, Rh, and Pd, totally about three times more than that of thermal reactor fuels for the given burn-up. These metals form intermetallic alloys among themselves and also with uranium and plutonium. These alloys are difficult to dissolve. The high plutonium content of the fuel gives rise to need for precise monitoring systems or analytical techniques, which can track the strategic nuclear materials in various streams. The insoluble residues lead to loss of plutonium and create problems in solvent extraction by accumulating in the interfaces, and these residues can be removed by high speed centrifuges.

7.9.1.2   Solvent Extraction

Plutonium Chemistry and Third-Phase Formation

Plutonium is known to polymerize under low acid conditions; hence stripping of plutonium from loaded organic phase must be done under highly controlled condition. Third-phase formation is another important phenomenon in solvent extraction process during the reprocessing of spent nuclear fuel containing a high proportion of plutonium. When the Pu(IV) concentration in the organic phase (extractant phase) increases beyond a particular level, the incompatibility of the plutonium solvate with the non-polar organic phase leads to the separation of a third phase containing almost all the plutonium solvate. This phenomenon is of particularly serious concern during the extraction of tetravalent actinides, e.g. Pu(IV) with TBP [14]. Third-phase formation can lead to phase inversion in the case of loaded organic phase with higher densities and could result in criticality-related issues with plutonium. Third-phase formation can also lead to flooding in organic/aqueous phase separators, e.g. mixer–settler, centrifugal extractor, etc. Due to flooding, organics can carry over to locations where aqueous solutions are collected, e.g. evaporator. Evaporation of aqueous products containing dissolved organics, e.g. TBP and metal nitrates generated from a solvent extraction process, can lead to "red oil formation", which can result in explosion-related issues. Third-phase formation is particularly important with respect to fast reactor fuel reprocessing, where the plutonium concentrations are much higher. To avoid this phenomenon, it is necessary to restrict the plutonium loading in the TBP-based organic phase, which affects the throughput of the process. Usually, feed solutions with U(VI) concentration of about 350 g/L and Pu(IV) concentration of ~1 g/L in 3–4 M HNO_3 medium are used in PUREX process for thermal reactor fuel reprocessing. Therefore, third-phase formation is not expected to occur in TRFR.

The Limiting Organic Concentration (LOC) for third-phase formation—the limit above which third-phase formation could occur—is a function of several parameters including the concentration of aqueous nitric acid, concentration of TBP, the diluent used, the temperature, etc. [14]. With increase in aqueous nitric acid concentration, the LOC of plutonium initially decreases, and after passing through a minimum at around 2 M HNO_3, it increases with nitric acid concentration. It is also reported that the LOC value for Pu(IV) decreases with increase of uranium loading and reduction in free TBP concentration in the organic phase. The LOC for third-phase formation in the extraction of Pu(IV) from 4 M HNO_3 by 1.1 M TBP/n-DD is about 50 g/L. In general, co-extraction of U(VI) decreases the LOC of Pu(IV) at acidities ranging from 2 to 5 M HNO_3. The effect is more pronounced at higher acidities (4–5 M) than at lower acidities (2–4 M).

Though the organic loading of heavy metals in TRFR could be as high as 119 g/L based on the stoichiometry, typical organic loading of heavy metals in TRFR plant is generally in the range of 80 g/L; however, in FRFR, it is difficult to achieve such a loading of 80 g (U + Pu)/L with a Pu(IV) loading of 25 g/L during the extraction from nitric acid medium (3–4 M) due to the possibility of third-phase formation. Hence for the fast reactor fuel reprocessing, the desired extractant should have higher capacity to load plutonium without third phase formation.

The high radiation levels associated with fast reactor fuels demand a careful optimization of process parameters to ensure a high level of decontamination from fission products. The high radiation levels also cause degradation of extractants and diluents, leading to the formation of a number of products, which are deleterious to the solvent extraction/stripping process. The buildup of degradation products also leads to retention of plutonium in the organic phase and hence incomplete stripping. The fission products, e.g. zirconium, also have "enhanced retention" in the degraded organic phase. To minimize radiation degradation, the solvent extraction process with short cooled/high burn-up fast reactor fuels can be carried out using centrifugal extractors that reduce the contact time between the aqueous and organic phases, thereby reducing degradation of the organic phase. The treatment of the solvent to remove degradation products is an important step in aqueous reprocessing of fast reactor fuels to reduce waste volumes.

Solvent/Diluent Cleaning

To remove the degradation products of TBP, solvent is treated with sodium carbonate, which generates the major portion of medium level aqueous waste from a reprocessing plant. This waste volume can be reduced by the use of hydrazine carbonate instead of sodium carbonate since the spent hydrazine solution can be degraded in an electro-oxidation cell.

Partitioning

The presence of higher concentrations of plutonium in the extracted phase would require proportionately larger quantities of U(IV) for the reduction of Pu(IV) to Pu(III).

Equipment

Special equipment, such as, dissolver and extractor have to be designed and developed for use in FRFR. General principles behind the design of solvent extraction contactors were described in Sect. 7.4.1.2. Since exotic conditions are required for dissolving the mixed oxides or other ceramic fuels, the conventional thermosyphon dissolver is not adequate for FRFR. Electrolytic dissolver is one possible candidate. For solvent extraction in TRFR, pulse columns are used in most of the cycles, to meet the demand of practically no maintenance during active operations; however, in the uranium separation cycle, either pulse columns or conventional pump-mix type mixer–settlers can be used. In FRFR, however, requirements for mitigation of ill effects of acid/radiation induced solvent/ diluent degradation call for the use of centrifugal extractors, which have low residence times.

7.9.2 International Experience in Fast Reactor Fuel Reprocessing

The status of the development of fast reactor fuel reprocessing in various countries (as of 2011) is covered in detail in the IAEA Technical Report NF-T-4.2 [15]. No country is presently engaged in regular reprocessing of fast reactor fuel except India. However, France had reprocessed fuel discharged from RAPSODIE and PHOENIX reactors at its pilot facilities in Marcoule and La Hague. Development of the processes for reprocessing as well as waste management for fast reactor fuel cycle has been actively pursued by several countries (China, Germany, Japan, Korea, Russia, and USA) [15].

The Fast Breeder Test Reactor (FBTR) at Kalpakkam, Tamil Nadu, employs uranium–plutonium mixed carbide fuel with a composition of 70 as well as 55% Pu. The irradiated fuel is reprocessed in pilot plant facility, CORAL (Compact Reprocessing of Advanced Fuels in Lead Mini Cells) which was set up at Kalpakkam to validate the flow sheet for reprocessing fast reactor fuels with high plutonium content [12]. Several batches of irradiated fuel pins with varying burn-up, e.g. 25, 50, 100, and 155 GWd/t, have been reprocessed in this facility. The fuel was reprocessed by modified PUREX process, and the process flow sheet followed was similar to that of oxide fuel.

7.9.3 Alternate Extractants for Fast Reactor Fuel Reprocessing:

The phenomenon of "third-phase formation" was discussed in Sect. 7.9.1.2. The importance of this phenomenon, combined with a few additional limitations of the existing extractant (tri-n-butyl phosphate), has provided the motivation for development of alternate extractants.

7.9.3.1 Limitations of TBP

TBP meets most of the requirements of an ideal extractant. It has been the solvent of the reprocessing industry since last seven decades. However, there are limitations/drawbacks which raise the need to find alternate extractant to TBP.

The limitations of TBP are.

1. Significant aqueous solubility (~0.4 g/L).
2. The selectivity of U and Pu over Zr and Ru is low, and hence, decontamination factors (DF) are low.
3. TBP has the tendency to form third phase during extraction of tetravalent actinides at macro level from nitric acid medium. The Limiting Organic Concentration of

Pu(IV) by 1.1 M TBP/n-Dodecane at 2 M nitric acid is ~40 g/L. In the context of reprocessing fast reactor fuels with high plutonium content, these drawbacks of TBP are of major concern.

4. Chemical and radiolytic degradation products of TBP, viz. monobutyl phosphoric acid and dibutyl phosphoric acid, are responsible for the deterioration of the DF values of plutonium from fission products, and hence, it poses problems in the back extraction of uranium and plutonium.

5. Large volumes of the secondary (phosphate) waste due to incomplete incineration.

7.9.3.2 Alternate Extractants

There exists a huge database for TBP and seven decades of accumulated experience. Any alternate extractant must have all the desirable qualities of TBP without the associated disadvantages.

Higher Homologues of TBP

The higher homologues of TBP [16] are obvious choice as an alternate to TBP. The higher homologues are suitable candidates for the fast reactor fuel reprocessing (where plutonium content is high) from the point of view of aqueous solubility and third-phase formation. Studies on a large variety of extractant systems seem to indicate that tri-iso-amyl phosphate could be a suitable extractant for the fast reactor fuel reprocessing [16].

N,N-dialkyl amides:

A class of amides, known as N, N-dialkyl amides [17], has also been extensively used as an alternative to TBP. The salient features of amides as extractants are.

1. Low volume of secondary waste generated (completely incinerable).
2. Innocuous nature of chemical and radiolytical degradation products (better decontamination from fission products).
3. Low aqueous phase solubility.
4. Final U and Pu product streams are free of phosphorus contamination.

However, these compounds are capable of forming third phase with plutonium under certain operational conditions.

7.9.4 Innovations for Fast Reactor Fuel Reprocessing

The need for handling high levels of radioactivity, high concentrations of plutonium, and corrosive media also demand development of innovative equipment, such as contactors. In the reprocessing plant, especially dissolver solutions arising out of fast reactor, remote operation and maintenance are of major importance while designing and development of equipment, since the solutions processed are highly radioactive. Hence, development and commissioning of remotely operated processes are crucial. The application of remote technologies in reprocessing facility is to minimize radiation exposure to the personnel, safety, better throughput, and enhanced quality assurance. In addition, towards process and inventory control of reprocessing operation, it is important to analyse samples from the various process vessels to assess the plant performance. In view of the very high radioactivity associated with dissolver solution, the liquid sampling has to be carried out using remote techniques. The development of in-service inspection devices for assessing the healthiness of equipment and process vessels is also required. These considerations apply to TRFR as well as FRFR.

With regard to FRFR, there are additional requirements for innovation. The dissolution of mixed oxides or carbide fuels subjected to high burn-up in FBR needs appropriate dissolver, e.g. electrolytic dissolver. The presence of higher levels of plutonium and higher radiation levels lead to challenges in the selection of dissolver vessel material; dissolver vessel should have provisions to facilitate dissolution of fuel under harsh or reactive conditions. Since the components are prone to corrosion-related issues, material research is crucial to ensure and achieve the desired plant life. Development of superior austenitic stainless steels and alternate materials, e.g. titanium-based alloys are required, especially for making dissolver and evaporator materials for minimizing corrosion.

Product Storage:

Independent of the fuel-type processed, the purified finished oxide products are stored in a shielded vault. The radiations emitted from the products and from the residual fission products demand appropriate level of shielding, for the safe storage. Lead shielding is often used in a variety of applications for handling, processing, and storage of nuclear materials. The high density of lead (11.34 g/cm^3) makes it a useful material for shielding against X-ray and gamma radiation. Concrete material also has good shielding properties with respect to neutrons and gamma rays. Concrete is used as a radiation shielding material in some applications as it is cheap, strong, and easily moldable. Criticality analysis is generally carried out for nuclear material vault, and appropriate shielding is ensured for its safe storage.

7.10 Non-aqueous Reprocessing

An attractive feature of pyrochemical processes [18] is their use of liquid metal and salt solvents that are resistant to radiation damage. It is feasible to process high burn-up, short cooled fuels after they are discharged from reactor and thereby avoiding the out-of-reactor fuel-inventory costs associated with long cooling periods. This advantage can be exploited most effectively in a closed, on-site fuel cycle, where long cooling periods would not be dictated by long-distance shipping requirements. Because the liquid metal and salt streams can accommodate high concentrations of the fuel constituents, pyrochemical processes are compact. The non-volatile fission product wastes are in the form of solidified metal and salt, and the gaseous fission products are collected in a small volume of inert gas.

Advantages of pyrochemical processes

1. Suitability for treatment of highly refractory materials.
2. A wide temperature range for amplifying the differences in thermodynamic stability that control the separation factors among compounds.
3. Satisfactory irradiation resistance of the inorganic reactants used in the process, allowing treatment of materials with minimum prior cooling.
4. Good fuel solubility in molten salts.
5. Compact process equipment, in which several process steps can be performed in a single device.
6. Suitability for recycling; the process output product is generally of the same nature as the input stream (metal/metal, oxide/oxide, nitride/nitride).
7. Lower criticality hazard due to the absence of water and thus of a neutron moderator in the process.
8. Minimum aqueous waste is generated.

A few of the techniques studied are outlined below:

(a) **Halide volatility**: The higher volatilities of the hexafluorides of U and Pu compared to those of fission products are utilized for achieving separation among them. The fuel elements are dissolved in a eutectic molten fluoride salt mixture at about 450 °C in the presence of HF. When this mixture is heated in the presence of fluorine, UF_6 is formed and is distilled; it is possible to distil PuF_6, which is less stable.

(b) **Molten salt extraction**: The fuel is dissolved in a salt melt. A heat-resistant solvent of low volatility, like ionic liquids, with affinity towards actinides can be used analogous to solvent extraction. This technique can be useful in case of molten salt reactor fuel where reprocessing is continuous.

(c) **Molten salt transport**: The fuel is dissolved in a metallic melt, e.g. a molten Cu-Mg alloy, which is in contact through a stirred molten chloride salt at about 800 °C with another metallic melt containing a reductant, e.g. a molten Zn–Mg alloy. Noble metal fission products are retained in the Cu-containing melt, whereas U and/or Pu are collected in the Zn-containing melt. The less noble fission products concentrate in the molten salt.

(d) **Molten salt electrorefining**: The spent fuel acts as anode in a molten salt which also contains metal or liquid cathode. By applying an electric field between the anode and cathode, material is dissolved at the anode and deposited on the cathode. Careful control of the applied voltage makes it possible to obtain pure cathode product. The process was developed for purification of plutonium metal alloys. Currently, this is considered as the most potential technique for the reprocessing of high burn-up metallic spent fuels from fast reactors and is being developed into a commercial process in laboratories across the world.

The electrorefining technique involves chopping of the spent metal fuel, e.g. U–Pu–Zr, which will be taken at the anode basket; a eutectic mixture of LiCl–KCl (containing about 3 wt% UCl_3) is used as the molten electrolyte; the spent metal fuel is dissolved in an eutectic salt mixture; during electrorefining, actinide metals are recovered at a solid cathode, or a liquid Cd cathode. The process utilizes oxidation–reduction potentials of relevant elements; i.e. the species present in spent fuel which have lower standard potentials than zirconium, e.g. actinides and less noble fission products such as alkali metals (Rb, Cs), alkaline-earth metals (Ba, Sr), and lighter lanthanide fission products (La, Ce, Pr, Nd, Sm) are electrochemically dissolved at the anode, while elements having higher standard potentials, i.e. zirconium, iron (cladding), cadmium, and noble elements (e.g. ruthenium, rhodium, palladium, and molybdenum) remain undissolved in the anode basket. At solid cathode, uranium is preferentially reduced and collected since it is the most easily reduced element among the dissolved materials; plutonium is recovered with uranium in the liquid cadmium cathode because the reduction potentials of it are close to that of uranium. Cathode processing involves deposits being mechanically removed from the solid cathode rod or from cathode crucible which will be used for the electrorefining process; the cathode deposit is heated in a cathode processing step to separate the actinide metals by distillation of adhering salt and also that of cadmium. The actinide metals obtained during the cathode processing steps are further processed, e.g. using an Injection Casting technique with appropriate addition of uranium and zirconium to maintain the composition of recycled spent fuel.

(e) Molten metal purification: Metallic fuel elements can be dissolved in molten metals (e.g. a zinc alloy). In deficient amounts of oxygen, strongly electropositive fission elements form oxides, which float to the surface of the melt as slag and can thus be removed, while volatile FPs get distilled. The residual melt will contain mainly U, Pu, Zr, Nb, Mo, and Ru, known as "fissium alloy" and can be reused as new fuel elements. This technique was tested on metallic breeder reactor fuel elements.

The above techniques are well suited for metallic fuels. However, for application to ceramic fuels, a reduction step is needed prior to pyroprocessing.

The following are the limitations of pyrochemical processing:

1. Low separation factors compared to aqueous process. This implies that the product cannot be handled in glove boxes as in case of TRFR, and refabrication also needs to be carried out in hot cells.
2. Process media (molten salts and liquid metals) are highly aggressive to process equipment.
3. High reactant melting points, resulting in a risk of fouling of transfer systems and damage to process reactors in the event of an accidental temperature drop.
4. Sophisticated technology required by the high process temperatures and the need for operating under controlled atmosphere.
5. Processes difficult to develop for continuous operation.
6. Technology and process waste requiring suitable treatment.

Despite the above limitations, pyroprocessing seems to be the most suitable process for reprocessing irradiated spent metallic fuels.

Nuclear fuel reprocessing is the most vital link for the better utilization of uranium and to extract fissile materials for recycling and to reduce volume of high-level waste.

Fuel reprocessing, aqueous as well as non-aqueous, needs very complex facilities and expertise, and it is not surprising that very few countries are actually engaged in reprocessing. However, reprocessing does offer challenging problems for research, and a number of research groups in several countries are therefore actively engaged in studies related to reprocessing and associated separations technologies.

Questions

1. Why is solvent extraction preferred method of separation over precipitation and ion exchange process?
2. Why are the earlier members of the actinides series behave like transition metals and the later members like lanthanides?
3. Name three reducing agents used in the partitioning step of PUREX process.
4. Compare the extractants, amides, and phosphates for the nuclear fuel reprocessing?
5. What are the limitations of tri-*n*-butyl phosphate (TBP)?
6. What is hydrolysis and discuss its importance with respect to plutonium chemistry?
7. What is third-phase formation phenomena in solvent extraction and explain methods to overcome the issue?
8. What are the desirable properties of diluents in reprocessing?
9. Name some troublesome fission products, which are present in the product PuO_2 leading to lower decontamination of the product.
10. What is the form in which Tc is present in the dissolver solution?
11. Name two methods for the removal of degradation products from the solvent before reuse.

References

1. R. Taylor (ed.), *Reprocessing and Recycling of Spent Nuclear Fuel* (Woodhead Publishing, 2015).
2. J.J. Katz, G.T. Seaborg, L.R. Morss, *The Chemistry of the Actinide Elements*, vol. II, 2nd ed. (Chapman and Hall, London, 1986).
3. D.D. Sood, S.K. Patil, Chemistry of nuclear fuel reprocessing: current status. J. Radioanal. Nucl. Chem. **203**(2), 547–573 (1996)
4. G.M. Ritcey, A.W. Ashbrook, *Solvent Extraction: Principles and Applications to Process Metallurgy* (Elsevier Science Publishing Co., Inc., Part I, 1984)
5. J.D. Law, T.A. Todd, *Liquid-Liquid Extraction Equipment* (Idaho National Laboratory, United States, 2008). INL/CON-08-15151.
6. M. Benedict, T.H. Pigford, H.W. Levi, *Nuclear Chemical Engineering* (McGraw Hill Book Company, New York, 1981), pp.172–174
7. P.K. Dey, N.K. Bansal, Spent fuel reprocessing: a vital link in Indian nuclear power program. Nucl. Eng. Des. **236**, 723–729 (2006)
8. W.D. Bond, Thorex process, in *Science and Technology of Tributyl Phosphate (Applications of Tributyl Phosphate in Nuclear Fuel Reprocessing)*, vol. III, ed. by W.W Schulz, L.L. Burger, J.D. Navratil, K.P. Bender (CRC Press, Boca Raton, Florida, USA, 1990)

9. N.K. Pandey, N. Desigan, A. Ramanujam, *PUREX and THOREX Processes* (Aqueous Reprocessing), ed. by S. Hashmi. Materials Science and Materials Engineering (Oxford, Elsevier; 2016), pp. 1–15.
10. D. Das, S.R. Bharadwaj (Eds.), *Thoria-based Nuclear Fuels: Thermophysical and Thermodynamic Properties, Fabrication, Reprocessing, and Waste Management* (Springer-Verlag, London, 2013).
11. K. Anantharaman, P.R.Vasudeva Rao, Global perspective on thorium fuel, in *Nuclear Energy Encyclopedia: Science, Technology, and Applications*, ed. by S.B. Krivit, J.H. Lehr, T.B. Kingery (Wiley, 2011), pp 89–100.
12. R. Natarajan, Baldev Raj, Fast reactor fuel reprocessing technology: successes and challenges. Energy Procedia **7**, 414–421 (2011)
13. J. Bourges, C. Madic, G. Koehly, M. Lecomte, Dissolution of PuO_2 in HNO_3 medium with electro generated Ag(II). J. Less-Common Metals **122**, 303–311 (1986)
14. P.R. Vasudeva Rao, Z. Kolarik, A review of third phase formation in extraction of actinides by neutral organophosphorus extractants. Solv. Extr. Ion Exch. **14**(6), 1996, 955–993.
15. IAEA Nuclear Energy Series; Status of Developments in the Back End of the Fast Reactor Fuel Cycle; No. NF-T-4.2; 2011, pp. 38–64
16. A. Suresh, C.V.S. Brahmmananda Rao, B. Srinivasulu, N.L. Sreenivasan, S. Subramaniam, K.N. Sabharwal, N. Sivaraman, T.G. Srinivasan, R. Natarajan, P.R. Vasudeva Rao, Development of alternate extractants for separation of actinides. Energy Procedia **39**, 120–126 (2013).
17. V.K. Manchanda, P.N. Pathak, Amides and diamides as promising extractants in the back end of the nuclear fuel cycle: an overview. Sep. Pur. Technol. **35**, 85–103 (2004)
18. K. Nagarajan, B. Prabhakara Reddy, Suddhasattwa Ghosh, G. Ravisankar, K.S. Mohandas, U. Kamachi Mudali, K.V.G. Kutty, K.V. Kasi Viswanathan, C. Anand Babu, P. Kalyanasundaram, P.R. Vasudeva Rao, B. Raj, Development of pyrochemical reprocessing for spent. Energy Procedia **7**, 431–436 (2011).

Further Reading

1. K.E. Holbert, R.L. Murray, *Nuclear Energy, An Introduction to the Concepts, Systems, and Applications of Nuclear Processes*, 7th ed. (Butterworth-Heinemann, Elsevier, 2015).
2. G. Choppin, J.-O. Liljenzin, J. Rydberg, C. Ekbery, *Radiochemistry and Nuclear Chemistry*, 4th ed. (Elsevier, 2013).
3. Modern Nuclear Chemistry, *Water Loveland*, 2nd edn. (David J Morrissey, G T Seaborg, Wiley-Interscience, 2017)
4. I. Crossland (ed.), *Nuclear Fuel Cycle Science and Engineering*. Series in Energy (Woodhead Publishing, 2012).
5. K.H. Leiser, *Nuclear and Radiochemistry: Fundamentals and Applications*, 2nd ed. (Wiley-VCH, 2001).
6. K.L. Nash, G.J. Lumetta, *Advanced Separation Techniques for Nuclear Fuel Reprocessing and Radioactive Waste Management*, 1st ed. Series in Energy (Woodhead Publishing, 2011).
7. K.D. Kok (ed.), *Nuclear Engineering Handbook* (CRC Press, 2017).
8. S.B. Krivit, J.H. Lehr, T.B. Kingery (eds.), *Nuclear Energy Encyclopedia* (Wiley, 2011).
9. B. Raj, P. Chellapandi, P.R. Vasudeva Rao, *Sodium Fast Reactors with Closed Fuel Cycle* (CRC Press, 2015).
10. B. Raj, P.R. Vasudeva Rao, *Nuclear Fuel Cycle: Closing the Fuel Cycle* (BRNS, 2006).
11. D.D. Sood, A.V.R. Reddy, N. Ramamoorthy, *Fundamentals of Radiochemistry*, 4th ed. (Indian Association of Nuclear Chemists and Allied Scientists, 2010).

Chapter 8
Radioactive Waste Management

Smitha Manohar, G. Sugilal, R. K. Bajpai, C. P. Kaushik, and Kanwar Raj

8.1 Introduction

The nature of waste from nuclear fuel cycle depends upon various factors such as its source of generation, type of fuel and its cladding material, process used in purification/fabrication of fuel, type of nuclear reactor, burnup of the fuel, off-reactor cooling period, process flow sheet used in SNF reprocessing and techniques adopted in decontamination/decommissioning. In view of large variety of radioactive waste streams generated in the nuclear fuel cycle, the processes and technologies used in their treatment and disposal are equally diverse [1–3]. The techniques and technologies used in radioactive waste management are based on the concepts of [4]:

(a) delay and decay of short-lived radionuclides,
(b) concentration and confinement of radioactivity as much as practicable and

Authors C. P. Kaushik and Kanwar Raj: Formerly at Nuclear Recycle Group, Bhabha Atomic Research Centre.

S. Manohar (✉) · G. Sugilal · R. K. Bajpai · C. P. Kaushik · K. Raj
Nuclear Recycle Group, Bhabha Atomic Research Centre, Mumbai 400085, India
e-mail: smanohar@barc.gov.in

G. Sugilal
e-mail: gsugilal@barc.gov.in

R. K. Bajpai
e-mail: rkbajpai@barc.gov.in

C. P. Kaushik
e-mail: cpk.1962@gmail.com

K. Raj
e-mail: rajkanwar50@gmail.com

B. S. Tomar et al. (eds.), *Nuclear Fuel Cycle*,
https://doi.org/10.1007/978-981-99-0949-0_8

(c) dilution and dispersion of the resultant effluents of very low-level radioactivity
     to the environment as per the permissible levels, set by the national regulatory
     authority, which need to be in line with the international practices.

The underlying objective governing the management of radioactive waste is
protection of human beings and environment, now as well as in future. To meet
this objective, the necessary codes and guides have been framed by international
bodies like International Commission on Radiation Protection (ICRP) and International Atomic Energy Agency (IAEA) [1, 2]. One of the basic requirements is to set
up a national regulatory framework in every country. As an example, Atomic Energy
Regulatory Board (AERB) has been set up in India by an act of the Parliament,
Atomic Energy Act 1962. Under this Act, AERB is entrusted with the responsibility
of enforcement of safe disposal of radioactive waste in the country [3].

The challenges in management of radioactive waste, as compared to the traditional industrial waste, are basically on account of radiation associated with the
chemical species present in the radioactive waste. Therefore, in many radioactive
waste treatment facilities, the operation and maintenance (O&M) activities are to be
carried out completely remotely, and process equipment is housed inside hot cells
having thick biological shielding for the protection of operation and maintenance
(O&M) personnel. All gadgets and control equipment must have high resistance
to radiation. O&M personnel must be protected against radiation hazards by suitable plant and equipment design as well as health physics safety monitoring and
procedures. On the other hand, while handling low-level radioactive effluents, technological challenges are faced due to very low concentration of specific radionuclides
to be removed. The solidified waste product also must have excellent chemical durability to prevent leaching of radionuclides from a waste disposal system, thus putting
stringent demands on process design, control and quality assurance. The positive
aspect of radioactive waste is that radionuclides present in the waste decay over time
and become stable, thus, losing their radioactivity.

Details of various types of radioactive wastes and their characteristics, steps
in radioactive waste management, treatment of aqueous and organic liquid waste,
recovery of valuable radionuclides waste, development of matrices for immobilization of waste, vitrification process and technology, interim storage and disposal in
near surface and geological disposal facility are discussed in this chapter. Practices
adopted in radioactive waste management, and the operational experience, in various
countries are also presented for illustration.

In view of the limitations on the topics that can be covered, aspects of management
of radioactive wastes from use of isotopes in healthcare, research and industry are not
discussed in this chapter. Similarly, details of decontamination and decommissioning
processes and treatment of resultant wastes are also not covered here.

8.2 Radioactive Waste Classification

The radioactive waste streams are classified on the basis of their radiological, physical and chemical properties. The most common method of categorization of solid radioactive wastes is on the basis of their physical properties like combustibility and compressibility. This helps to select the volume reduction process/technique either by incineration/pyrolysis or by compaction. Solid radioactive wastes are also characterized by the radiation field on the surface of the waste package which helps to determine their further handling and disposal. Another method of categorization of solid waste is based on the half-lives of the radionuclides present in the waste whether short- or long-lived which determines the required period for isolation of waste and in selection of disposal site, backfill material and disposal concept [4–6].

Similarly, liquid radioactive waste streams are categorized based on various characteristics of wastes, e.g. concentration of the radionuclides and their half-lives, aqueous/organic nature, etc. Radioactive liquid wastes are commonly classified as low-level waste (37–3.7×10^6 Bq/l), intermediate-level waste (3.7×10^6–3.7×10^{11} Bq/l) and high-level waste (above 3.7×10^{11} Bq/l). This categorization of waste fulfils many objectives, viz. basis of segregation of waste in the waste generating facilities and in design of relevant storage area/tanks and selection of waste treatment process, selection of matrices for immobilization of waste. It also helps in proper documentation and communication with respect to various categories of radioactive waste among waste generators, managers and regulators.

Low- and intermediate-level liquid radioactive wastes (LILW) are normally generated in every stage of nuclear fuel cycle. These are classified primarily based on their activity level. In general, liquid effluents that would (i) normally not require shielding during storage/processing, etc., and (ii) may or may not require treatment would be classified as low-level waste (LLW). The wastes above LLW limits and requiring shielding would be commonly regarded as intermediate-level liquid waste (ILW). A commonly used classification for LILW is given in Table 8.1 for illustration.

Table 8.1 Classification of low- and intermediate-level waste (LILW)

Category	Activity (m^{-3}) mixed β/γ emitters[a]	Remarks
Low-level waste (LLW)	(1) <37 kBq	No treatment required; released after measurement[b]
	(2) 37 kBq–37 MBq	Treated, no shielding required
	(3) 37 MBq–3.7 GBq	Treated, shielding sometimes required according to radionuclide composition
Intermediate-level waste (ILW)	(4) 3.7 GBq–370 TBq	Treated, shielding necessary in all cases

[a] Concentration of alpha activity is negligible
[b] Related to the release rates, licenced by the respective competent authority

8.3 Characteristics of Radioactive Wastes

8.3.1 Types and Characteristics of LILW

Low- and intermediate-level radioactive waste (LILW) streams are generated from various nuclear facilities. LILW can be aqueous or organic in nature. Organic liquid waste (OLW) is treated by processes which are different from those used for aqueous streams. OLW treatment is discussed separately in Sect. 8.6.

There is significant variation in the composition of LILW originating from different facilities and even within a facility. Typically, LILW streams can be further categorized into three streams for ease of segregation and treatment. These are (i) process wastes, (ii) decontamination wastes and (iii) potentially active waste. Examples of process wastes are solutions from regeneration of ion exchange columns used for polishing of reactor coolants/spent fuel storage pools, vapour condensates from evaporation of reprocessing streams, chemical wastes generated in various conversion processes, etc. Decontamination wastes, as the name suggests, arise from decontamination of in-service equipment and piping and during decommissioning of nuclear facilities. This stream typically contains significant concentrations of complexants. Potentially active waste streams usually emanate from the change rooms of nuclear facilities, floor drains and other cleaning activities. This stream usually would require minimal or no treatment, if appropriately segregated at source.

Further, there are special wastes which are facility-specific. For example, tritiated wastes are almost entirely of nuclear reactor origin. Another example of special waste is alkaline liquid waste generated during chemical de-cladding of spent nuclear fuel during its reprocessing.

Therefore, LILW streams have a wide range of characteristics ranging from highly acidic to highly alkaline conditions, spectrum of total solids concentration ranging ppm to gm per litre levels. One factor that characterizes LILW is the presence of ^{137}Cs and ^{90}Sr activity. Other radionuclides are usually found in significantly lesser concentration which include alpha activity, ^{60}Co, ^{106}Ru, ^{125}Sb and ^{99}Tc. Their generation rates are remarkably diverse depending upon the source.

8.3.2 Fuel Reprocessing Waste

The radioactive liquid wastes generated during fuel reprocessing are categorized into three steams, viz. high-level liquid waste (HLLW), intermediate-level liquid waste (ILLW) and low-level liquid waste (LLLW) depending on their radioactivity levels. HLLW accounts only about 3% by volume and is acidic in nature. It contains more than 99.5% of the radioactivity that gets generated in the entire nuclear fuel cycle. The generation of ILLW is about 7% of the volume and it contains about 1–2% radioactivity whereas LLLW accounts for less than 0.5–1% of radioactivity but is generated >90% of the total volume. HLLW contains several valuable radionuclides and is, therefore, considered as a resource rather than the waste.

8.3.3 High-Level Liquid Waste (HLLW)

During reprocessing, the spent nuclear fuel is first dissolved in the nitric acid and the plutonium and unused uranium are recovered in the chemical separation plant. The remaining solution, containing about 99% of the dissolved fission products, together with some impurities primarily from cladding materials and inactive process chemicals, traces of unrecovered plutonium and most of the transuranic (TRU) elements, constitutes HLLW.

8.3.3.1 Major Components of HLLW

The waste contains both long- and short-lived fission products, minor actinides (americium, neptunium and curium), unrecovered Pu/U, processing chemicals like nitric acid and sodium nitrate, dissolved cladding material like aluminium, and zirconium fines and corrosion products of material of construction of storage tanks and piping.

- **Un-recovered U and Pu**: Although reprocessing facility aims at the extraction of maximum U and Pu from dissolved spent fuel, small amounts of U and Pu are always present in raffinate stream along with fission products. This could be due to the limitation of separation system, degradation of solvents, etc.
- **Fission products**: The majority of fission products, present in spent fuel during reprocessing, appear in HLLW. The radioactivity associated with HLLW depends upon the extent of burnup and cooling period. The longer off-reactor cooling period reduces the radioactivity due to decay of short-lived radioactive fission products. Major fission products present in HLLW are ^{90}Sr, ^{106}Ru, ^{137}Cs, ^{144}Ce, etc.
- **Minor Actinides**: The TRU elements neptunium, americium and curium are formed by neutron capture. The amounts formed are small in the low burnup fuels, particularly curium, but the yields become significant as the burnup increases in fast reactors and when plutonium is used as fuel. These elements are of particular concern in long-term waste management because of their long half-lives.
- **Fuel elements and clad material**: Spent fuel elements are decladded mechanically in a few reprocessing processes. The traces of clad material adhere to the fuel, pass into the dissolver and end up in the waste solution, e.g. traces of metallic fines of Zircaloy cladding report in HLLW. In certain processes, chemical dissolution of cladding material aluminium results in small amount of aluminium in HLLW.

- **Chemicals added during reprocessing**: Earlier reprocessing flowsheets involved use of many chemicals such as ferrous sulphamate and sodium nitrite for conditioning and adjustment of oxidation state. Hence, HLLW generated from such reprocessing facilities contains substantial amount of sodium, iron, sulphate, etc., in addition to the nitric acid. In some cases, fluoride or mercury (as mercuric nitrate) was being added to catalyse the fuel dissolution. They finally result in HLLW. In modern reprocessing flowsheets, addition of chemicals, except for nitric acid, is avoided to minimize the final waste volume.
- **Corrosion products**: Iron, nickel and chromium, arising from corrosion of process equipment, piping and tanks are also present in HLLW.
- **Traces of organic compounds**: Solvents used in reprocessing undergo degeneration due to prevailing high level of radiation. For example, PUREX-based spent fuel reprocessing involves recovery of U and Pu using tributyl phosphate (TBP) dissolved in an alkane diluent. As a result, traces of TBP and its degradation products are also present in HLLW.
- **Soluble poison**: The chemical processing of enriched uranium or plutonium-based spent fuel requires addition of soluble poison such as gadolinium or boron to avoid any possibility of criticality. Such soluble poison finally ends in HLLW.

The composition of HLLW depends on various factors, e.g. type of reactor, type of nuclear fuel and its burnup, off-reactor cooling period, reprocessing flowsheet, etc. Further, the composition of vitreous matrix, design of equipment involved in vitrification and vitrification parameters highly depend upon the composition of HLLW.

The fission product distribution in the nuclear fuel depends upon the energy of neutron utilized for fission of fissile material, as well as the fissile nuclide. The amount of fission products will be more in HLLW generated from reprocessing of PWR (burnup ~45,000 MWd/Te) as compared to that of PHWR (burnup ~6700 MWd/Te). Minor actinides will also be more in HLLW from processing of higher burnup spent fuel as compared to that of lower burnup. The presence of fission products in HLLW with shorter half-life will largely depend upon off-reactor cooling period of the spent fuel prior to reprocessing. Finally, the composition of HLLW, especially inactive components, depends also on the reprocessing flow sheet and inactive elements used during reprocessing. HLLW from PUREX process (Fig. 8.1) is acidic in nature and contains nitric acid and salts in nitrate form [7, 8].

The common constituents present in HLLW from reprocessing of spent fuel are shown in Table 8.2.

8.3.4 Intermediate-Level Radioactive Liquid Waste

Intermediate-level radioactive liquid wastes (ILLW) generated from reprocessing activities are alkaline in nature and contain radionuclide such as ^{137}Cs, ^{90}Sr and ^{106}Ru besides trace concentration of alpha activity due to actinides. These wastes

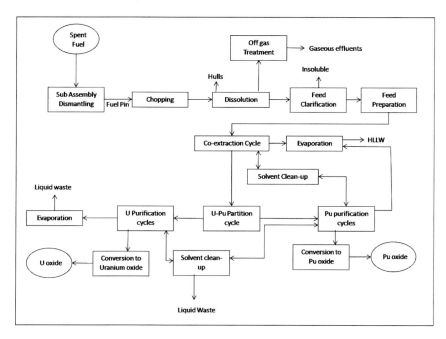

Fig. 8.1 Schematic flowsheet of PUREX process showing generation of radioactive wastes

Table 8.2 Common constituents of HLLW generated during reprocessing of spent fuel

Chemical nature	Acidic (mainly nitric acid ~1–4 M)
Major fission products	^{137}Cs, ^{90}Sr, ^{106}Ru, ^{144}Ce, etc.
Alpha emitters	^{241}Am, ^{237}Np, ^{242}Cm, U, traces of Pu
Process chemicals	Ions of Na, Fe, Al, Mn, Ca
Corrosion products	Ions of Fe, Cr, Ni, Mo
Organic products	Traces of TBP and its degradation products

are stored in mild steel tanks and due to addition of alkali for neutralization contain significant concentrations of sodium nitrates and carbonates. The present trend is to eliminate generation of ILLW by modifying the flowsheet whereby generation of salts is eliminated and by replacing carbon steel as material of construction with stainless steel for storage tanks.

8.3.5 Organic Liquid Waste

The intense ionizing radiation present during reprocessing process deteriorates the solvent due to oxidative degradation. Several degradation products of solvents and diluents are formed during processing, and many of them pose deleterious effects on the process performance.

8.3.6 Solid Waste

Solid radioactive waste is generated during maintenance and housekeeping, in laboratories and also in the form of protective wears. These are generally of low-level category except in case of mops used in decontamination of spills, etc. Another type of waste containing alpha radionuclides is generated from glove boxes, as used equipment and in the form of protective wears. These alpha-contaminated solid wastes generated from reprocessing facilities are classified as *metallic wastes* (e.g. used equipment, glove boxes, hull waste, etc.) and *non-metallic wastes* (e.g. cellulosic wastes, rubber and plastics).

8.4 Basic Steps in Radioactive Waste Management

The basic steps involved in radioactive waste management are pre-treatment, treatment, conditioning/immobilization, packaging and transportation, and storage/disposal. Depending on the nature of waste, some or all of these steps could be the part of the scheme for management for a particular waste stream [4–6].

8.4.1 Segregation/Minimization

Minimization of waste generation is an aspect to be taken care during design, construction, operation and decommissioning of nuclear fuel cycle facility. There are several good practices and examples in the nuclear fuel cycle facilities where waste generation has been reduced. Some of these examples are (a) proper segregation at the source of generation to avoid mixing of non-radioactive and radioactive materials, (b) suitable segregation of plant areas and good housekeeping to reduce contamination of protective wears, (c) proper selection of material of construction to minimize and reduce corrosion, (d) minimizing use of chemical reagents and salts to reduce the waste loading, (e) selection of process chemicals/solvents/materials of construction with high radiation resistance, etc. Minimization of radioactive waste helps in reducing the volume of final waste to be disposed which has very significant impact on the conservation of land required for waste disposal and on the safety of environment. At the waste management facilities, process and technologies are selected with the objective of reduction of the final waste volume. Some of these techniques are (a) recovery and recycle of valuables from waste; (b) volume reduction of solid waste by compaction, incineration/pyrolysis; (c) volume reduction of liquid wastes by thermal and solar evaporation; (d) decontamination of effluents; (e) concentration of chemical sludge, etc.

8.4.2 Pre-treatment

The waste received at the waste management facility is characterized with respect to various radiochemical properties. A few examples of pre-treatment are (a) removal of organics, which interfere with the separation process, from ILW prior to ion exchange treatment, (b) concentration of HLLW by evaporation to reduce thermal load on vitrification melter and (c) size reduction of disused equipment prior to compaction.

8.4.3 Treatment

The objectives of waste treatment are (a) decontamination, that is, to remove the radionuclides from waste stream, (b) reduce the volume, (c) remove harmful chemical species that could enhance corrosion, affect decontamination factor and/or properties of waste product. A few examples of treatment are chemical treatment of LLW to remove Sr, Cs and Ru; ion exchange treatment of ILW to remove Cs and Sr, denitration of HLLW to reduce corrosion due to nitric acid as well as volatilization of Ru and Cs.

8.4.4 Conditioning/Immobilization

The concentrate, chemical residue/sludge and other compacted solid radioactive wastes (viz. spent ion exchange resin) are immobilized in inert stable matrices, in order to immobilize radionuclides, present in the waste, fill the voids and obtain a monolithic waste product of desired characteristic. The most significant property of conditioned waste product is chemical durability defined by leachability. Matrices commonly used for immobilization are cement, polymer and glass depending on the nature and concentration of radionuclides present in the waste besides few other parameters.

8.4.5 Packaging/Transportation

Waste packaging is usually done at the waste generating facility to meet radiological requirements of transportation, storage and disposal. The external surfaces should be free of transferable contamination, and shielding design must satisfy the regulatory limits of maximum permissible radiation field. Similarly, design of shielding cask must meet the requirements of safety during handling and transportation.

8.4.6 Storage/Disposal

The storage/disposal of waste packages is usually the last step in management of radioactive waste. Depending on the management strategy, waste packages are sometimes stored for an interim period before final disposal, e.g. vitrified HLLW may be stored under cooling for about thirty years which facilitates reduction in decay heat to about half. The waste storage facility must be equipped with material handling gadgets and equipment like crane for safe de-casking and placement of waste package at the desired storage locations.

Either after interim storage or directly from the waste generating facility, the waste packages are transferred to the disposal facility. Similar to the storage, the waste disposal facility is also equipped for safe handling of waste packages, backfilling, closure, monitoring and surveillance. Depending upon the nature of disposed waste, the waste disposal modules are located either near the land surface or underground at varying depths.

8.5 Management of LILW

The treatment of LILW can be done at source itself or can be performed in a centralized waste management facility if there are number of co-located waste generating facilities at a given nuclear site. Four treatment processes, viz. *chemical treatment, ion exchange/sorption, evaporation and reverse osmosis*, are normally used for treatment of waste. The treatment leads to generation of a small volume of secondary waste in the form of *concentrates* which contain bulk of radioactivity. These waste concentrates are immobilized usually in an inert and stable matrix like cement. The treated aqueous stream, which forms major part in terms of volume, is further treated as per the requirements [9–12].

Various treatment methods for LILW steams are described in the following sections.

8.5.1 Chemical Treatment Process

This is the most common treatment method employed for LILW streams. The process involves removal of radionuclides from LILW streams by creating precipitates by externally added chemicals and then subjecting them to settling. For example, radioactivity due to Cs is removed by co-precipitation with potassium copper ferrocyanide by the following reaction:

$$CuSO_4 + K_4Fe(CN)_6 \rightarrow K_2CuFe(CN)_6 + K_2SO_4 \qquad (8.1)$$

In the above case, Cs is carried with potassium of potassium copper ferrocyanide at pH 8 efficiently. Sr is removed by co-precipitation with precipitate of barium sulphate or calcium phosphate at pH 9.5.

The actinides commonly encountered in LILW, viz. U and Pu, are separated by co-precipitation with precipitate of ferric hydroxides at pH 8–9. Radionuclides of Ru, Sb, etc., can be separated from LILW by co-precipitation using different reagents. Decontamination factor (DF) between 10 and 100 is usually obtained in the process depending on the feed characteristics. After the chemical addition, LILW stream is sent to settling equipment for separation of precipitates as radioactive sludge. The settling equipment is usually an equipment with low H/D ratio. Photograph of a typical settling equipment is shown in Fig. 8.2.

Fig. 8.2 Chemical treatment of LILW: settlers used for precipitation of sludges

The sludge has low total solids concentration which is further reduced in volume by means of centrifugation, filtration, or evaporation. The concentrated sludge is then immobilized in cement matrix. Bituminization is also an immobilization process, used at La Hague [10].

8.5.2 Ion Exchange/Sorption Process

Ion Exchange/Sorption process is used extensively in nuclear industry for separation of specific radionuclides from LILW. This process is normally used for relatively clean LILW streams such as primary coolant in a nuclear reactor and water used in spent fuel storage pool, and for waste streams rich in radionuclides of Cs and Sr (e.g. de-cladding waste from fuel reprocessing and supernatant of legacy neutralized HLLW). In this process, Resorcinol Formaldehyde Polycondensate Resin (RFPR) is used in repeated loading–elution-regeneration cycles for removal of ^{137}Cs, the major radionuclide present in ILW [11, 12]. The separation mechanism is as follows:

$$RFPR\text{-}O\text{-}H \xrightarrow{\ NaOH\ } RFPR\text{-}O\text{-}Na + H_2O \xrightarrow{\ Cs\ } RFPR\text{-}O\text{-}Cs \qquad (8.2)$$

$$\underset{HNO_3}{\xleftarrow{\hspace{5cm}}}$$

The process splits ILW into two streams: (1) a small volume of eluted Cs-rich stream and (2) large volume of LLW. Cs-rich stream is further concentrated and is a source of ^{137}Cs for various applications like blood irradiation. The second stream, i.e. LLW from RFPR treatment, contains various radionuclides, viz. ^{90}Sr, ^{125}Sb, ^{106}Ru and ^{99}Tc besides large concentration of salts, mainly sodium nitrate. A reductive co-precipitation process using FeS and Fe(OH)$_2$ is employed for removal of Tc and Ru. Ruthenate and technetate ions are reduced to lower oxidation state using Na$_2$SO$_3$ in acidic condition and then precipitated along with FeS and Fe(OH)$_2$. The precipitation of Ru, Tc and Cs is effective in acidic/neutral pH and that of Sr and Sb in alkaline pH. The two-step chemical treatment process is shown in Fig. 8.3.

A list of typical selective sorbents used in LILW treatment is given in Table 8.3.

The resins are mostly used in fixed bed mode in columns. DF obtained in the process is highly variable depending on the objective but is usually between 10 and 1000.

Ion exchange process is also employed for treatment of contaminated water. Use of zeolites for treatment of contaminated water from Three Mile Island incident is an example [14–16]. Another example of multi-nuclide removal by ion exchange/sorption is the treatment of large volumes of contaminated water at Fukushima site in Japan [17–20]. Here, the mixing of seawater and fission products produced a particularly challenging waste stream due to its high salinity and very high radioactivity (caesium: 5×10^6 Bq/ml). Another source of contamination was groundwater finding its way to parts of nuclear power plant, thus, necessitating

Fig. 8.3 Two-stage
chemical treatment process
for removal of [106]Ru [13]

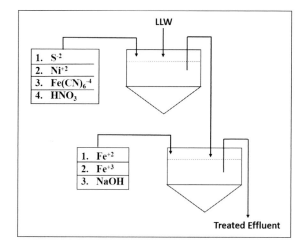

treatment and removal. In view of large volume of waste, various water treatment systems were deployed at the site progressively. Initial efforts involved treatment of water with the aim to recycle treated water for cooling the reactors. The herschelite-type zeolite in combination with a sedimentation and reverse osmosis-based system was used to decontaminate the water and recycle the same. Another system, called as Simplified Active Water Retrieve and Recovery System, treated the accumulated water in an ion exchange system comprising of engineered zeolite and crystalline sili-cotitanate columns. The magnitude of the contaminated water treatment is succinctly captured by the statistics that as of Aug 20, 2020, about 12,26,400 m^3 of water has been treated through these systems. The problem of presence of tritium in the treated stored water still persists.

8.5.3   Evaporation Process

In the evaporation process, LILW is boiled to drive off water at pressures marginally below atmospheric pressure. Since most of the radionuclides present in LILW streams are non-volatile, high DFs (>1000) are obtained as compared to other processes. From among the wide variety of evaporators used in the chemical process industry, use of pot, thermosiphon and forced circulation type evaporators is more common. The choice of evaporator type is decided on the processing rate required, salts present in the waste and desired DF. A train of evaporators is also used to obtain overall desired DF, e.g. for conversion of ILW into LLW. A novel variant of evaporation process is the *solar evaporation* of effluents in dry regions.

Evaporation process is tolerant to the variations in feed conditions. The major disadvantage is the use of steam which makes the evaporation process energy intensive in comparison with other ambient temperature processes. The performance gets

Table 8.3 Types of resin/sorbents used for LILW treatment

Radionuclide to be removed	Acidic LILW	LILW in pH range	Alkaline LILW
Cs	Ammonium molybdophosphate impregnated/coated on various substrates	Copper-ferrocyanide impregnated on various substrates like zeolites, resins, etc., crystalline silicotitanate and natural zeolites	Resorcinol formaldehyde polycondensate resin, crystalline silicotitanate
Sr	–	Zeolite 4A	Iminodiacetic acid

affected while handling LILW having foaming or scaling tendencies. Thus, the use of evaporation is usually limited to concentration of LILW effluents in nuclear power plants, reprocessing facilities and vitrification plants for *limited volumes* of waste.

8.5.4   Reverse Osmosis Process

Reverse osmosis (RO) process is used for concentration of LILW effluents using a semi-permeable membrane. The membrane allows the water to pass through it and retains bulk of the salts. It is generally a high pressure process with pressures ranging from 10 to 30 bars depending on the total salt concentration of input LILW stream. The process gives good decontamination factors, which lie in between those offered by chemical treatment/ion exchange processes and evaporation process. The commonly used membranes are the composite polyamide and thin film composite membranes. The membranes are usually deployed in spiral-wound modules, due to the high surface area to volume ratio available in this type of configuration. The modules are then connected in various ways to meet the processing rate and to satisfy membrane limits like minimum/maximum flow and pressure drop. The RO system consists of pre-filtration units, high pressure pump (multistage centrifugal or double/triple acting plunger pump), membrane modules and storage tanks. The process is relatively cheaper than the evaporation process. One major advantage of the process is that system can be readily built as modular unit and deployed at site, minimizing installation time. The major disadvantage of the process is that volume reduction obtained in the process is lower as compared to other processes. Hence, this process is normally used in conjunction with other treatment processes mentioned previously.

8.5.5   Auxiliary Processes

To ensure effective performance of main four processes described in preceding sections, several auxiliary processes are employed. These include pH adjustment, filtration and advanced oxidation process. *pH adjustment* is required for most of LILW treatment processes, except evaporation, as a pre-treatment step. pH adjustment is carried out with mineral acids, strong alkali, weak alkali and even with CO_2 depending on the initial and final desired conditions for further processing. *Filtration* is another common step used in various stages of LILW processing. Selection of filter depends on the size of suspended solids, their concentration and nature. Commonly used filters include vertical pressure candle and bag filters and variations of these types of filters. Ultrafiltration membranes can separate suspended solids up to 0.1 micron and can also tolerate a wide range of pH. Ceramic membranes are also available, which extend the scope of pH ranges considerably into acidic/alkaline regions.

Advanced oxidation processes provide a solution for such organic bearing wastes. One such process is photo-fenton reaction which involves irradiation of the reaction solution in presence of hydrogen peroxide and iron salt catalyst using ultraviolet/visible light as per Eq. (8.3).

$$Fe(II) + H_2O_2 \rightarrow Fe(III) + OH + OH^- \qquad (8.3)$$

The hydroxyl radicals oxidize the organic compounds. Many organic substances have been reported to undergo mineralization by this reaction. This destruction is essential as presence of these organic compounds interferes in further processing by ion exchange and chemical treatments. Ozone oxidation reactions can be used to decompose dissolved organics and render the alkaline ILW suitable for further treatment by ion exchange. Another important advantage of these processes is that they do not introduce any new hazardous substances in the process [21].

8.6 Management of Organic Liquid Waste

Various organic materials are used in the nuclear industry, such as specific organic reagents in separation technologies which are the mainstay of nuclear fuel cycle. Due to repeated use in the prevailing radiation environment for a prolonged time, these organic compounds degrade and lose their desirable properties and are to be managed appropriately. Organic wastes deserve special attention on account of their nature. The organic wastes are generally not amenable to treatment and immobilization processes developed for aqueous and inorganic waste streams.

8.6.1 Types of Organic Wastes

Ion exchange resins: Ion exchange resins are used to remove contaminants from water in water-purification circuits of nuclear reactors, reprocessing plants and effluent treatment plants prior to discharge of effluents to the environment. Typically, a 220 MWe nuclear power plant generates annually 20 cubic metre of spent ion exchange resins which are normally polystyrene resin with divinyl benzene cross-linkages.

Lubricants: Lubricants include both oils and greases used in water circulators in power reactors, and in smaller applications such as vacuum pumps and experimental facilities. Radioactive oil waste produced in nuclear power plants, reprocessing and waste management plants consists of lubricating oils from primary heat transport pumps, hydraulic fluids from fuelling machines and turbine oils. These are normally categorized as low-level wastes.

Organic solvents: A large variety of organic solvents are used in nuclear research centres and medical establishments. The solvent extraction systems employed in fuel reprocessing generate organic radioactive liquid wastes. Solvents include tributyl phosphate (TBP) and diluent, usually a saturated hydrocarbon such as dodecane or a mixture of paraffins. With use and exposure to prevailing radiations, solvents, their modifiers and diluents get degraded by hydrolysis and radiolysis, and the performance is significantly reduced. These are then declared as spent solvents and become radioactive organic waste. Such waste also contains heavy metals and fission products.

Scintillation liquids: Liquid scintillators are used in routine radiochemical analysis of radionuclides with low energy emissions. The scintillators typically consist of mixtures of an organic solvent such as dioxane, toluene or xylene with the scintillating compound such as PPO(2-phenyl dioxazole). After use, these are stored as waste and require treatment before disposal.

Aqueous streams containing organics: At times, radioactive aqueous waste streams containing organics are generated. These streams require treatment to remove radioactivity as well as destruction of organic molecules to satisfy COD limiting criteria for discharge to environment. These streams include decontamination liquids, solvent wash solutions, detergent wastes generated during washing of used personal clothing and oxalic acid solution generated in the reconversion cycle of PUREX flow sheet.

8.6.2 Organic Waste Treatment Processes

Thermal treatment by way of incineration or pyrolysis to treat organics has been industrially practiced for many decades. However, in view of the emission norms getting more and more stringent, other advanced treatment options have also been developed. Often, a combination of processing techniques could lead to a more optimum solution. Whatever be the adopted treatment process, radioactivity associated with the waste stream does get consolidated in one or the other streams and is required to be immobilized into stable matrix. Some of the organic wastes, which emanate in small amount, are many times directly immobilized in appropriate matrix. Various processes for management of some of the principal organic wastes are discussed in following paragraph:

Incineration process: Incineration process yields high volume reduction factor. The products of complete combustion are the oxides of elemental constituents, carbon dioxide and water. Incinerator for solid organic waste can be classified as excess air or starved air (pyrolysis) types [22]. Incinerators for liquid waste are generally excess air type. Incineration is being practiced for complete destruction of diluent recovered from alkaline hydrolysis of TBP and after successive reuse in reprocessing plants.

Wet oxidation process: Wet oxidation technique uses soluble salts of redox-sensitive elements with hydrogen peroxide or air/oxygen to initiate the chain reaction for oxidation of organic materials in the presence of catalyst at 100 °C. A plant was constructed and operated in Winfrith, UK based on this process [23]. In India, R&D has been focussed on wet oxidation, photo-oxidation and wet air oxidation which

Fig. 8.4 Process block diagram of wet oxidation of organic waste

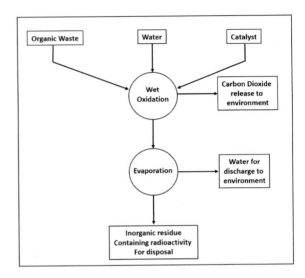

helped in conversion of organic waste into inorganic form. Wet air oxidation of divinyl benzene polystyrene resin by 50% H_2O_2 in presence of copper sulphate catalyst at 50–60 °C was demonstrated and a flow sheet was developed (Fig. 8.4).

Electrochemical oxidation process: This process is based on the creation of highly reactive oxidant ions such as Ce(IV) or Ag(II) and their subsequent use to oxidize the organic bonds compounds. The process operates at room temperature and pressure. The feed stock is fed to the reaction vessel containing a solution of silver nitrate and nitric acid as anolyte of an electrochemical cell [24]. The catholyte is nitric acid and the compartments are separated by a semi-permeable membrane which allows selective migration of ions but prevents mixing of electrolytes. The reactive Ag(II) species formed electrochemically reacts with organic waste and destroys it and, in the process, gets reduced to Ag(I). The Ag(II) is regenerated electrochemically. The advantage of electrochemical oxidation process is that it can treat a range of liquid organic wastes, solid as well as liquid wastes with very high content of organic material (as much as 100%).

Acid digestion process: The process uses a mixture of nitric and sulphuric acids at 250 °C. Materials of construction with high corrosion resistance are essential for the equipment used in this process. Extensive off-gas scrubbing is also required since sulphur dioxide and nitrogen dioxides are formed during the process. Organic solvents such as hexone and TBP can be efficiently treated in this process. This technology was pioneered in Belgium and operated in large-scale plants in Germany and USA [24]. The process has technologically matured with TBP, but not successfully demonstrated with other organic liquids.

Phase separation process by adduct formation: Phase separation by adduct formation can be considered as a pre-treatment step prior to destruction or disposal of solvents like TBP-diluent mixtures. This is normally done by mixing spent solvent with concentrated phosphoric acid. TBP is solubilized quantitatively in phosphoric acid and forms TBP-phosphoric acid polar adduct $3TBP \cdot H_3PO_4 \cdot 6H_2O$ and $TBP \cdot 4H_3PO_4$. Most of the radioactivity and degradation products get transferred to adduct phase, which can be either incinerated or recycled. The adduct can be split by dilution with water.

Alkaline hydrolysis process: Alkaline hydrolysis is a process to treat used TBP-diluent mixture waste generated in fuel reprocessing. TBP-diluent mixture is first washed with sodium carbonate solution as a pre-treatment step. This removes heavy metal, which may cause precipitation and emulsion during hydrolysis. Hydrolysis is conducted by heating the mixture of TBP-diluent with 50% sodium hydroxide solution to 125–130 °C [25].

$$(C_4H_9O)3PO(TBP) + NaOH \rightarrow (C_4H_9O)2POONa + C_4H_9OH \tag{8.4}$$

The principal reaction product is dibutyl phosphate (DBP) with small quantity of mono-butyl phosphate (MBP) and inorganic phosphates. The resultant aqueous waste contains reaction products and bulk of radioactivity associated with the waste. This is immobilized in cement matrix. The recovered hydrocarbon diluent contains very low level of radioactivity and is suitable for reuse. The flowsheet of alkaline hydrolysis process is shown in Fig. 8.5.

Distillation process: Distillation process was adopted in USA and Brazil for managing contaminated scintillation fluids. It is carried out in conventional distillation tower which is an established technology. The active residue could be either immobilized or destroyed by incineration. Vacuum distillation process based on short path distillation followed by fractional distillation has been demonstrated in engineering scale [26]. Vacuum distillation can be used as pre-treatment step for separation of large volume of diluent and modifier. The stream containing organic extractant with concentrated activity can be taken to solvent clean-up or destruction. This waste stream can be either immobilized in an inert matrix or can be incinerated in alpha contained specially designed incineration system.

Cementation Process: Immobilization in cement matrix is a process for solidification of organic wastes like spent ion exchange resin [22]. Compatibility of Ordinary Portland Cement (OPC) with spent IX resins is an issue in view of swelling pressures of wet IX resins which lead to cracking of the cement waste form. This has been circumvented by adopting a slag-based cement formulation with appropriate admixtures. In addition to the matrix formulation, technology development of 'cement in drum' process can be carried out to eliminate the requirement of dry cement addition in radioactive operating area.

Advanced oxidation process: Advanced oxidation process is a class of waste treatment method that includes the use of ultraviolet light and an oxidant, e.g. hydrogen peroxide or ozone, to destroy organic materials producing carbon dioxide and water. Sometimes, catalyst is also used in combination. The potential streams for this treatment are DBP containing aqueous stream after solvent clean-up; organic solvents like D2EHPA, TEHDGA; stream containing DTPA and lactic acid; oxalic

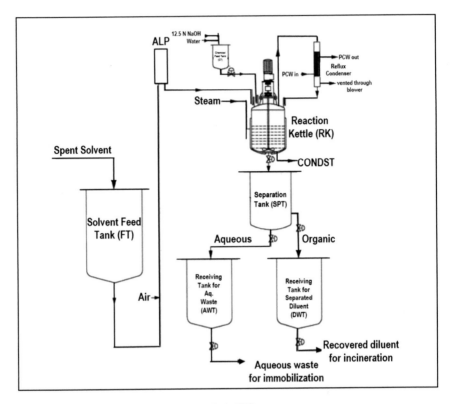

Fig. 8.5 Process schematic of alkaline hydrolysis [25]

acid after precipitation of heavy metal; and laundry effluents containing soap and detergent.

Pyrolysis Process: Pyrolysis is thermal decomposition process for organic waste in an inert or oxygen deficient atmosphere. Incineration is used for LLW of non-alpha type, and pyrolysis is considered for ILW as well as alpha-contaminated waste. The operating temperature is 500–550 °C which is lower than incineration temperature. At this temperature and reduced oxygen level, volatile radioactive components, such as ruthenium and caesium, are largely retained in the pyrolysis reactor. Organic matters convert to syngas which subsequently gets oxidized to carbon dioxide and water vapour. Off-gas passes through a system of high temperature filters, scrubber and finally HEPA filters before discharge to environment. This process is suitable for treatment of radioactive solid wastes like cellulose, plastic and rubbers. This process can also be adopted as pre-treatment stage for spent ion exchange resin.

8.7 Recovery of Useful Radionuclides from Waste and Their Utilization

With the development of novel extractants and advanced separation processes, it is now possible to achieve selective separation of radionuclides from radioactive waste and subsequently use them in various medical and industrial applications (Due to off-reactor cooling of spent fuel for 1–3 years, short-lived radionuclides produced in fission are not present in HLLW).

8.7.1 HLLW: Resource Material for Societal Benefit

The inventory of any radioisotope in HLLW can arrive from the knowledge of fission yield, burnup, cooling period and the volume of the generated waste. For example, around 240 g of ^{137}Cs (fission yield: 6%) is present in one tonne of fuel irradiated in PHWR to a burnup 6700 MWD/tonne and cooled for 5 years. This corresponds to about 20,000 Ci of ^{137}Cs which is useful in radiation technology applications. Other useful radioisotopes present in the HLLW and their potential areas of application are given in Table 8.4. In view of the application potential of each of the useful radioisotope and its inventory, HLLW can be considered as an important resource material of societal significance in terms of benefits towards serving the mankind.

8.7.2 Steps Involved in Recovery of Useful Radionuclides

Towards utilization of a radioisotope, first and foremost step is its separation from HLLW in radiochemically pure form. The purity requirement, though application-specific, should be in conformity with the stringent regulatory limits for impurities. Further, excluding the therapeutic applications, the radioisotope is to be converted in

Table 8.4 A short list of valuable radionuclides present in HLLW

Radioisotopes	Half life	Radiation type	Energy (MeV)	Major area of application
^{137}Cs	30 y	Gamma	0.66	Blood irradiation Food irradiation
^{90}Sr/^{90}Y	28 y	Beta	0.5 and 2.7	Power source
^{90}Y	64 hr	Beta	2.70	Bone pain palliation Radio-pharmaceutical
^{106}Ru	365 d	Beta	3.54	Treatment of eye cancer (brachytherapy)
^{238}Pu	88 y	Alpha	5.59	Power source
^{241}Am	433 y	Alpha	5.48	Power source

non-leachable form, so that release of radioactivity does not occur during its lifetime use. For therapeutic applications, the recovered radioisotope is tagged with suitable carrier molecule before applications.

Details of processes for the recovery of three important radioisotopes, viz. ^{137}Cs, ^{90}Sr and ^{106}Ru, from HLLW followed by their conversion in suitable form for end use applications are discussed sharing international as well as Indian experience in this field. Initially, chemical precipitation-based processes have been used for the large-scale separation of ^{137}Cs from highly radioactive acidic solutions. Later, improved separation techniques like ion exchange and solvent extraction-based processes have evolved.

In chemical precipitation-based processes, separation of the traces of radioelements from large volumes of waste solution is accomplished by formation of an insoluble compound, which entraps the element of interest by the mechanisms like (i) ion exchange, (ii) co-precipitation and (iii) sorption. Regarding recovery of ^{137}Cs, a process involving precipitation of ^{137}Cs by zinc hexacyanoferrate was adopted at US DOE laboratory in 1959 for the large-scale recovery of ^{137}Cs from PUREX raffinate at pH ~ 2 [27]. Separation of Cs$^+$ by the precipitate occurs by ion exchange mechanism with potassium as a carrier for ^{137}Cs:

$$K_4Fe(CN)_6 + ZnSO_4 \leftrightarrow K_2ZnFe(CN)_6 \downarrow + K_2SO_4 \tag{8.5}$$

After separation of the precipitate from the bulk solution, it is washed with water and then hydrolysed using steam at 250–300 °C. The leached Cs solution is processed further to obtain ^{137}Cs product in the form of CsCl powder [28].

Other precipitation-based processes such as use of phospho-tungstic acid, $H_3[PW_{12}O_{40}]$, for selective precipitation of ^{137}Cs from acidic raffinate were adopted at US DOE laboratory. In acidic solution, the replacement of one or more hydrogen of $H_3[PW_{12}O_{40}]$ by Cs$^+$ leads to the formation of an insoluble precipitate. The precipitate after separation from bulk liquid was dissolved in NaOH and processed further for the recovery of ^{137}Cs.

The use of precipitation-based processes has been discontinued in recent times due to their inherent disadvantages like bulkiness of process pot, higher shielding requirements and complexity in use of filtration/centrifugation in high radiation fields. Ion exchange-based process is a better option mainly due to ease of solid–liquid separation and better separation performance. An ion exchange-based process can be represented as

$$R - O^-X^+(Solid) + Cs^+(Sol) = R - O^-Cs^+(Solid) + X^+(Sol) \tag{8.6}$$

where R is the polymer matrix, X$^+$ is the exchangeable ion and Cs$^+$ is the ion of interest to be separated from the waste solution.

This mass transfer technique is simpler to operate imparting better separation. The efficient separation of the ion of interest, or the performance of an ion exchanger, will depend on various factors such as nature of ion exchanger and functional group, nature and composition of the waste including concentration of the metal ion as well as competing ions in the solution. HLLW has low concentration of Cs^+ and significantly higher concentrations of competing Na^+ ions. Hence, ion exchangers possessing very high affinity to Cs^+ are selected. Also, efficient elution of Cs^+ is desirable when recovery of Cs^+ is the main objective.

The first successful application based on the fixed bed use of ion exchange column was reported from Russia for recovery of mega curies of ^{137}Cs from acidic solution [29]. In this process, ^{137}Cs in acidic waste stream was loaded on fixed bed column of copper hexa cyano ferrate on silica support and eluted by using 8.0 M HNO_3. The sorbent was then regenerated by passing a reducing mixture of $NaNO_2$ and KNO_3. During industrial-scale use of the process at Mayak (Russia), about 7 MCi of ^{137}Cs was recovered using a 120 L column in sorption–desorption-regeneration cycles, for loading of ^{137}Cs from alkaline solution.

Solvent-extraction processes, based on a variety of extractants, have been evaluated in recent years for the separation of ^{137}Cs from acidic waste streams. Among others, two extractants, viz. (i) chlorinated cobalt dicarbollide (CCD) and (ii) calix-crown, are widely popular for their high selectivity for extraction of Cs from acidic solution. The process based on use of CCD in polar diluent (metanitrobenzo trifluo-ride) was developed at the Nuclear Research Institute in Czechoslovakia and success-fully adopted in the commercial separation plant at Mayak PA in Russia [30]. The loaded Cs on CCD is stripped using 5–6 M HNO_3. A total of about 15 MCi of ^{137}Cs–^{90}Sr were separated from 400 m^3 of HLLW, using CCD along with polyethylene glycol.

The discovery of crown ethers by Pederson attracted researchers' interest in devel-oping suitable extractants for selective separation of alkali and alkaline earth metal, notably caesium and strontium. Various crown ethers have been identified as effec-tive extractants for separation of caesium and strontium [31–33]. A process (CSEX) based on the use of di-benzo-18-crown-6 has been developed at ANL and tested with simulated waste along with SREX (Sr extraction by using di-t-butylcyclohexano-18-crown-6) in an iso-paraffinic hydrocarbon diluent with TBP solvent phase modifier. The extraction mechanism for Cs with dibenzo 21-Crown-7 can be represented as follows:

$$Cs^+(aq) + NO_3^-(aq) + HNO_3 + 21C7 = Cs21C7 \cdot NO_3 \cdot HNO_3(org) \quad (8.7)$$

$$Cs^+(aq) + NO_3^-(aq) + HNO_3 + 21C7 = Cs21C7^+ \cdot HNO_3(org) + NO_3^-(org) \quad (8.8)$$

The development of calixarenes by Izatt in mid-1980 heralded the next generation extractant for selective separation of Cs from acidic as well as basic solutions [34]. In this new class, several compounds have been evaluated as superior extractants for Cs separation. The common feature of this class of compound is that the base molecule, calix[4]arene, is anchored with one or two crown ether, specifically 18-Crown-6, moiety in 1,3-alternate conformation. In addition, several alkyl and aromatic groups are incorporated to enhance cation-binding ability of the compound. The attractive feature of the process is that Cs can easily be stripped from organic phase using dilute HNO$_3$ [35–38].

The first successful demonstration on Cs separation using bis (t-octylbenzo-18crown-6) was done in a small-scale centrifugal contactor using actual waste at Savannah River site. It has also been demonstrated at the Oak Ridge National Laboratory that the calix[4]arene-crown ethers are highly effective for selective separation of Cs from acidic solution.

Large amounts (more than 8 MCi) of Stronium-90 were separated by ORNL and Hanford Site of US DOE using precipitation and solvent extraction process for making Sr-90 fuelled thermoelectric generator for SNAP-7. In the StRontium EXtraction (SREX) process, solvent extraction using crown ether (di-t-butylcyclohexano-18-crown-6) in *n*-octanol medium has been employed for the extraction of Sr.

8.7.3   Recovery of Cs from HLLW for Cs-Glass Pencil

A typical multi-step process flow sheet for recovery of valuable radionuclides from HLLW is shown in Fig. 8.6.

After removal/recovery of residual uranium and plutonium from HLLW using 30% TBP in n-dodecane, the HLLW is subjected to recovery of Cs. Cs-selective extractant calix-crown-6 in IDA-dodecane solution is used for selective separation of Cs. Dilute nitric acid is used for the stripping of Cs from loaded organic phase.

The Cs-rich product stream is further concentrated and is used for making of Cs-glass pencils for gamma irradiation applications. A thumbnail view of Cs pencil-making process is shown in Fig. 8.7. Concentrated Cs solution is vitrified in an induction-heated metallic melter (IHMM) to produce Cs-rich glass. Special glass matrix has been developed and deployed for making Cs-glass accommodating up

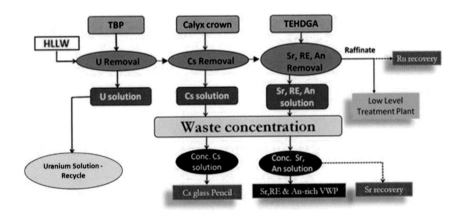

Fig. 8.6 Process flow sheet for recovery of valuable radionuclides from HLLW

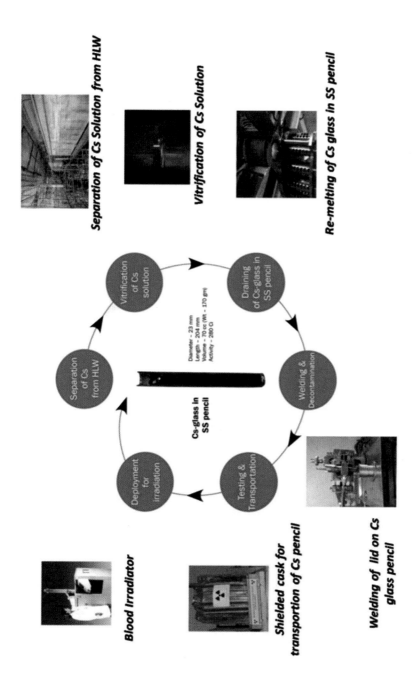

Fig. 8.7 Process steps for making of Cs-glass pencil for blood irradiation

to 5 Ci/gm. [55] The sealed pencil is subjected to leak tests and externally decontaminated using ultrasonics. The Cs-glass pencils, containing Cs with high specific activity, have potential to be used in various other gamma irradiation applications such as food irradiation and gamma chamber.

8.7.4 Recovery of $^{90}Sr/^{90}Y$

Though ^{90}Sr has several applications, the major demand is due to its daughter ^{90}Y. It is used as radiopharmaceutical in treatment of liver cancer, neuroendocrine tumours, etc. ^{90}Sr–^{90}Y are in secular equilibrium, and half-life of the daughter is only 64 h. Therefore, separation of ^{90}Sr is to be carried out first. Further challenges lie to meet stringent radiopharmaceutical purity requirement of ^{90}Y for use as nuclear medicine. Recovery of ^{90}Sr from HLLW is carried out by solvent extraction process. The raffinate stream of second cycle (obtained after separation of Cs) of solvent extraction process is subjected to co-extraction of Sr-actinides-lanthanides using extractant Tetra Ethyl Hexyl Di-Glyco Amide (TEHDGA) in Isodecyl alcohol-Dodecane solution. The stripping of the radionuclides from loaded TEHDGA phase is carried out using dilute HNO$_3$. The product, Sr–An–Ln rich, is further concentrated and used for recovery of bulk amount of Sr. Multi-step separation processes are deployed involving ion exchange process, membrane process, chemical precipitation, etc., to recover purified form of Sr avoiding contamination of other radionuclide as per desired product quality. The recovered Sr product is treated in two-step supported liquid membrane (SLM) system for milking out Yttrium-90 (^{90}Y).

The SLM-based milking process for generation of pure ^{90}Y is carried out using a three-chamber glass assembly cell as presented in Fig. 8.8 [39]. The mixture of ^{90}Sr and ^{90}Y, adjusted to pH 1–2, is placed in the first chamber (feed chamber). The second chamber, also called intermediate compartment, contains 4 M HNO$_3$, and the third chamber (receiver chamber) contains 1 M CH$_3$COOH. Between the feed and intermediate chamber, 2-ethylhexyl phosphonic acid (KSM-17)-based SLM has been inserted, whereas CMPO-based SLM is placed between intermediate and receiver chamber. Under these conditions, selective transport of ^{90}Y from feed chamber to aqueous phase containing acetic acid via 4 M HNO$_3$ is feasible.

8.7.5 Separation of ^{106}Ru

^{106}Ru is one of the important fission products produced in thermal fission of uranium. It is a low energy beta emitter. However, its daughter, ^{106}Rh, emits high energy beta radiation, which is especially useful for brachytherapy applications, particularly for the treatment of cancer of eyes.

Separation of ^{106}Ru is carried out from raffinate obtained after three cycle solvent extraction process as discussed previously. The process utilizes oxidation of Ru

Fig. 8.8 Schematic diagram of ^{90}Sr–^{90}Y supported liquid membrane generator [39]

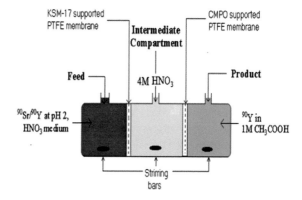

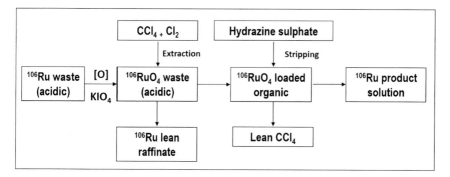

Fig. 8.9 Flowchart for separation and purification of ^{106}Ru from HLLW [40]

nitrosyl nitrate to RuO_4 by addition of KIO_4 followed by extraction of RuO_4 in chlorinated CCl_4. Finally, the extracted Ru is stripped from CCl_4 using acidic hydrazine solution [40]. A schematic of the process flow sheet is presented in Fig. 8.9. The purified ^{106}Ru solution is used electro-deposit ^{106}Ru on a silver substrate followed by preparation of sealed source in form of plaque. ^{106}Ru bearing plaques are in use at different hospitals IN India for treatment of cancer of the eye.

In summary, utilization of the radioisotopes separated from HLLW clearly validates that it is a valuable source of wealth for societal benefits. There are many more valuable radioisotopes, and some non-radioactive isotopes present in HLLW such as Platinum Group Metals (PGM).

8.8 Actinide Partitioning

The objective of closed fuel cycle is to enhance resource utilization by reprocessing, recovering and recycling the remaining Pu and U in the spent fuel and render the waste suitable for conditioning for interim storage and long-term disposal in geological repository. Twin objectives of the actinide partitioning of HLLW are (a) to reduce the radiotoxicity and, thereby, the potential impact of disposal of radioactive wastes on the environment, and (b) to optimize the cost of the disposal in the geological disposal facility by reducing the volume and also heat load of the disposed wastes [41, 42]. The HLLW contains minor actinides (Np and transplutonium elements, which are not extracted by TBP) that have long half-lives and high radiotoxicity, and therefore, separation of minor actinides (MA) from HLLW is an important step.

During the initial 100 years after discharge of the spent fuel from a reactor, major thermal load of HLW is on account of ^{90}Sr and ^{137}Cs. This restricts the geological disposal facility design conditions, and consequently, removal of these heat-generating nuclides from the HLLW permits the effective utilization of geological repository.

8.8.1 Process Development for Minor Actinides (MA) Separation

The separation of the trivalent MA and lanthanide elements is a challenging task. Since TBP is not suitable for extracting the trivalent MA ions, a range of candidate ligands or molecules-based around oxygen and nitrogen donor atoms have been developed. Among the various parameters to consider, while developing a new separation system for An(III)/Ln(III) separation, the most important ones are the following [43–46]:

- High separation factor or selectivity towards the desired MA(III) versus the metal ions and nitric acid present in the feed to ensure high decontamination.
- High radiolysis and hydrolysis stability of the organic extractant, the diluent and the phase modifier.
- The secondary waste generation from the process should be as low as possible.

8.8.2 Approach to MA Partitioning

The actinide partitioning process is based on a multi-step solvent extraction process using individual solvents to perform specific tasks. Removal of residual uranium and plutonium from HLLW is performed using PUREX solvent. Since the amount of MA(III) is much smaller than that of the fission products contained in HLLW, the selective extraction of MA(III) directly from HLLW is practically difficult. Most of the processes described in the literature follow a two-step approach for separation of MA(III). In the first step, bulk separation of trivalent MA and lanthanides from HLLW is carried out by solvent extraction process taking advantage of similarity of chemistry of these two groups. This separation is performed using hard donor ligands like CMPO (the TRUEX process) or malonamides (the DIAMEX process) or TODGA/TEHDGA (the DGA process). In the second step, MA(III) are separated from an MA(III) + Ln(III) product solution either by preferential complexation in aqueous phase or by selective extraction in organic phase [46–50].

8.8.3 Bulk Separation of MA Along with Lanthanides

Several processes based on range of candidate ligands/molecules around oxygen donor atoms have been developed to perform this step [43, 46, 50]. The US researchers used CMPO (n-octyl(phenyl)-N, N-diisobutyl carbamoyl methyl phosphineoxide) in the TRUEX process to separate the actinide along with the lanthanides from acidic aqueous waste streams. However, CMPO-metal complex has limited solubility in n-dodecane and hence TBP as phase modifier has been used to mitigate third-phase formation at higher waste loadings. Use of ~1 M TBP leads to a large uptake of nitric acid from HLLW feed solution. This extracted acid hinders stripping operations. Moreover, the degradation products of both CMPO and TBP, i.e. phosphinic acid and DBP, respectively, are known to take up actinides irreversibly. Use of complexant-based stripping is required to mitigate this problem.

In recent years, trend is to use molecules that fulfil the CHON principle, whereby the molecules contain only carbon, hydrogen, oxygen and nitrogen atoms. This approach is beneficial as it permits direct incineration of used process solvent. An example is use of malonamide-based molecule in the DIAMEX process of France that co-extracts An and Ln ions [43, 46, 50].

More recently, DiGlycolAmide (DGA) molecules have been developed for the co-extraction of trivalent actinides and lanthanides, which demonstrated significant promise for use in MA separation processes. The researchers in Japan and France have used TODGA as their reference molecules [43, 46, 50]. In India, TEHDGA has been selected as ligand molecule for the development and plant scale deployment of partitioning processes [53, 54]. However, to utilize the DGAs in practical solvent extraction processes, phase modifiers are required to be used along with the diluents to achieve higher metal loading without third-phase formation. Structures of various candidate molecules for bulk separation of MA along with lanthanides are shown in Fig. 8.10.

8.8.4 Group Separation of MA from Lanthanides

On account of the chemical similarity of MA(III) and Ln(III), this separation is practically not feasible with common extracting ligands which coordinate via relatively hard oxygen atoms. Most of the solvent extraction methods for their separation rely on the slightly greater covalency of MA(III) bonding interactions and, hence, slightly greater strength of the interaction that trivalent minor actinides exhibit with softer atoms (e.g. as a donor atom in a ligand (N, S) [44, 45, 47, 48]. Separation of trivalent

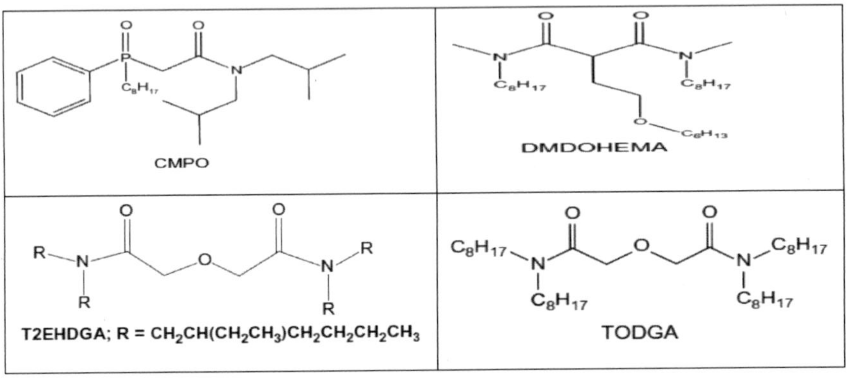

Fig. 8.10 Structures of various candidate molecules for bulk separation of MA with lanthanides

minor actinides from lanthanides is based on the manipulation of this chemistry, using soft-donor extractant molecules (actinides are selectively extracted into the organic phase) or soft-donor complexants (actinides are preferentially retained in the aqueous solution). While most processes are applicable to aqueous solutions in the pH range of about 2.5–3.5, some are applicable at acidity in the range of 0.5–1.0 M.

One of the extensively tested approaches of employing soft-donor aqueous complexants is the TALSPEAK process [48, 49]. This process is based on the partitioning of lanthanides and minor actinides between an organophosphorus extractant and an aqueous phase containing a high concentration of a carboxylic acid buffer and a polyaminopolycarboxylate complexant. The separation is achieved by the balance between the relative affinity of cation-exchanging solvent extraction reagents and comparably powerful aqueous complexants for the two classes of metal ions. The latter reagent is principally responsible for holding back the trivalent minor actinides, allowing the selective transfer of the lanthanides into the organic phase. The most thoroughly studied TALSPEAK separations are based on bis-(2-ethylhexyl) phosphoric acid (HDEHP, organic extractant), diethylenetriamine pentaacetic acid (DTPA, holdback reagent) and lactic acid (carboxylic acid buffer).

In the SANEX process and its variants of the European advanced fuel cycle research programme, separation has been approached primarily through the design of soft-donor extractant molecules, of which the bistriazinyl-pyridine and bistriazinyl-bipyridine (BTP and BTBP) derivatives have undergone the most extensive development [46, 47, 51, 52]. The tridentate BTPs and the tetradentate BTBPs belong to the most promising nitrogen donor extracting agents for the MA(III)–Ln(III) separation as they are able to directly extract MA(III) from solutions containing up to molar concentrations of nitric acid with separation factors for Am(III) over Eu(III) between 100 and 300. Structures of various candidate molecules for group separation of MA from lanthanides are shown in Fig. 8.11.

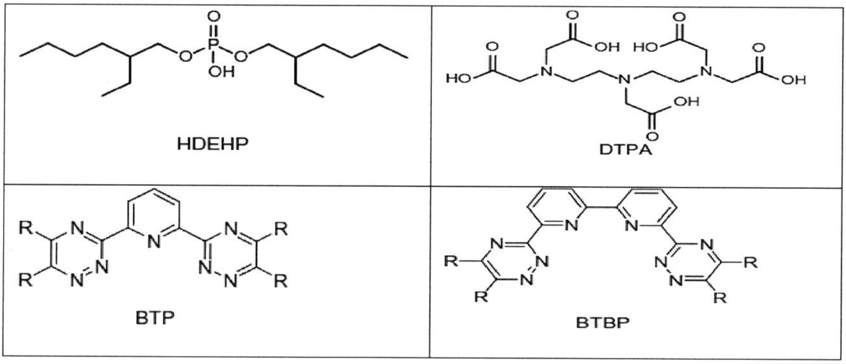

Fig. 8.11 Structure of molecules developed for group separation of MA from lanthanides

8.9 Vitrification of High-Level Liquid Waste

HLLW is generally concentrated by evaporation and stored as an aqueous acidic solution in high integrity stainless-steel tanks. The storage tanks are equipped with cooling arrangements (coils and/or an outer jacket) to remove decay heat. The tanks are provided with agitation system to keep the solids, if any, in suspension. In most storage systems, multiple containment is provided to guard against the escape of radioactivity.

8.9.1 Conditioning of HLLW—A Necessary Processing Step

Conditioning of HLLW is an important step to minimize the migration of radionuclide into human environment in the longer time span. HLLW must be immobilized to solidified/conditioned waste form so that its interim and long-term storage followed by its ultimate disposal is technologically feasible, economical and environmentally safe. The conditioned waste product should possess the desirable properties, viz. chemical durability, radiation stability, good thermal conductivity, high waste loading, high volume reduction and amenability for cost-effective production process.

8.9.2 Glass—A Desired Matrix for Conditioning

One of the major reasons for selection of 'glass' as a matrix is its stability over a long period of time [56]. Other favourable features of glass are as follows: it can accommodate a large variety of radionuclides and other elements present in the waste due to its random three-dimensional polymeric structure; it has excellent chemical durability, thermal and radiation stability as well as thermal conductivity; and it provides high volume reduction factor. In addition to the above, the technologies based on glass matrix can meet demands of the vitrification process which are amenability for remotely controlled processing in shielded hot cells; ability to retain to some extent the problematic components (sulphates, phosphates, etc.) present in the waste which are likely to create phase separation or release due to volatilization (radionuclides of Ru, Cs, etc.); and low secondary waste generation.

8.9.3 Matrix Design Criteria

This aspect will be illustrated by taking example of India where borosilicate glass matrix has been adopted for vitrification of HLLW. Initially, glass formulations were developed based on glass-forming regions of a three-component phase diagram involving SiO_2, B_2O_3 and Na_2O. Suitable modifications have been made in order to accommodate compositional changes in the waste to accommodate chemical species like sulphate, sodium, aluminium, thorium and fluoride. Waste loading, glass additives and the processing temperature are the essential parameters to be considered for development of suitable glass formulation. SiO_2 and B_2O_3 are the basic glass-forming oxides while Na_2O, Cs_2O, SrO, BaO and MnO_2 are glass modifiers. The glass formulations are designed to obtain maximum waste loading considering processing temperature limitations and desired product characteristics. The waste loading is constrained either by solubility limit of any waste component in the matrix or limit imposed due to decay heat associated with heat-generating radionuclides of HLLW.

8.9.4 Vitrification Process

HLLW is converted into inert glass matrix using the vitrification process. Vitrification process involves feeding of metered quantity of pre-concentrated waste along with glass-forming additives in the form of slurry or glass frit into specially designed melter equipment [57, 58]. It consists of major six processing steps, namely evaporation of volatile components like nitric acid and water, drying of solids, calcination of nitrate salts to respective oxides, fusion of oxides to make glass, soaking of glass to homogenize the product and its casting into the canister. Each processing step occurs in a definite temperature range. The details of steps are as follows:

Evaporation: In evaporation step, the liquid waste is heated to its saturation temperature. Majority of water and nitric acid are converted into vapour phase and discharged along with the off-gas stream. The salts of active and inactive elements remain inside the evaporation vessel. Typical temperature range of evaporation process is 100–150 °C.

Drying: Further heating evaporates the bound moisture from the salts and dries the salts which are heated to raise the temperature. The process usually occurs at temperature range 150–250 °C.

Calcination: Most of salts present in HLLW are in nitrate form. These salts thermally decompose to respective oxides at their decomposition temperature. Generally, calcination process starts after 250 °C and all nitrate salts are converted into their respective oxides up to 800 °C.

Fusion: On further raising of the temperature, oxides of different elements get fused along with oxides of glass-making additives to form a glassy mass. Fusion of oxides to make glass takes place between 800 and 900 °C for salts normally present in HLLW.

Homogenization: The vitrified mass is kept at appropriate temperature for about 4–6 h to reduce the viscosity of vitrified mass below 50 poise to ensure good homogeneity.

Casting of vitrified mass: After adequate soaking of the vitrified mass, it is drained into the stainless-steel container called 'canister'.

8.9.5 Energy Requirement

Vitrification is an energy-intensive process and requires substantial amount of energy to convert HLLW into vitreous waste product (VWP). All process steps involved in vitrification are endothermic and require heat/energy. Major proportion of energy, about 60%, is utilized during evaporation step in the form of latent heat of vaporization. Around 1000–1500 kcal energy is required to convert a litre of HLLW into vitreous mass based on factors such as salt concentration and acidity of HLLW. Normally, energy requirement increases with increase in salt concentration.

8.9.6 Single-Step Vitrification Process

In this process, both streams, i.e. pre-concentrated HLLW and glass additives, are directly added to the melter. All the steps of vitrification process are carried out in single equipment, viz. melter, with varying operating temperature ranging from 100 °C to melting temperature. *Pamela process*, developed by France, is an example of single-step process in which HLLW is fed directly, together or separately with the glass forms, into the vitrification melter where the process steps of evaporation, calcination and melting occur simultaneously [59–62]. The Pamela process is shown in Fig. 8.12.

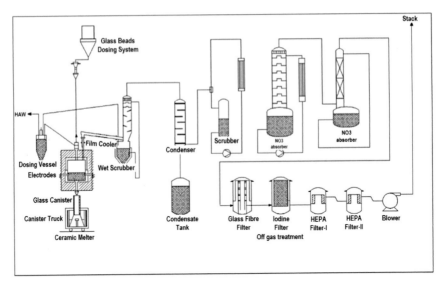

Fig. 8.12 Single-step vitrification process of Pamela plant [60]

8.9.7 Multi-step Vitrification Process

Multi-step process consists of a combination of a rotary kiln and an induction-heated metal glass-melting crucible. Evaporation, drying and calcination take place inside the rotary calciner. The calciner is a slightly inclined tubular kiln which rotates at slow speed. The tube is heated externally by an electric resistance furnace divided into four zones. The first two zones are devoted to the evaporation having a higher heating capacity than other two zones which are for drying and calcination. The temperature varies from 225 °C at the feed point to a maximum of 600 °C. The calcined products flow by gravity into a melter, which is heated by medium frequency induction to about 1150 °C, along with glass-making additives. The process has been successfully operated in France at Marcoule vitrification plant and is named *AVM* (Atelier de Vitrification de Marcoule) process which is illustrated in Fig. 8.13.

8.9.8 Product Quality Assurance in Vitrification

The quality of vitrified waste product made by the vitrification process is ensured by qualifying the process and not by sampling during radioactive operations. In particular, control of the composition and properties of the vitrified waste product is achieved by sampling and analysis of HLLW prior to feeding, control of the base glass matrix and other additives and careful control of process conditions. Research and development work, performed before hand, with inactive simulants at laboratory and

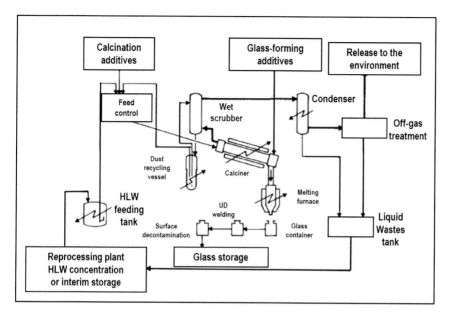

Fig. 8.13 Two-step AVM vitrification process [60]

pilot plant scale is used to finalize and standardize the parameters of feed chemistry and process operating conditions that yield desired good product quality. The rigorous approach to process control and feed analysis provides the confidence that the final product does not require routine sampling as part of the quality assurance process [61].

8.9.9   Treatment of Melter Off-Gas

Off-gases generated in the vitrification process contain water and nitric acid vapours, NOx, volatile radionuclides like Cs and Ru. Besides, it also contains leakage air since operation is carried under sub-atmospheric pressure and particulate carry-over having constituents from both waste and glass-forming additives. Extensive off-gas cleaning system is provided for minimizing the atmospheric release of radioactivity from the vitrification facility. Off-gases from melter are initially passed through a dust-scrubber wherein carried-over particulates are stripped off by direct-contact scrubbing and temperature is also lowered resulting in partial condensation of vapours. Resulting off-gases are subjected to condensation in heat exchanger wherein majority of water vapour and nitric acid vapour get condensed giving rise to acidic secondary aqueous waste stream. Off-gases then enter 'NOx scrubbers' wherein majority of NOx is removed by direct-contact scrubbing with water or alkaline medium. Off-gases are finally subjected to filtration. Filtration is usually by 'High Efficiency

Particulate Air Filters' (HEPA) which have an efficiency of $\geq 99.97\%$ for submicron particulates. Additional clean-up steps, which are usually radionuclide specific, may be introduced in between, to trap volatile components depending on the composition of the waste being handled. Treated off-gases are ultimately mixed with plant ventilation air and subjected to one more cycle of HEPA filtration before their monitored release to atmosphere through a tall stack.

8.9.10 Management of Secondary Waste

Vitrification of HLLW generates secondary liquid as well as solid wastes. The major sources of secondary liquid waste are vapour condensate collected from condenser and scrub solutions. The secondary wastes are acidic in nature and specific activity ranging from 0.1 mCi/L to few Ci/L depending upon activity of HLLW and stream of secondary waste. The radioactivity is mainly due to ^{137}Cs. However, ^{106}Ru and other radionuclides of HLLW, such as ^{90}Sr and alpha radionuclides, may also be present in traces. The secondary liquid waste is treated using multistage evaporation cycles to achieve required volume reduction as well as higher decontamination factor. The overall decontamination factor achieved after multiple evaporation cycles is in the range of 10^9, thus achieving near zero discharge level of radioactivity.

The secondary solid waste from vitrification process are mainly off-gas filters, filters from exhaust/ventilation system, mops and tools from decontamination activities, used protective wears, contaminated tools, machines, equipment and pipes. The solid wastes are segregated based on their physical and chemical characteristics, radioactivity content and nature of radionuclides present. These are properly categorized and packed for further treatment/disposal.

8.9.11 Vitrified Waste Canister Handling

After the canister is filled with vitrified waste product (VWP) during processing, it is allowed to cool to its steady state temperature before it is removed from the filling station. A lid is placed on the top opening of canister and welded remotely. The external surface of canister is decontaminated with water using ultrasonic effect and checked for surface contamination. If the contamination level of canister surface is found sufficiently low, it is remotely placed inside the shielded cask and is transported to interim storage facility.

8.10   Matrices for Vitrification

The term vitrification implies immobilizing radionuclides in a suitable glassy matrix. The selection of glass as the matrix is motivated by a variety of factors which include:

- High chemical and radiation stability: required due to the high level of radioactivity incorporated
- Compositional flexibility: ability to reliably immobilize a range of elements
- Simple and robust production technology: ideally adapted from industrial glass making
- Small volume of the waste form: high volume reduction factor in view of geological disposal requirement.

8.10.1   The Glassy State: A Brief Introduction

One of the main reasons for selection of glass matrix for immobilization of radioactive elements is its long-term stability. Naturally occurring Obsidian glasses have been dated to about 20 million years and have been used by humans as arrow heads and blades since the dawn of humanity. In the simplest terms, glass is a solid with no long-range order in atomic arrangement and exhibiting glass transition behaviour. The glass transition phenomenon is not one of simple dynamic arrest caused by increasing viscosity of the glass-forming melt. Rather, it is a complex interplay of viscosity and thermodynamics, combining the concept of dynamic arrest with the divergence of time scales in the vicinity of the glass transition temperature, T_g [63–65]. The glass transition region upon cooling from the melt is presented in Fig. 8.14.

Fig. 8.14 Enthalpy versus temperature curve for a melt upon cooling [65]. Note the discontinuous change in enthalpy (volume behave similarly), at melting point, indicating the formation of a crystalline solid, implying a first order transition. The glass forms at a comparatively rapid cooling rate by imposition of thermodynamic and viscosity constraints on the supercooled melt

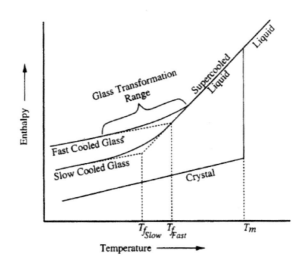

The higher the cooling rate, the higher the glass transition temperature. From a glass matrix perspective, T_g is an important parameter. Above this temperature, the glassy material behaves like a viscoelastic liquid and will deform under its own weight. Below T_g, a glass behaves like a solid that is familiar at room temperature. Another important aspect is that the dynamic arrest referred to earlier progressively relaxes as we approach T_g making movement of atoms easier, which in turn makes unintentional crystallization more likely. Therefore, it is imperative that the maximum temperature of the waste form remains well below the T_g as crystallization can deleteriously impact the properties of the waste form, particularly chemical durability.

8.10.2   Glass Formation: Structural Theories

Common oxides in glasses can be classified into:

1. Network formers: e.g. SiO_2, B_2O_3, P_2O_5 or GeO_2
2. Network modifiers: e.g. R_2O, $R^{''}O$ (where R = Alkali element; $R^{''}$ = Alkaline earth element)
3. Conditional network formers: e.g. Al_2O_3, ZnO.

8.10.3   Glass Formation: Kinetic Theories and the TTT Curve

The variation of T_g with the cooling rate (refer Fig. 8.15) indicates a kinetic factor in the formation of a glass. The estimation of the cooling rate is then predicted on the following basis:

- What is the volume fraction of crystals in a glass that can be detected and identified?

Fig. 8.15 TTT curve for cooling rates A and B, with A < B [65]

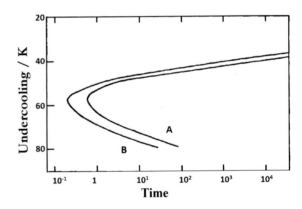

- Relating the volume fraction of crystals to the kinetic constants describing the nucleation and growth processes.
- Correlation of these kinetic constants to readily measurable experimental parameters.

Under the assumption that the crystals are uniformly distributed in the liquid, 10^{-6} is a reasonable volume fraction of crystals that can just be detected.

The cooling rate required to prevent the formation of a given fraction of crystals can be estimated by construction of a TTT curve, an example of which is presented in Fig. 8.16 for two different cooling rates. The nose in a TTT curve is the minimum time required for the crystallization of a given volume fraction. The shape of the TTT curve indicates a competition between the driving force for nucleation, which increases with decreasing temperature and increasing viscosity of the melt, which increase as the temperature decreases. TTT curves can be plotted for various crystalline fractions or different temperatures to understand ease of glass formation in a system. The cooling rate required to avoid a given fraction of crystals can be obtained from these curves.

The development of glass matrix for vitrification involves a judicious optimization of matrix properties with process stability. Therefore, glasses used for vitrification are optimized based upon the waste stream under consideration. Glasses being used for waste immobilization are either borosilicate-based glasses or phosphates [66]. Borosilicates are more ubiquitous. In Russia, aluminophosphate-based glasses are in use for immobilization since 1987. Phosphate melts are more aggressive towards

Fig. 8.16 Induction-heated metallic melter (IHMM)

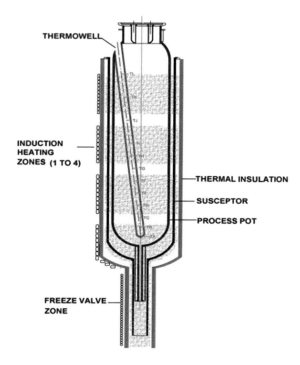

refractory linings, and this has limited their utility. However, phosphate glasses exhibit higher solubility of cations problematic to borosilicates such as Mo and elements of platinoid groups. A list of globally used glass compositions is presented in Table 8.5.

It may be noted from Table 8.5 that SiO_2 content of waste immobilization borosilicate glasses is lower than that in natural analogues such as obsidians where it can exceed 65 wt%. However, the melting point of such high SiO_2 compositions is likely to exceed 1300 °C, at which temperature, volatility of species such as Cs is significant. Additionally, the temperature range may damage the melter and its components. This illustrates the point that although a given composition may be ideal as a matrix, process conditions may not allow such a composition to be taken up on the plant scale. Typical compositions used at WIP, Trombay and Tarapur in India are collected in Table 8.6.

8.11   Melter Technologies for Vitrification

Vitrification of HLLW is a challenging technology due to high temperature corrosion by molten glass. Very high radiation field in the vitrification cell calls for demanding requirements of remote operation and maintenance leading to complexities in design and construction. To meet these challenges, various technologies have been developed in different countries and demonstrated for robust operations. Melter is the most critical equipment of vitrification process. A brief description of various melters for vitrification is as follows [67–71].

8.11.1   Induction-Heated Metallic Melter

In *pot vitrification*, HLLW and glass-forming additives are fed to a metallic vessel housed inside an electric furnace. In the early sixties, *FINGAL* (Fixation In Glass of Active Liquors) process was developed in the UK. In this process, the process vessel also served as the final container. *PIVER* (Pilote Verre) process was subsequently developed in France by incorporating a freeze valve in the bottom of the process vessel for casting the vitrified waste into a canister. In India, pot vitrification in an *Induction-Heated Metallic Melter* (IHMM) was developed initially to carry out vitrification process [5, 68].

IHMM is essentially a multizone induction-heated furnace housing two concentric metallic pots as shown in Fig. 8.16. The outer pot is used as a susceptor for induction heating and the inner one as the process pot. Both process and susceptor pots are made of high Ni–Cr alloy so as to withstand high temperature and corrosive conditions. HLLW stream fed to the process pot enters top of the working zone above the molten glass. The working zone consists of three subzones in which evaporation, calcination and glass formation take place. The molten glass after homogenization is drained

Table 8.5 Compositions of various matrices used for vitrification in various plants

(All constituents are expressed in wt%)

Plant, waste, country	SiO_2	P_2O_5	B_2O_3	Al_2O_3	CaO	MgO	Na_2O	Misc	Waste Loading
R7/T7, HLW, France	47.2	–	14.9	4.4	4.1	–	10.6	18.8	≤28
DWPF, HLW, United States	49.8	–	8.0	4.0	1.0	1.4	8.7	27.1	≤33
WVP, HLW, UK	47.2	–	16.9	4.8	–	5.3	8.4	17.4	≤25
PAMELA, HLW, Germany—Belgium	52.7	–	13.2	2.7	4.6	2.2	5.9	18.7	<30
Mayak, HLW, Russia	–	52.0	–	19.0	–	–	21.2	7.8	≤33
Radon, LILW, Russia	43	–	6.6	3.0	13.7	–	23.9	9.8	<35

Table 8.6 Glass compositions in wt% used in Tarapur and Trombay WIP

Composition	Tarapur		Trombay	
	Basic sodium borosilicate IR110	Modified sodium borosilicate IR111	Lead-based borosilicate WTR-62	Barium-based borosilicate SB-44
Glass formers (SiO_2 + B_2O_3)	46	46	50	50.5
Glass network intermediate (TiO_2)	7	7	–	–
Glass modifiers (Na_2O + MnO + PbO + BaO)	26	16	30	28.5
Waste oxide	21	31	20	21

out from the process pot through a bottom freeze valve to a stainless-steel canister, allowing reuse of the process pot.

IHMM is a very compact and simple system which can be used for a wide range of glass compositions. It has, however, low throughput on account of limited melter diameter and shorter melter life due to high temperature molten glass corrosion. The *AVM* process developed in France has two steps—calcination in a rotary calciner followed by melting in an IHMM. An ovoid metallic pot is used in their vitrification facilities at La Hague, in place of circular cross-section of AVM melter, for enhancing the melter capacity. UK has adopted this technology at its vitrification facility at the Sellafield site.

8.11.2 Joule-Heated Ceramic Melter

Work on all-electric direct-heated glass melter for vitrification was initiated by USA in 1973. In this design, Joule heat generated by passing a large alternating current through the molten glass contained in a ceramic tank is used for vitrification. In parallel, this technology was developed by INE/KfK at Karlsruhe Germany and demonstrated by radioactive solution vitrification at the Mol plant in Belgium. *Joule-Heated Ceramic Melter (JHCM)* was adopted by USA, Germany, Russia, Japan and China for the vitrification of HLLW [67, 69, 70]. Advanced vitrification systems employing JHCM are also under operation at Tarapur and Kalpakkam in India [5, 62, 68, 72].

In JHCM, the heat for vitrification is provided by passing an alternating current through the molten glass itself employing submerged, air-cooled high Ni–Cr electrodes. The molten glass is contained in a tank constructed of fused-cast alumina–zirconia–silica blocks backed by insulating layers as shown in Fig. 8.17. In a liquid-fed JHCM, the incoming waste stream undergoes evaporation, drying and calcination; resulting in a crust, known as cold cap, which floats on the pool surface. The waste feed rate is controlled to prevent the cold cap from direct contact with the hot wall. The crust gradually fuses with the glass frits to form vitrified waste product. Since glass is non-conducting when cold, auxiliary heating by radiant heaters housed in the melter-super structure is used for starting up JHCM. Pouring of the glass from the melter into the canister is done using an electrically regulated freeze valve at the melter bottom.

Availability of unrestrained heat transfer area and amenability to continuous mode of operation of JHCM facilitate larger processing capacity. Use of sufficiently thick glass corrosion-resistant refractory wall enhances the life of JHCM. Natural convection currents prevailing in the electrically conducting molten glass pool improve the product quality. By virtue of the large thermal inertia of the glass pool, JHCM can accommodate variations in the feed streams to a great extent. However, the major operating constraint for JHCM is that the metallic electrodes are ideally not

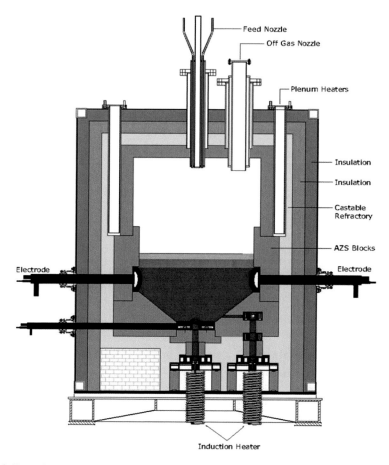

Fig. 8.17 Joule-heated ceramic melter (JHCM)

to be continuously exposed to temperatures higher than 1000 °C in order to ensure reasonably long melter life.

8.11.3 Cold Crucible Induction Melter

The globally emerging vitrification technology, based on the *Cold Crucible Induction Melter (CCIM)* (Fig. 8.18), offers several advantages such as compact size, long melter life, high temperature availability, high waste loading and high specific capacity [71, 73]. CCIM is also more tolerant to presence of noble metals than the traditional JHCM because of the heat release in the melt by direct induction. The CCIM technology can effectively be used to condition hard-to-process waste streams. It is also a promising technology for adopting glass ceramic for HLLW immobilization [74–79].

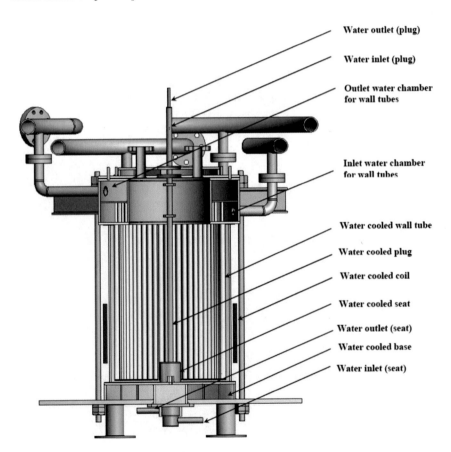

Fig. 8.18 Cold crucible induction melter

In cold crucible induction glass melting, glass is directly heated by electromagnetic induction employing a segmented crucible which is manufactured from contiguous segments forming a cylindrical volume but separated by a thin layer of electrically insulating material. An inductor surrounding the segmented crucible induces eddy currents in each segment creating an oscillating electromagnetic field, which in turn generates induction heat in the molten glass inside the crucible. Direct heating of glass facilitates heating the process material to high temperatures. In order to avoid corrosion of the metallic crucible by the molten glass, internal cooling of the segments is provided. This cooling produces a solidified glass layer, which acts as a protection against glass corrosion along the melter inner wall. High temperature availability without substantial corrosion of the melter makes the CCIM a promising technology for vitrification.

Russia employs CCIM for vitrification of LILW since 1998. CCIM was deployed in France in 2010 for vitrifying HLW from reprocessed U–Mo–Sn–Al spent fuel. Pilot-scale facilities based on CCIM technology are operational in India and USA.

8.12 Vitrification Experiences on the Industrial Scale

Several countries have been operating facilities for vitrification of HLLW. The experience of vitrification of some of these is presented here [57, 58].

8.12.1 Vitrification Experience in France

The two-stage vitrification process developed and deployed in France involves evaporation and calcination of HLLW in a rotary calciner. The calcined mass is fed to a metallic melter along with glass formers in the form of primary glass (frit) simultaneously. The melter is made of a metallic pot heated by induction. The process has been utilized at the Atelier Vitrification de Marcoule at Marcoule, France, and vitrification plants R7 and T7 located at La Hauge, France. The vitrification plants R7 and T7 at La Hague are scaled up versions of AVM with increased through put and enlarged melter capacity [57, 59].

8.12.2 Vitrification Experience in UK

In the UK, reprocessing of spent fuel from the UK's civil nuclear power stations is performed at Sellafield [61]. The resultant HLLW is converted into borosilicate glass at the Sellafield Waste Vitrification Plant (WVP). The Sellafield WVP uses a two-stage design based on the continuous French AVH (Atelier de Vitrification de la Hague). In the two-stage vitrification process used in WVP, HLLW is first mixed with sugar solution to reduce ruthenium volatilisation and enhance denitration. It is then fed to an electrically heated rotary calciner. This calcine is discharged by gravity along with a metered quantity of base glass frit directly into an induction-heated melter crucible and heated to ~1050 °C. The calcine reacts with the molten base glass and the resulting homogeneous melt is then periodically poured into product containers. Once filled, the product canisters are allowed to cool for at least 24 h before having a lid welded in place. They are then decontaminated, checked for any activity on the outer surface and then transferred to the interim storage facility.

8.12.3 Vitrification Experience in Germany/Belgium

The Federal Republic of Germany has successfully operated a Pamela process at the Belgium Eurochemic reprocessing plant [57]. The Pamela vitrification plant is a single-step process. It is based on a liquid-fed Joule-heated ceramic melter in which the high-level fission product solution is fed directly—together or separately with the glass forms—into the glass melter where the process steps of evaporation, calcination and melting occur simultaneously. These melters are constructed from refractory materials with high corrosion resistance. The power input is obtained by four pairs of Inconel-690 plate electrodes, placed in two levels of the melter pool at 1150–1200 °C.

8.12.4 Vitrification Experience in USSR

A radioactive facility at Kyshtym was put into operation during 1987 using a single stage melter system. The melter was made of a high-alumina zirconium refractory, was heated using molybdenum electrodes and used orthophosphoric acid as a fluxing agent with molasses added to reduce radionuclide entrainment to the off-gas system. The phosphate glass is poured into canisters. The pour vessels (canisters) are sealed with welded lids [60].

8.12.5 Vitrification Experience in India

The Waste Immobilization Plant at Tarapur, the first vitrification plant in India, is based on a pot glass process—a single-step process with metallic melter. Pre-concentrated waste and glass-forming additives in the form of slurry are metered as separate streams into the process vessel located in a multizone furnace. The susceptor temperature is initially maintained at 600 °C. The process vessel which is made of Inconel-690 incorporates a freeze valve section operable by an independent induction coil. With simultaneous concentration and calcination of waste, the solid–liquid interface moves vertically upwards. The level of liquid waste is indicated by the temperature sensed by thermocouples located at different heights. The feed is stopped when the vessel is about 75% full of calcine. At this stage, the furnace temperature is raised to 950 °C, the calcined mass is fused into glass and it is soaked at 950–1000 °C for six hours to achieve homogenization. The molten mass is then drained into storage canister by operating the freeze valve. The *Waste Immobilization Plant at Trombay* is similar as WIP, Tarapur, but with improvised design feature based on operational feedback of WIP, Tarapur. The *Advanced Vitrification System (AVS), Tarapur*, is the first vitrification facility based on ceramic melter in India. Based on operational feedback of JHCM at Tarapur, another vitrification facility based on JHCM was constructed at *Waste Immobilisation Plant, Kalpakkam*. The HLLW is subjected to separation of residual uranium, using pulse column-based solvent extraction system using by TBP to enhance the waste loading in vitrified mass. The HLLW is pre-concentrated and fed to JHCM for vitrification [5, 62, 68].

8.13 Vitrified Waste Product Characterization

The borosilicate glass system is the preferred matrix for immobilization of HLLW. The confidence that this matrix will adequately contain various radionuclides and isolate them from the biosphere for long period of time is derived from several scientific studies carried out independently by various groups around the globe. These studies, which involve matrix characterization, also provide the basis for qualifying waste forms and for quality assessment of the waste product. This section contains details of various matrix characterization techniques used to characterize the properties of product [56, 66, 67].

8.13.1 Acceptance Criteria

To establish that a waste form meets the acceptance, the following criteria are used:

1. the uniformity of the waste form,
2. the amorphous nature of the waste form,
3. chemical durability and radiation stability,
4. optimum concentration of various chemical species in the product (waste loading),
5. thermal stability of waste form.

These aspects of product characteristics are discussed here in brief.

(a) **Uniformity/homogeneity**

The uniformity of the waste form is checked in terms of the distribution of various elements in the matrix, using either energy dispersive spectroscopy (EDS) or electron probe microanalysis (EPMA). Alternately, the Z contrast afforded by backscattered electron (BSE) imaging is used to distinguish clusters of high Z elements. Figure 8.19 shows an example of a uniform sample under optical microscopy and BSE. The images in the figure appear featureless, indicating good homogeneity.

Superimposition of X-ray image on the SEM image can yield information about the distribution of elements. The X-ray image of a sample of barium borosilicate-based glass, developed for sulphate-bearing wastes, superimposing Ba and S Kα lines is presented in Fig. 8.20. This image shows uniform distribution of $BaSO_4$ over a micron size scale. This finding was important since high density $BaSO_4$ could settle to the bottom of the melter and clog the drain port.

De-vitrification, i.e. crystallization in glasses, is generally considered deleterious since migration of atoms may take place during the crystallization process will lead to inhomogeneous distribution of glass constituents and, hence, may affect its chemical durability. There is also uncertainty about redistribution of radionuclides among the glassy and crystalline regions. For example, in Mo-containing sodium borosilicate glass, mixed Cs-Na molybdates are formed when the MoO_3 concentration exceeds 2 wt%. These phases are more vulnerable to leaching, with the possibility of Cs remobilization. Another example is nepheline crystallization during vitrified product homogenization in the vitrification melter. This phenomenon is responsible for the problem of clogging of the pouring spout as this phase is highly refractory and settles down. Also, when nepheline is formed during the cooling of the melt, Al_2O_3 and

Fig. 8.19 Optical microscope image (**a**) and BSE (**b**) of a surrogate borosilicate glass from WIP, Trombay, showing no segregation of elements

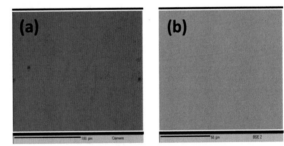

Fig. 8.20 B and S Kα X-ray image showing uniform distribution of S and Ba in the glass

Fig. 8.21 X-ray diffractogram of barium borosilicate glass (Note the absence of crystalline reflections) [80]

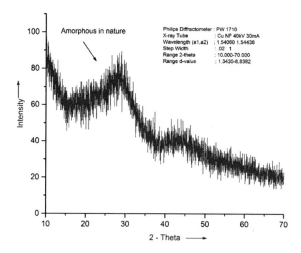

SiO$_2$ are depleted from the residual glass which renders the waste product more vulnerable to degradation. In contrast, wastes such as the defence waste at Hanford, upon vitrification yield small crystallites of spinels distributed in the glass, possibly arising from corrosion products such as Fe, Mn and Ni. If these crystals are small (less than a few μm) and less than 1–2% by volume in the waste form, there are no adverse effects on the performance of waste form. It is, therefore, important to identify crystalline phases formed and confirm amorphous nature of the glass. For this, the most common technique used is powder X-ray diffraction. In Fig. 8.21, the X-ray diffractogram of a glass is presented where the absence of any sharp peaks is indicative of the amorphous nature of the sample.

(b) **Waste loading**

Waste loading is the amount of waste that can be immobilized in a fixed amount of glass without any phase separation. The waste form which provides higher waste loading is preferred to minimize the final waste volume for storage and disposal. It is imperative to establish the waste loading limit that can be achieved before the commencement of formation of crystalline phases. To illustrate this aspect, the case of immobilization of sulphate-bearing HLLW stream is presented which is a challenge due to limited solubility of sulphates in borosilicate glasses. A new barium borosilicate-based composition was developed for the same at WIP, Trombay. A waste loading of up to 23 wt% of waste oxides was found acceptable, but at higher waste loading, a yellow phase is formed which contains sodium sulphate and chromate as well as barium sulphate and chromate (Fig. 8.22). The yellow phase acts as a sink for radioisotope of Cs and Sr, and the waste product exhibits poor aqueous leach resistance behaviour. Therefore, the appearance of this phase signifies the waste loading limit in to the glass. The identity of the crystalline regions is further ascertained using EDS/EPMA.

Fig. 8.22 Representative X-ray diffractogram of the yellow phase. The sharp reflections indicate crystalline phase formation [80]

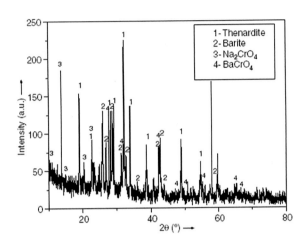

(c) **Chemical durability**

Chemical durability is one of the most important properties of the vitrified waste product (VWP). It is assumed that, over a long period of time, VWP may encounter the ground water of the disposal repository. Leaching of radioactivity from the waste and its subsequent migration, therefore, is the principal mechanism through which radionuclides of waste can be released to the human environment. The leaching of the glass is closely related to the dissolution of matrix constituents which is a reasonable measure of the release of radioactive species and, hence, of the chemical durability of the waste form.

Leaching of VWP depends on a large number of factors like temperature, radiochemical compositions of VWP, leachant and its change of characteristics due to radiolysis, composition of the repository system including material of canister, backfill and host rock/clay. Leaching involves several sequential processes like hydration, ion exchange, network dissolution and precipitation of phases as surface layers. Sodium is chosen as an index element to measure the leach rate because of its chemical similarity with radiocaesium. The nature of the surface species formed and the morphology of the glass surface after leaching are investigated by XRD, scanning electron microscopy (SEM) and energy dispersive X-ray spectroscopy (EDS) techniques. Structural studies are carried out by infrared (IR) absorption to confirm the integrity of the borosilicate network.

(d) **Radiation Stability**

The major source of radiation in VWP is alpha decay of the actinides and beta decay of fission products such as ^{90}Sr and ^{137}Cs. In the repository, beta decay of fission products is predominant in the first thousand years and alpha decay of long-lived radionuclides will predominate thereafter. Radiation may induce change in volume, leach rate, stored energy, microstructure and mechanical properties of VWP.

(e) Effect due to alpha decay

The alpha decay of actinides results in energy dissipation of high energy alpha particles (of few MeV) by elastic collisions with matrix atoms and also by ionization.
Such collisions, ionization and accumulation of helium may result in change of the
structure of VWP. The simulation of effects due to alpha decay has been carried
out by various methods such as doping with a short-lived actinide isotope such as
^{238}Pu (T½ = 86.4 years), ^{242}Cm (T½ = 163 days) or ^{244}Cm (T½ = 18.1 years). Other
methods of simulation are heavy-ion irradiation with lead and utilization of the ^{10}B(n,
α) ^{7}Li reaction by subjecting borosilicate waste form to slow neutron irradiation in
a reactor.

The atomic displacements induced by alpha decay may lead to changes in volume
or density. Volume changes in glass of about 1% have been observed over a very
large period of time, which can be tolerated very well by the waste container design.
Displaced atoms have higher energy than those in equilibrium positions, and the
difference is generally referred to as stored energy. Saturation values in waste glasses
of up to about 400 J/g have been found, and if this were released instantaneously, the
rise in temperature would be approximately 130 °C for a waste glass with a specific
heat of 3.0 J/g/°C. However, in practice, the energy release occurs over a long period
of time and would result in only a small rise in temperature.

Studies on actinide-doped glass by XRD and electron microscopy indicate that
radiation has no significant impact on the microstructure of the high-level waste
glasses. No significant change in leach rate was observed, when boron-containing
glasses were irradiated in a slow neutron flux to simulate a dose equivalent to 10^{22}
alpha particles/kg produced by ^{10}B(n, α) ^{7}Li.

(f) Effect due to decay of fission products

Decay of fission products yields high energy beta particles with energies up to 2 meV,
accompanied by gamma radiation. Both beta and gamma radiation dissipate their
energy mainly by ionizing process. Studies have been conducted using a high energy
electron beam to simulate a beta gamma dose in a glass waste form. Irradiation up to
a cumulative dose of 10^8 Gy on glasses was found not to have any measurable effect
on the leach rate. Gamma radiation of the glass waste form itself was not found to
increase the leachability.

(g) Thermal properties

The knowledge of thermal properties of glass, like glass transition, conductivity and
viscosity as a function of temperature as well as coefficient of thermal expansion,
is desirable for various purposes including design of melter and storage canister,
assessment of stresses in waste product during cooling, control of melt draining,
etc. Crystallization may take place in waste glass, if sufficient time is allowed in the
temperature region above T_g. The presence of heat-generating radionuclides such
as ^{137}Cs and ^{90}Sr leads to rise in temperature of the waste form. The dimensions of
the waste canister are so designed that the centre line temperature does not exceed
T_g. T_g is measured using a differential scanning calorimeter (DSC) or a differential

Fig. 8.23 Representative
DTA plot of a barium
borosilicate glass [80]

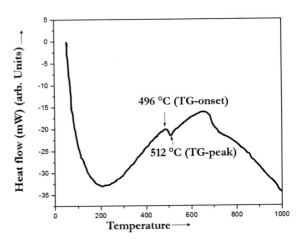

thermal analyser (DTA). A DTA plot of a barium borosilicate glass indicating the glass transition temperature is presented in Fig. 8.23.

(h) Coefficient of thermal expansion

Thermal expansion of VWP is a property of interest for evaluating the extent of thermal stress induced in VWP during pouring/cooling of melt. Thermal expansion of the glass samples of barium borosilicate glass is measured using a Thermo mechanical Analyzer. The mean thermal expansion coefficient between 100 and 400 °C ($\alpha_{100-400}$) is 98×10^{-7} °C^{-1}.

(i) Viscosity of glass melt

Viscosity of VWP has impact on homogenization process in the melter as well as design of drainage of melt since it affects the flow characteristics. Viscosity values of barium-based glass product measured at different temperatures using Brookfield Viscometer were found to be varied from 156 dPa s at 850 °C to 50 dPa s at 925 °C. Viscosity of the vitreous melt increases with fall in temperature. This large variation of viscosity within this temperature range hinders the crystallization of the melt and at the same time facilitates easy pouring of molten glass into the canister.

8.14 Radioactive Gaseous Waste Management

The gaseous radioactive waste from nuclear fuel cycle and radioactive waste processing facilities is a major source for potential direct environmental impact. Therefore, effective control and management of gaseous waste in both normal and accidental conditions is an important aspect in design and operation of nuclear fuel cycle and waste processing facilities [81–83]. Due to its very nature and large volumes, it is not feasible to store the gaseous waste and it is processed in the generating facility itself.

The objective of gaseous waste treatment system is to retain the gaseous contaminants present in the waste. The design of this system is based on several parameters like source of the waste; nature and concentration of various contaminants, generation rates; physical and chemical properties of gases, discharge limits for radioactivity as well as chemical species present. Gaseous streams from highly contaminated areas, e.g. equipment and process vessel, are called off-gas streams. These may contain higher concentrations of radionuclides. Therefore, off-gas streams are treated prior to mixing with the ventilation air. The secondary wastes produced in solid or liquid waste form are processed for further storage and/or disposal.

8.14.1 Ventilation and Air Cleaning Systems

Ventilation and air cleaning systems are a vital part of nuclear fuel cycle facilities. The objectives of providing ventilation and air cleaning systems are to control airborne contamination below safe working levels. For this purpose, the air is supplied after filtration on a once-through basis, with the following objectives:

(a) to maintain directional flow from the point of least contamination potential to the point of greatest contamination potential,
(b) to clean the exhaust air before discharge to the atmosphere,
(c) to monitor contaminants in the working areas and releases to the environment.

In nuclear facilities, the ventilation and air cleaning systems are usually designed to serve for both normal and accidental conditions. The supply air is filtered, and at times washed, to reduce the dust load. The exhaust air is filtered using high efficiency particulate air (HEPA) filters. Wherever necessary, additional clean-up is provided. Typical containment and ventilation system components include hot cells, fume hoods, glove boxes, filters, fans and control dampers. Enclosures such as glove boxes and hot cells are maintained at negative pressure to ensure containment of radionuclides within the process enclosures.

8.14.2 Off-Gas Treatment System

Gaseous streams from highly contaminated areas, e.g. equipment and process vessel, are called off-gas streams. These may contain higher concentrations of radionuclides. Therefore, off-gas streams are treated prior to mixing with the ventilation air. The objective of off-gas waste treatment system is to retain the gaseous contaminants present in the waste. The design of this system is based on several parameters such as source of the waste; nature and concentration of various contaminants, generation rates; physical and chemical properties of gases, discharge limits for radioactivity and chemical species present. The secondary wastes produced in solid or liquid waste form are processed for further storage and/or disposal.

Off-gas systems of ambient temperature waste processing facilities

The off-gas systems for radioactive waste ambient temperature processing plants are typically rated for general area containment and the operator comfort. For example, the handling of the waste in compaction and cementation plants are essentially mechanical processes. These produce solid particulates which are considered as potentially radioactive. These particles can remain suspended long enough to become caught in off-gas system. The ventilation system provides air condition control, general building ventilation with low-level monitoring. The ventilation system for potentially active areas, e.g. low-level liquid waste treatment facility, is equipped with HEPA filtration and monitoring. Such facilities handling LLW with significant concentration of alpha radionuclides are provided with separate off-gas system.

Off-gas system of radioactive waste treatment facilities operating at high temperatures

The off-gas system of radioactive waste treatment facilities operating at high temperatures, e.g. vitrification and incineration/pyrolysis, is elaborate and complex. Such off-gas system consists of several devices. Absorbers and scrubbers are used for removal of gaseous contaminants, e.g. ^{14}C oxides, iodine and noble gases. To obtain desired level of decontamination, several stages of filters are employed, namely dry filters operating at high temperature and wet filters operating with aqueous solutions. To remove sulphur and nitrogen oxides from gases, scrubbers and catalytic reactors are used. Chillers as well as dilution are used to decrease the temperature of off-gas streams and to facilitate removal of contaminants from gaseous streams by condensation. This also helps in protection of off-gas filters. Finally, the gases are routed through absolute filters, i.e. HEPA filters. The details of off-gas system of vitrification facilities are discussed in Sect. 8.9.9. The off-gas cleaning systems for incineration follows in many aspects a similar approach.

8.14.3   Gaseous Waste Treatment Options

Various treatment methods used for decontamination of gaseous waste are described here briefly highlighting their features, limitations and aspect of secondary waste generation.

8.14.3.1   High Efficiency Particulate Air (HEPA) Filtration

The HEPA filters can retain solid submicron particles with high efficiency ($\geq 99.97\%$). These are made of glass fibre filter media and are used extensively in nuclear fuel cycle facilities. Control of humidity present in input off-gases is required to prevent damage to filter media as well as to restrict pressure drop across filters. Moisture separators are used online before HEPA filter. Pre-filters are provided to extend life of HEPA filters [84].

8.14.3.2 Sorption

This is used for removal of inorganic and organic iodine in reactors and reprocessing plants. The sorption media includes chemically impregnated charcoal or zeolites. In this process, humidity control is required. Spent sorption media is the secondary waste.

8.14.3.3 Cryogenic Method

This method is useful for trapping ^{85}Kr from off-gases by sorption on solid sorbent, e.g. charcoal. The system requires elevated pressure and reduced temperature operating parameters. The loaded ^{85}Kr can be recovered and sorbent reused multiple times. Further processing and packaging for long-term storage is essential. The secondary waste generated consists of spent/degraded sorption media.

8.14.3.4 Delay/Decay

This process is used for decay of short-lived noble gases, namely ^{133}Xe, ^{135}Xe, ^{87}Kr, ^{88}Kr and ^{41}Ar. Large beds are required to provide for long residence times.

8.14.3.5 Wet Scrubbing

Scrubbers of various types are used to provide the contact of target compounds or particulate matter with the scrubbing solution. These are commonly used for treatment of process off-gases. Scrubber solution used is demineralized water or aqueous solution of reagents that specifically target certain compounds. Used scrubber solution constitutes liquid waste.

8.15 Management of Radioactive Solid Waste

Solid radioactive wastes are mainly plant protective equipment (PPE) from routine O&M of nuclear facilities, decommissioned equipment, piping and gadgets, secondary process solids (e.g. chemical sludges, filter cake, concrete debris), etc. The wastes are segregated as per their physical and chemical properties. To minimize the volumes of the waste destined for disposal, processes as applicable to the waste characteristics are deployed, viz. compaction, incineration, acid digestion and electrochemical processes [85].

8.15.1 Compaction

Compaction is the mechanism of volume reduction by the application of external force (about 200–2000 Tones) to the waste container where the voids are minimized, and the trapped air is vented through dedicated ventilation system for confinement of the radioactivity. The waste materials amenable for compactions are small pieces of pipes, plates and clads, rubber and plastic-type PPEs and filters. Volume reduction factor achieved by compaction varies from 3 to 10 depending on type of waste being compacted.

8.15.2 Incineration

Incineration is a thermal decomposition process for volume reduction of combustible waste forms. The incineration system consists of a primary chamber where waste is decomposed with the supply of heat. The decomposed waste form gets incinerated in excess air supplied in the chamber and subsequently in the secondary chamber [86]. The ratio of air to primary/secondary chamber may vary from 80/20 to 60/40 depending on the process requirement. Conventional incinerators are either furnace oil- or diesel-fired. Primary and secondary chambers are maintained at about 800 and 900 °C, respectively. The volume reduction factor achieved in the incinerator is in range of 5–50 depending on the waste and its inorganic contents. A stationary chamber (fixed grate) type incinerator is generally used for radioactive wastes. In this incinerator, waste is fed to the primary chamber having fixed grit where decomposition and primary combustion of flue gases take place. Residual ash remains in the same chamber while the flue gases pass through the secondary chamber for complete combustion in the presence of excess air. Incombustible or residue is removed from bottom or side openings and immobilized in cement matrix before transfer for disposal in NSDF. The flue gases are given extensive treatment by using a combination of various off-gas treatment equipment before discharge to the environment.

   In advanced incinerator systems, plasma is used as heat source. Due to high plasma temperature (>2000 K), various types of waste can be processed, including organic wastes, viz. rubber and plastic with negligible releases of toxic gases, like dioxins and furans [87].

8.15.3 Acid Digestion

Acid digestion is a process of converting solid waste to liquid using strong acidic solutions. It is a chemical decomposition process, with little volume reduction, by which all toxic constituents get dissolved in the solution while only small or no residue is left behind. In general, for organic matter dissolution, a combination of acids is used considering vigorous reaction hazards with HNO_3 type reactant with various organics. The resultant solution needs further treatment as per radioactive contents.

8.15.4 Electrochemical Process

Electrochemical process is a method by which the solid wastes are treated with the application of the direct current through an electrolyte and two electrodes. Treatment of the wastes is carried out by two methods: (a) electrolytic dissolution and (b) mediated electrolytic dissolution. Nitric acid is a preferred electrolyte from the point of further processing of secondary wastes.

8.15.5 Mediated Electrolytic Dissolution

This treatment method similar to discussed in Sect. 8.6.2 involves generation of highly oxidizing ions like Ag(II)/Ag(I) and Ce(IV)/Ce(III) at anode by passing direct current. Ag(II) is found to be effective for the treatment of cellulosic wastes. It produces highly reactive OH and NO_3 radicals in nitric acid, which attack organic compounds converting them into CO_2, water and inorganic ions. This method is a safer alternative to direct incineration as it operates at room temperature and does not produce toxic gases like dioxins. However, the ions generated at anode and cathode are reactive to each other leading to low efficiency. Efficiency can be increased substantially by providing porous divider between anolyte and catholyte to prevent bulk mixing but allowing passage of current through it. Ce(IV)/Ce(III) couple is also found to be more suitable for decontamination of metallic components. The metal components undergo uniform corrosion by this method. Agitation may be required for complex geometries for uniform reaction and approach.

8.15.6 Electrolytic Dissolution

This method is used for the decontamination of metallic surfaces using a movable cathode. The metal surface to be decontaminated is made anode. The cathode is

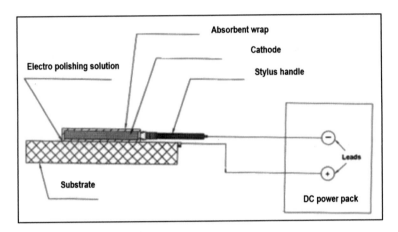

Fig. 8.24 Schematic for movable cathode decontamination

wrapped with a porous sponge material to prevent the electrodes coming in direct
contact which can cause a short circuit. The electrolyte is recirculated through the
cathode which keeps the sponge wet. Current passes through the electrolyte in sponge
which is kept wet by electrolyte recirculation. The schematic of movable cathode
method is provided in Fig. 8.24. The metallic ions remain dissolved in the electrolyte. With these ions, radioactivity also gets leached, rendering the surface free of
contamination.

8.16 Storage of Vitrified Waste

Vitrified HLW has substantial decay heat which needs to be removed for ensuring
the centre line temperature of canister well within the limit determined by the glass
transition temperature. Therefore, the main objective of air-cooled interim storage
of vitrified waste product is to remove the decay heat sufficiently so as to make it
amenable for transport to and disposal in GDF.

8.16.1 Requirement of Storage Facility for Vitrified Waste

The transport of VWP canisters without interim storage to the ultimate disposal
site would lead to complexity in design of transport cask and result in inefficient
utilization of the repository space. The most important factor that influences the
footprint, as well as design of GDF, is the initial decay heat associated with VWP
canister. Hence, it is necessary to store the solidified radioactive wastes for an interim

period of about 30 years before disposal [88–90]. A few countries are planning an even longer period, about 60 years, of interim storage.

8.16.2 Heat Removal Methodology at Storage Vault

(a) Vitrified waste can be stored in water pool or air-cooled vault. Storage of vitrified waste in a water pool will be an extension of the existing mode of storing spent fuel from the nuclear reactors. As a heat transfer media, water has higher thermal conductivity and heat capacitance as compared to air and also ensures better containment in case of failure of VWP canister. However, corrosion due to interaction of water, at elevated temperature, with canister material and pool lining is a major drawback of use of water as coolant considering the fact that integrity of canister is required to be maintained for a very long time. Besides, requirements of make-up water, maintenance of water chemistry and need for continuous operation of pump, etc., for three to six decades of interim storage are costly as compared to air cooling. Hence, storage of VWP canisters in water pools is not being used on industrial scale.

(b) Air-cooled storage vault is designed based on natural or forced convection of air flow for removal of decay heat. Forced convection requires continuous operation of large capacity fans resulting in additional cost of operation and maintenance. Therefore, passive design of natural convection air cooling, assisted by induced draught due to stack, is preferred. It utilizes decay heat and a suitably designed stack to provide the driving force for the movement of air through the storage vault. The decay heat increases the air temperature, causing an upward movement of air due to buoyant forces. The driving force due to buoyancy is balanced by the friction effect of air passage through the system to establish equilibrium for the loading condition of each compartment of the vault. Natural convection-based air-cooled vault can be designed with cross-flow of air across the vitrified waste canisters.

8.16.3 Optimization of Canister Dimensions

Optimization and standardization of dimensions of vitrified waste canister are essential to meet the objectives of (1) efficient removal of decay heat present in VWP canisters within the limits of maximum centre line temperature and (2) facilitate design of remote handling system at vitrification plant, interim storage and DGR. Design of canister storage unit is conceptualized based on multiple-barrier philosophy. In the storage unit, there are three barriers for any outward movement of radioactivity; first, the glass matrix in which the waste oxides are incorporated; second, the high integrity all welded stainless-steel canister and third, the stainless-steel over-pack/thimble. As indicated earlier, dimensions of canisters and over-pack/thimble are selected in such a way that the maximum temperature of vitrified waste product does not exceed the safe limit, derived from glass transition temperature of glass matrix, at any point of time, during transportation inside shielded cask and in the storage facility [91, 92].

8.16.4 Illustrations of VWP Interim Storage Facilities

(a) India

Solid Storage Surveillance Facility (SSSF) is operational at Tarapur since more than two decades. Another such facility, Vitrified Waste Storage Facility (VWSF) has been constructed at Kalpakkam. SSSF, Tarapur, is constructed based on passive design considering natural convection of air flow draught assisted by stack effect. Cross-flow of air across the vitrified waste containing over-pack is adopted for removal of decay heat. To ensure optimal heat transfer efficiency, coolant air distribution, storage unit array and filling pattern are optimized. The storage vault is divided into two blocks, and each of the blocks is further divided into three compartments. This also helps in isolation of any particular compartment, if necessary. SSSF has 'double vault design' to facilitate the requirements of both structural stability and thermal expansion. The 'inner storage vault' is designed on thermal consideration, and the design of the 'external vault' is based on structural and biological shielding considerations. The external vault is also designed to isolate any seepage of ground water from the immediate environs of the storage units. The thermal vault is supported on specially designed bearings from the outer vault enabling the free sliding of the thermal vault, thereby relieving temperature stresses on the thermal vault [92].

Vitrified Waste Storage Facility (VWSF), Kalpakkam, the second Indian facility for interim storage of vitrified HLW, is designed to store vitrified waste canisters in four independent above-ground storage vaults. The vault is designed to receive and store canisters containing vitrified waste as 'single containment concept'. Each location has thimble tube, in which the vitrified waste canister is placed. The thimble acts as secondary containment. The channelized axial flow of air is adopted for the removal of heat. The vault has an additional safety feature in the form of induced draft air cooling with HEPA filtration, which can be brought in line as and when airborne activity is detected in the cooling air. Besides this, each location has been provided with monitoring of temperatures and air sampling. The system is self-regulating and can compensate for changes in heat load or seasonal/weather variation conditions. As decay heat reduces, both the air flow and air temperatures decrease. The schematic and actual views of VWSF at Kalpakkam are shown in Fig. 8.25.

(b) France

Vitrified waste from the earliest French reactor—the natural uranium-graphite-gas (Uranium Naturel Graphite Gaz—UNGG) spent fuel reprocessing is stored at the Marcoule facility. HLW from this process was calcined and embedded in a glass matrix by a processing line in the vitrification plant (Atelier de Vitrification de

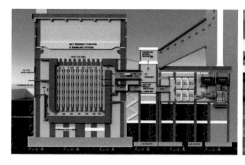

Fig. 8.25 Schematic and actual view of VWSF, Kalpakkam, India

Marcoule—AVM) since 1978. The vitrified waste containers are stored in pits at the Marcoule site.

A cross-section of the vitrified waste store in the La Hague Plant illustrates the principle of harnessing natural convection to cool vitrified waste packages located in shafts. As the stack creates an air draft, no mechanical ventilation system is required. The shafts are located under the floor of a large hall featuring multiple 'manhole covers', as the shaft cover plates (Fig. 8.26).

Table 8.7 provides information relating to vitrified waste interim storage facilities in La Hague. These facilities are cooled by natural or mechanical air circulation. The shafts are sealed and a partial vacuum created, and the air in them is treated and conditioned.

(c) United Kingdom

Vitrified HLW at Sellafield UK is stored in stainless-steel canisters. Some 840 m^3 of vitrified high-level waste has been produced and is stored within 5,600 canisters in an engineered air-cooled facility [94]. Current practice in UK is for the waste to be stored for at least 50 years before disposal.

Fig. 8.26 Cross-section of a vitrified waste storage building [93]

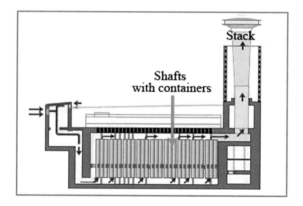

Table 8.7 Interim storage of vitrified waste at the La Hague site [93]

Facility	La Hauge (R7 T7)	La Hauge (EEE and T7)
Date commissioned	1989 and 1992	1995
Number of shafts	500 and 400	360
Capacity in containers	4500 and 3600	4320
Convection type	Forced	Natural

The engineered stores for vitrified HLW are fully equipped with remote devices for package handling and sufficient shielding provisions inside the building structure. Stores are usually designed on a modular basis to allow extension when required.

8.17 Disposal of Radioactive Waste

Disposal of radioactive waste is defined as the emplacement of waste in approved specified facilities without the intention of retrieval. In comparison with other wastes, viz. municipal waste, hospital waste, industry waste and chemical waste, radioactive wastes contain radionuclides which decay with time and emit radiation. The primary and the most important objective of disposal of radioactive waste is thus to provide confinement and isolation of these wastes from accessible environment for sufficiently long time period which is stipulated by the regulatory bodies to ensure that the radiation doses received by member of public even in distant future are less than the permissible limits [95–98]. The permissible doses to occupational worker and member of public are set by the national regulators.

Several varieties of radioactive waste disposal facilities are operational worldwide. The design, depth and location of these facilities vary from country to country and are essentially controlled by type of waste, their radioactivity, geological conditions and safety requirements. These facilities can be grouped into six classes based on their design, depth and type of waste they accommodate. These include landfills, near-surface disposal facilities, intermediate depth facility, geological disposal facilities, borehole disposal and tailing ponds.

8.17.1 Near-Surface Disposal Facility

These facilities comprise engineered trenches and vaults constructed below the ground surface in the depth range of five to few tens of metres and are meant for LLW and short-lived ILW with half-lives of up to 30 years. The development of these facilities proceeds in stages that include siting, site characterization, design, construction, operation, closure and post-closure monitoring. Site selection of a NSDF is mainly controlled by geological and hydro-geological characteristics of the site. Among these, important parameters are thickness of soil, groundwater flow direction and their flow velocity, porosity and sorption parameters for various radionuclide. These parameters in combination with other components of disposal system, i.e. design, waste form, type and quantity of waste, other engineered barriers like buffers and backfills, disposal trenches/vaults, together insure adequate radiological protection in accordance with safety requirements set by national regulations [98–101].

Some of the important features that control the choice of a site include absence of active tectonics, seismicity and very low probability of potentially disruptive natural events like volcanism, uplift, subsidence, igneous intrusion, faulting, climate change and alteration of topography results in longer structural integrity of disposal modules. The absence of major surface and subsurface water bodies reduces the groundwater-assisted radionuclide transportation from the site to biosphere. The important and desirable features of a good host rock or soil are manifested by lower porosity, permeability, solubility and sorption capacity. Besides these technical requirements, several other important considerations involved in the process of site selection are general lack of economic mineral or other natural resources in the area, low land value, sparse population, easy accessibility and transport facilities, nearness to waste-producing centres, etc. Generally, the required period over which these facilities are designed to provide isolation and confinement to the disposed waste is of the order of 300 years driven by the half-life of isotopes of prominent fission products present in the waste, e.g. ^{137}Cs and ^{90}Sr.

The Oak Ridge facility, a simple trench-type disposal module built in 1944 at Tennessee USA for disposal of contaminated broken glassware and other materials, is considered as the first near-surface disposal facility [102]. Presently, about 150 NSDF sites are operational worldwide. The Drig facility UK, Centre de La Manche and Aube Facility France, LLW disposal facility US, El Cabril Facility Spain, Rokkasho Facility Japan, NSDFs at Trombay, Tarapur, Kalpakkam, Kota, Narora, Kakrapar and Kaiga in India are examples of such facilities.

8.17.2 Geological Disposal Facilities

Geological disposal facilities are meant for disposal of heat emitting long-lived highly radioactive materials, viz. spent fuel and vitrified HLLW from open and closed fuel cycles, respectively. In addition to these, other types of wastes such as TRU waste, long-lived ILW and spent radiation sources are also planned to be disposed in these facilities. Geological disposal facilities primarily comprise specifically designed and constructed tunnels, vaults and silos in a particular geological host rock in the depth range of few hundred metres below the ground level. These facilities are widely referred as Geological Disposal Facilities (GDF). GDF concepts under development worldwide aim to meet the following objectives [103–106].

(i)   During the safety period considered in the design and construction of GDF, the release of the radioactivity to the biosphere shall not exceed the concentration limits set by the regulatory authority.

(ii) GDF system shall protect the waste from inadvertent human intrusion and would also protect it from disruptive natural events like earthquakes, tsunamis and flooding.

(iii) To incorporate concept of retrieval of certain type of disposed waste in design to provide opportunities to future generations to take decision about the disposed waste in the light of evolving technologies.

An example of GDF facility is WIPP in USA constructed in bedded salt at a depth of 700 m for non-heat-emitting long-lived transuranic waste [107]. Similar facilities for disposal of spent fuel and vitrified high-level waste are at various stage of development in countries like Finland, Sweden, France, Belgium, United Kingdom, USA, India, Japan and China.

8.17.2.1 Types of Geological Disposal Facilities

There are several bases for classification of GDF. In terms of chosen host rock, they can be classified as (a) crystalline, (b) sedimentary and (c) volcanic host rock-based systems. Crystalline host rock-based GDF are at different stages of development in Sweden, Finland, India, China, Canada, France, Belgium, Switzerland, etc., are evaluating sedimentary host rock-based GDF. Yucca Mountain-based GDF explored in USA is an example of unsaturated zone-based GDF. Similarly, in terms of mode of disposal they can be grouped into vertical pit mode disposal systems and tunnel mode disposal systems. Pit mode disposal systems are being considered in countries like Sweden, India and Finland whereas horizontal tunnel-based GDF are under consideration in France.

8.17.2.2 Safety Features of GDF

Safety indicators represent the radiation risk resulting from both occupational exposures and exposures of the general public. Effective dose constraint is one of the widely used safety indicator. In most of the countries, the regulatory recommended value of dose constraint is between 0.1 and 0.3 mSv/year for the critical group whereas 0.5 mSv/year is desirable limit for the member of public.

Geological barriers under consideration as host rock for GDF are invariably heterogeneous in terms of their structural elements, chemical, mineralogical, thermal, hydraulic and rock mechanical characteristics. This introduces a large-scale uncertainty in assessment of their capacity to provide desired level of safety overlong time periods. The designs of GDF have, therefore, provisions for both natural/geological

Fig. 8.27 The concept of
multi barrier for GDF

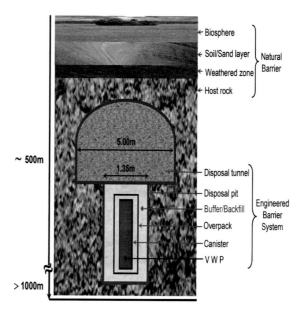

and engineered barriers. The engineered barriers essentially comprise materials like immobilizing matrix (viz. specially formulated glasses), canister/overpacks, clay buffers, seals and grouts. Figure 8.27 illustrates the concept of multi-barriers for GDF [96, 108]. Host rock or geological formation, in which a GDF is to be constructed, constitutes the natural barrier. Several types of geological formations have been investigated worldwide regarding their suitability for a hosting GDF. Among these, crystalline rocks such as granite, gneiss, basalts, salt domes, bedded salt formations, sedimentary formations (particularly mudstone or shale), basalt, tuff and unconsolidated ocean sediments are noteworthy. To serve as a good host rock, these geological formations need to provide long-term stability and mechanical integrity.

Additional favourable conditions, for GDF within such geological formations, include long flow times for radionuclides to reach the biosphere and additional retardation of migrating radionuclides by the rock matrix. Therefore, geological formations are studied for their chemistry, particularly their capability and capacity for buffering the pH and Eh (redox potential) of co-existing pore waters. Both containment and controlled-release performance of engineered barriers, radioactive waste forms and radionuclides are sensitively related to these chemical variables [108–110]. In particular, the solubility of many radioelements especially actinides are extremely low under low Eh and/or high pH conditions. Furthermore, retardation of radionuclides within the geological formations is strongly affected by these same chemical variables.

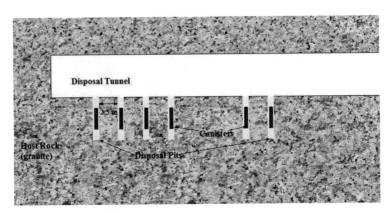

Fig. 8.28 In-tunnel emplacement concept for GDF

8.17.2.3 Criteria for Selection of Site for GDF

The sites meant for hosting a GDF are expected to lie in seismically stable areas with occurrence of large volumes of homogeneous rock mass with minimum potential for groundwater as well as surface water infiltration. Absence of structural discontinuities and very low uplift and erosion rates together with a very low potential for flora and fauna are other desirable features. The host rock for heat-generating radioactive waste also needs favourable thermal characteristics to facilitate dissipation of heat without undue influence on the performance of the disposal system.

There are three principal designs for a repository: in-tunnel emplacement, borehole emplacement and cavern emplacement. In in-tunnel emplacement, waste containers are emplaced in tunnels excavated in the rock, as given in Fig. 8.28.

As each waste container is emplaced, the container is backfilled, and the tunnel sealed. Alternatively, borehole emplacement involves excavating additional boreholes within the floor or walls of tunnels. One or more waste containers would be loaded into these holes and backfilled. For repository closure, the tunnels themselves would be backfilled and sealed. Typical horizontal tunnel disposal concept pursued by India is shown in Fig. 8.29. In some countries, cavern emplacement is under consideration. This involves multiple waste containers or boxes stacked together within one large silo of engineered barriers. In the 'WP-cave' concept, HLLW containers are placed within a radial set of tunnels connected by inner and outer shafts. Individual containers are not specifically backfilled. This central storage excavation would, in turn, be surrounded by a thick (several metres) bentonite–sand backfill.

In Indian GDF *concept*, preliminary design of disposal facility considers disposal of 10,000 waste canisters in an area of 4 sq. km. Each canister is expected to have heat generation rate of 0.5 kW at the time of emplacement with skin temperature of about 80 °C. Four disposal areas are planned. Each disposal area can accommodate 2,500 canisters and is planned to be developed in stages.

Fig. 8.29 GDF layout under
consideration in India

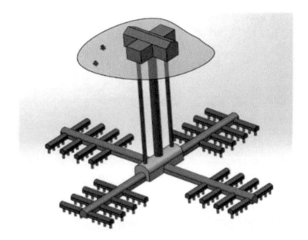

8.17.3 Other Types of Disposal Facilities

Landfill disposal Facility resembles typical municipality landfill facility and is used
in some countries for very low-level radioactive waste with low concentration
of radioactive elements. For example, National Research Council in the State of
Michigan USA is using land disposal facility for certain very low activity wastes.

Intermediate-level waste (ILW) disposal facilities are being used in a few countries
for disposal in mined cavities or in caverns, vaults or silos constructed in the depth
range varying from few tens of metres to few hundred metres. The Swedish final
repositories (SFR) built in granitic rock at depth of 60 m under the Baltic Sea bottom
for such type of waste is an example of such facilities (Fig. 8.30). In Czech Republic,
a similar facility has been built in Richard II mine cavern in the depth range of
70-80 m. Figure 8.31 depicts a schematic of ILW/LLW disposal facility in Finland
[111].

Fig. 8.30 Swedish final
repository for ILW [111]

Fig. 8.31 ILW/LLW
disposal facility in Finland
[111]

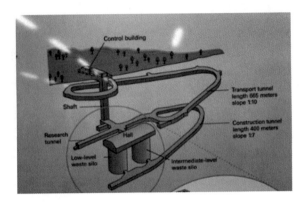

Borehole disposal concept is in use in a few countries like South Africa for disposal of spent radiation sources. Disposal in boreholes is in consideration in many countries like Denmark, Switzerland, Sweden and USA. This involves drilling of large diameter deep boreholes up to a depth of 5000 m and emplacing waste canisters in lower 2000 m depth followed by backfilling of the remaining 3000 m. A single borehole thus can accommodate 400 canisters of 5 m length each.

Tailing Ponds for mining and mineral processing waste are essentially near-surface disposal facilities co-located with the mines. These have large storage capacity to accommodate such type of waste. Universally, such facilities are integral part of all uranium mining projects world over. Moab uranium mill tailings pile is one such facility near largest uranium deposit in the United States in Utah. Sillamäe uranium tailing dam is located in Russia. In India, such tailing ponds are operational at uranium mining sites located in the states of Jharkhand and Andhra Pradesh.

8.18 Summary

Radioactive waste is generated in almost every step of the nuclear fuel cycle. The volume and characteristics of waste vary depending on nature of step and activities involved with respect to radioactive operations. In reality, radioactive waste management starts at the generating facility itself in the form of minimization of waste generation which includes selection of quality process and materials, good housekeeping and maintenance practices, segregation of waste material at the point of generation, etc. Awareness among O&M personnel about the significance of reduction in waste and its benefit in environment protection plays an especially important role in the minimization of radioactive waste generation.

The radioactive waste generating and treatment facilities are provided with waste storage areas/equipment for solid and liquid wastes. This, like transit storage before shipment, solves many problems that may arise due to break down in transportation, outages in downstream facility due to maintenance, etc. Solid waste storage is provided with material handling equipment and biological shielding as per the waste to be stored. Similarly, above surface storage tanks with dykes are provided for low-level liquid waste having liquid-level monitoring and transfer systems. Storage of ILW and HLLW is done in underground high integrity tanks with provisions for liquid-level/density/temperature monitoring, mixing and transfer systems. Storage before treatment helps in decay of short-lived activity as well as ensures uniform composition of waste feed which is essential for obtaining good quality of waste product, e.g. in vitrification. Depending upon the scale of operations, liquid and solid storage waste facilities are at times quite large for a centralized waste treatment facility.

Transportation of radioactive waste is required for various purposes, e.g. from waste generating facility to waste storage area, waste generating to waste treatment facility, waste treatment to waste disposal facility, etc. Liquid waste is transferred through specially designed piping system, if distance involved is not large. Road transportation of liquid waste by tanks is also used, though not quite common. Transportation of solidified (vitrified) HLLW and spent fuel has been performed though sea route, e.g. from France to Japan. In future, requirement will arise for regular shipment of solidified HLLW from generating/storage facilities to geological disposal facility. Transportation system for radioactive waste follows internationally accepted rules and regulations in which elements of transport are designed and qualified by testing to meet stipulations of radiological and industrial safety. In India, NSDFs are co-located with nuclear complexes to minimize the need of transportation of solidified LILW over long distances through densely populated territories.

The technologies for treatment of radioactive waste streams are diverse in view of different types of waste to be processed. The wastes differ in physical properties (physical state, compressibility, combustibility), chemical nature (organic/inorganic, pH, volatility), radionuclide content (concentration, half-lives), etc. Also, from same step of nuclear fuel cycle, composition may vary with time. Being a downstream facility, radioactive waste treatment system should be robust enough to accommodate minor variations in upstream process/system.

The future challenges in radioactive waste management are also the opportunities for adoption of newer and improved processes as well as development of better materials and gadgets. Some of the areas are:

- There is always demand for further research for reduction of waste volumes, improvement in decontamination factor and reducing the radioactive discharges. The work in these areas would lead to less demand of land required for waste disposal and reduced impact on environment due to radioactive discharges through terrestrial, aquatic and aerial routes.
- Developments in the areas of next generation reactor system, change in nuclear fuels from oxide to metallic/carbide, etc., are expected to put new demands on radioactive waste management. In India, utilization of thorium resources has opened new area of research in radioactive waste from Th fuel cycle.
- The emphasis on recovery and recycle of valuables from HLLW has led to development of novel solvents and related process developments, a trend which is likely

to continue in future as well to meet the ever growing need of radionuclides for societal benefits. This would also help in reducing the requirement of isolation of waste from the environment for very long time periods by reducing the load of long-lived radionuclides in the disposed waste. Another benefit is reduction in heat load on the repository and possible reduction in the size of geological disposal facility.

- As the present nuclear facilities grow in age, the need for dismantling and decommissioning is expected to increase requiring processing of waste from decontamination of facilities as well as disposal of disused equipment and hardware. Experience has been gained in various countries with respect to these areas and further development would be required to improve decontamination and volume reduction of such wastes. Final objectives are to reduce waste volumes and radioactive contents of disposed wastes.
- Another important area of future research and development is geological disposal facility. Research and development work are continuing globally on development of methodology for safety assessment, repository design, radioactive waste package handling, performance assessment of waste matrices and backfill materials, etc. These studies are also in the form of pilot repositories in various countries. These would help demonstrate safe disposal of HLLW as the ultimate step in the nuclear fuel cycle.

Questions:

1. What are the advantages of following closed fuel cycle?
2. How does radioactive waste differ from conventional chemical waste?
3. What are the sources of radioactive waste?
4. What are the steps to minimize the radioactive waste?
5. Why segregation of radioactive waste is important?
6. What are the parameters on which radioactive waste can be segregated?
7. What are the sources of gaseous radioactive waste?
8. What radionuclide will be present in gaseous waste during mining/reactor operation/reprocessing of spent fuel/high-level waste vitrification process?
9. What are the treatment mechanisms for gaseous radioactive waste?
10. What treatment mechanism is used for management of gaseous waste containing particulate matter?
11. What is efficiency of HEPA filters? Which filters are normally used in gaseous stream prior to their discharge to environment from nuclear fuel cycle facilities?
12. What are sources of radioactive solid waste?
13. Explain the categorization of radioactive solid waste.
14. Explain the differences between low-level liquid waste, intermediate-level liquid waste and high-level liquid waste.
15. What are the sources of low-level liquid waste? What are the salient features of low-level liquid waste involving chemical nature, radioactivity, major radionuclides present and extent of volume?
16. What are the treatment methods of low-level liquid waste?
17. How chemical precipitation method helps in treatment of low-level liquid waste?

18. What is the source of intermediate-level liquid waste? Explain the character-istic of intermediate-level waste involving chemical nature, radioactivity, major radionuclides present and extant of volume.
19. How intermediate-level liquid waste is treated?
20. Why ion exchange is better treatment option then cementation?
21. What is high-level waste? Explain three stage management strategy for high-level waste.
22. Explain the criteria for selecting matrix for immobilization of high-level liquid waste. Why glass is the best suitable matrix for immobilization of high-level liquid waste?
23. Why vitrified HLW is interim stored in air-cooled vault prior to disposal in GDF?
24. Explain multibarrier concept of disposal of vitrified HLW in GDF.
25. Explain the site selection criteria for GDF.
26. What are the advantages of partitioning and transmutation of minor actinide?
27. Explain the waste from wealth concept.

References

1. Predisposal management of radioactive waste. IAEA Safety Standards Series No. GSR-5 (2009)
2. Disposal of radioactive waste. IAEA Safety Standards Series No. SSR-5 (2011)
3. Management of radioactive waste. AERB Safety Code No. AERB/NRF/SC/RW (2007)
4. P.K. Wattal, Indian programme on radioactive waste management. Sadhana **38**, 849–857 (2013)
5. K. Raj, K.K. Prasad, N.K. Bandal, Radioactive waste management practices in India. Nucl. Eng. Des. **236**, 914–930 (2006)
6. P.K. Wattal, Backend of Indian nuclear fuel cycle—a road to sustainability. Prog. Nucl. Energy **101**, 133–145 (2017)
7. R. Natarajan, Reprocessing of spent nuclear fuel in India: present challenges and future programme. Prog. Nucl. Energy **101**, 118–132 (2017)
8. P.K. Dey, N.K. Bansal, Spent fuel reprocessing: a vital link in Indian nuclear power program. Nucl. Eng. Des. **236**, 723–729 (2006)
9. Handling and treatment of radioactive aqueous wastes. TECDOC No. 654, (IAEA, 1992)
10. Waste Treatment at La Hague and Marcoule Sites, Document No. ES/WM-49, ORNL, USA, pp. 2–9
11. P.D. Ozarde, S.K. Samanta, K. Raj, Management of intermediate level waste from past repro-cessing using caesium specific resorcinol formaldehyde resin, in *Proc of the IAEA Intl Conf on issues and trends in radioactive waste management.* IAEA-CN-90/51 (2002)
12. K. Raj, C.P. Kaushik, R.K. Mishra, Radioactive waste management in U/Th fuel cycles, in *Thoria-Based Nuclear Fuels*, ed. by D. Das, S.R. Bharadwaj (Springer, London, 2013), pp. 357–358
13. Presented in lecture on "Management of Radioactive waste from reprocessing plants" in *BRNS Theme Meeting on Status and Trends in Thermal Reactor Spent Fuel Processing in India*, October 22, 2010
14. V.M. Efremenkov, Radioactive waste management at nuclear power plants, IAEA Bulletin, 4/1989, IAEA (1989), p. 38

15. Process Information, LERF & 200 Area ETF, Hanford Facility RCRA Permit Dangerous Waste Portion, Addendum C, p. C-67
16. New developments and improvements in processing of problematic radioactive waste. IAEA TECDOC No. 1579 (2007), p. 43
17. Fukushima Daiichi Accident, https://www.world-nuclear.org/information-library/safety-and-security/safety-of-plants/fukushima-daiichi-accident.aspx
18. Overview of the multi-nuclide removal equipment (ALPS) at Fukushima Daiichi Nuclear Power Station, https://www.tepco.co.jp/en/nu/fukushima-np/handouts/2013/images/handouts_130329_01-e.pdf
19. J. Braun, T. Barker, Fukushima Daiichi emergency water treatment. Nucl. Plant J. 36–37 (2012)
20. T. Tsukada et al., Early construction and operation of a highly contaminated water treatment system in Fukushima Daiichi Nuclear Power Station (I)—ion exchange properties of kurion herschelite in simulating contaminated water. J. Nucl. Sci. Technol. **51**(2014)
21. C. Srinivas, G.S. Singh, A review study on organic waste management by green chemistry. Green Chem. Technol. Lett. **06**, 4–13 (2020)
22. Treatment and conditioning of radioactive organic liquids. Technical Report Series No. 656, IAEA (1992)
23. Predisposal management of organic radioactive waste. Technical report Series No. 427, IAEA (2004)
24. Handling and processing of radioactive waste from nuclear applications. IAEA Technical Report Series No. 402 (2001)
25. S. Manohar, C. Srinivas, T. Vincent, P.K. Wattal, Management of spent solvents by alkaline hydrolysis process. Waste Manage. **19**, 509–517 (1999)
26. K.K. Halder, K.V. Ravi, M.M. Malusare, Sanjay Kumar, Development and demonstration of vacuum distillation process for recovery of pure TBP and n dodecane from simulated organic liquid waste, BARC Newsletter, Jan–Feb (2018)
27. B.F. Judson, R.L. Moore, H.H. van Tuyl, R.W. Wirta, Chem. Eng. Symp. Ser. (A.I.Ch.E) vol. 55 (1959)
28. V. Kourim, J. Rais, B. Million, Exchange properties of complex cyanides-I: ion exchange of cesium on ferrocyanides. J. Inorg. Nucl. Chem. **26**, 1111–1115 (1964)
29. V.N. Ramonovsky, R&D activity on portioning in Russia, in *fifth OECD/NEA information exchange meeting on actinides and fission product portioning and transmutation* (SCK-CEN, Mol, Belgium, 1998)
30. V.N. Ramonovsky, Management of accumulated high level waste at the amyak production Association in the Russian Federation' in issues and Trends in Radioactive waste management, in *Proc. of international conference* (IAEA, Vienna, 2003)
31. C.J. Pederson, The discovery of crown ethers. Science **241** (1988)
32. H. Gerow, G.E. Smith, M.W. Davis, Extraction of Cesium (+1) and Strontium (+2) from nitric acid solution using macrocyclic polyethers. Sep. Sci. Technol. **16**, 519–548 (1981)
33. J.W. McDowell, G.N. Case, Selective extraction of cesium from acidic nitrate solutions with didodecylnaphthalenesulfonic acid synergized with bis(tert-butylbenzo)-21-crown-7. Anal. Chem. **64** (1992)
34. S.R. Izatt, R.T. Hawkins, J.J. Christensen, R.M. Izatt, Cation transport from multiple alkali cation mixtures using a liquid membrane system containing a series of calixarene carriers. J. Am. Chem. Soc. **107** (1985)
35. Z. Asfari, C. Bressot, J. Vicens, C. Hill, J.F. Dozol, H. Rouquette, S. Eymard, V.L.B. Tournois, Doubly crowned calix[4]arenes in the 1,3-alternate conformation as cesium-selective carriers in supported liquid membranes. Anal. Chem. **67**(18), 3133 (1995)
36. J.F. Dozol, N. Simon, V. Lamare, H. Rouquette, S. Eymard, B. Tournois, D. De Marc, R.M. Macias, A solution for cesium removal from high-salinity acidic or alkaline liquid waste: the crown calix[4]arenes. Sep. Sci. Technol. **34**, 877–909 (1999)
37. M.A. Norato, M.H. Beasley, S.G. Campbell, A.D. Coleman, M.W. Geeting, J.W. Guthrie, C.W. Kennell, R.A. Pierce, R.C. Ryberg, D.D. Walker, J.D. Law, T.A. Todd, Demonstration

of the caustic-side solvent extraction process for removing 137cs from high-level waste at savannah river site. Sep. Sci. Technol. **38**, 2647–2666 (2003)

38. P.V. Bonnesen, T.J. Haverlock, N.L. Engle, R.A. Sachleben, B.A. Moyer, Development of process chemistry for the removal of cesium from acidic nuclear waste by calix[4]arene-crown-6 ethers, in *ACS Symposium Series 757, Calixarenes for Separations*, ed. by G.J. Lumetta, R.D. Rogers, A.S. Gopalan (American Chemical Society, Washington, DC, 2000)

39. P.S. Dhami et al., Studies on the development of a two stage SLM system for the separation of carrier-free 90Y using KSM-17 and CMPO as carriers. Sep. Sci. Technol. **42**(Issue 5) (2007)

40. M. Blicharska, B. Bartoś, S. Krajewski, A. Bilewicz, Separation of fission produced 106Ru from simulated high level nuclear wastes for production of brachytherapy sources. J. Radioanal. Nucl. Chem. (2013)

41. G.P. Salvatores, Radioactive waste partitioning and transmutation within advanced fuel cycles: achievements and challenges. Prog. Part. Nucl. Phys. **66**, 144–166 (2011)

42. M. Salvatores, Partitioning and transmutation of spent nuclear fuel and radioactive waste, in *Nuclear Fuel Cycle Science and Engineering* (Woodhead Publishing Series in Energy, 2012), pp. 501–530

43. S. Tachimori, Y. Morita, Overview of solvent extraction chemistry for reprocessing. Ion Exch. Solvent Extr. Ser. Adv. **19**, 1–64 (2009)

44. K.L. Nash, J.C. Braley, Chemistry of radioactive materials in the nuclear fuel cycle, in *Advanced Separation Techniques for Nuclear Fuel Reprocessing and Radioactive Waste Treatment* (Woodhead Publishing Series in Energy, 2011), pp. 3–22

45. Paulenova, Physical and chemical properties of actinides in nuclear fuel reprocessing, in *Advanced Separation Techniques for Nuclear Fuel Reprocessing and Radioactive Waste Treatment* (Woodhead Publishing Series in Energy, 2011), pp. 23–57

46. Hill, Development of highly selective compounds for solvent extraction processes: partitioning and transmutation of long-lived radionuclides from spent nuclear fuels, in *Advanced Separation Techniques for Nuclear Fuel Reprocessing and Radioactive Waste Treatment* (Woodhead Publishing Series in Energy, 2011), pp. 311–362

47. Hill, Overview of recent advances in An(III)/Ln(III) separation by solvent extraction. Ion Exch. Solvent Extr. Ser. Adv. **19**, 119–194 (2009)

48. K.L. Nash, The chemistry of TALSPEAK: a review of the science. Solvent Extr. Ion Exch. **33**, 1–55 (2015)

49. M. Nilsson, K.L. Nash, A review of the development and operational characteristics of the TALSPEAK process. Solvent Extr. Ion Exch. **25**, 665–701 (2007)

50. C.A. Sharrad, D.M. Whittaker, The use of organic extractants in solvent extraction processes in the partitioning of spent nuclear fuels, in *Reprocessing and Recycling of Spent Nuclear Fuel* (Woodhead Publishing Series in Energy, 2015), pp. 153–189

51. G. Modolo, A. Geist, M. Miguirditchian, Minor actinide separations in the reprocessing of spent nuclear fuels: recent advances in Europe, in *Reprocessing and Recycling of Spent Nuclear Fuel* (Woodhead Publishing Series in Energy, 2015), pp. 245–287. https://doi.org/10.1016/B978-1-78242-212-9.00010-1

52. B.A. Moyer, G.J. Lumetta, B.J. Mincher, Minor actinide separation in the reprocessing of spent nuclear fuels: recent advances in the United States, in *Reprocessing and Recycling of Spent Nuclear Fuel* (Woodhead Publishing Series in Energy, 2015), pp. 289–312. https://doi.org/10.1016/B978-1-78242-212-9.00011-3

53. S. Manohar, V.P. Patel, U. Dani, M.R. Venugopal, P.K. Wattal, Engineering scale demonstration facility for actinide partitioning of high level waste, BARC Newsletter 332, May–June (2013) 13–18

54. S. Manohar, V.P. Patel, U. Dani, M.R. Venugopal, P.K. Wattal, hot commissioning of an actinide separation demonstration facility, BARC Newsletter Founder's Day Special Issue (2015) 237–246

55. C.P. Kaushik, Amar Kumar, N.S. Tomar, S. Wadhwa, D. Mehta, R.K. Mishra, J. Diwan, S. Babu, S.K. Marathe, A.P. Jakhete, S. Jain, A. Gangadharan, K. Agarwal, Recovery of cesium from high level liquid radioactive waste for societal application: an important milestone, BARC Newsletter, March-April (2017) 3–4

56. K. Raj, C.P. Kaushik, Glass matrices for vitrification of radioactive waste—an update on R&D efforts, in *IOP Conference Series: Materials Science and Engineering*
57. Design and operation of high level waste vitrification and storage facilities. Technical Report Series No. 333 (IAEA, Vienna)
58. W. Baehr, Industrial vitrification processes for high-level liquid waste solutions, IAEA Bull. **4** (1989)
59. R.D. Quang, E. Pluche, C. Ladirat, A. Prod'Homme, review of the French vitrification program, in *WM'04 Conference*, February 29-March 4 (Tucson, AZ, 2004)
60. D.J. Bradley, K.J. Schneider, Radioactive waste management in the USSR: a review of unclassified sources, 1963–1990. PNL Report No. 7182 (Pacific Northwest Laboratory Richland, Washington), p 99352
61. T. Harrison, Vitrification of high level waste in the UK, in *2nd International Summer School on Nuclear Glass Waste form: Structure, Properties and Long Term Behavior, SumGLASS 2013, Procedia Materials Science*, vol. 7 (2014), pp. 10–15
62. G. Suneel et al., Experimental investigation and numerical modelling of a joule-heated ceramic melter for vitrification of radioactive waste. J. Hazard. Toxic Radioact. Waste **23** (2019)
63. L. Luezzi, Th. M. Nieuwenhuizen, *Thermodynamics of the Glassy State* (CRC Press, 2007)
64. A.K. Varshneya, *Fundamentals of Inorganic Glasses: Society of Glass Technology* (Sheffield, UK, 2006)
65. D.R. Uhlmann, A kinetic treatment of glass formation. J. Non-Cryst. Solids **7**, 337–348 (1972)
66. M.I. Ojovan, W.E. Lee, *New Developments in Glassy Nuclear Wasteforms* (Nova Science Publishers Inc., New York, 2007)
67. M.I. Ojovan, W.E. Lee, *An Introduction to Nuclear Waste Immobilization* (Elsevier, Oxford, UK, 2005)
68. C.P. Kaushik, Indian programme for vitrification of high level radioactive liquid waste. Proc. Mater. Sci. **7** (2014)
69. C.C. Chapman, Nuclear waste glass melter design including the process and control systems. IEEE Trans. Ind. Appl. **IA-18**(1) (1982)
70. S. Weisenburger, Nuclear waste vitrification in a ceramic lined electric glass melter. IEEE Trans. Ind. Appl. **IA-18** (1982)
71. D. Gombert, J.R. Richardson, Cold crucible induction melter design and development. Nucl. Technol. **141** (2003)
72. G. Sugilal, A. Thess, G. Weidmann, U. Lange, Chaotic mixing in a Joule-heated glass melt. Phys. Fluids **22** (2010)
73. G. Sugilal, Experimental analysis of the performance of cold crucible induction glass melter. Appl. Therm. Eng. **28** (2008)
74. R.D. Quang, V. Petitjean, F. Hollebecque, O. Pinet, T. Flament, A. Prod'homme, Vitrification of HLW produced by uranium/molybdenum fuel reprocessing in COGEMA's cold crucible melter, in *Proc. Waste Management Symposium* (Tucson, USA, 2003)
75. R.A. Day, J. Ferenczy, E. Drabarek, T. Advocat, C. Fillet, J. Lacombe, C. Ladirat, C. Veyer, R. Do Quang, J. Thomasson, Glass-ceramic in a cold crucible melter: the optimum combination for greater waste processing efficiency, in *Proc. Waste Management Symposium* (Tucson, USA, 2003)
76. Jouan, R. Boen, S. Merlin, P. Roux, A warm heart in a cold body—melter technology for tomorrow, in: *Proceedings of the Spectrum 96, International Topical Meeting on Nuclear and Hazardous Waste Management* (Seattle, USA, 1996)
77. F.A. Lifanov, I.A. Sobolov, S.A. Dimitriev, S.V. Stefanovsky, Vitrification of low and intermediate level waste: technology and glass performance, in *Proc. Waste Management Symposium* (Arizona, USA, 2004)
78. R. Didierlaurent, E. Chauvin, J. Lacombe, C. Mesnil, C. Veyer, Cold crucible deployment in La Hague facility: the feedback from the first four years of operation, in *Proc. Waste Management Symposium* (Arizona, USA, 2015)
79. S.V. Stefanovsky, A.G. Ptashkin, I.A. Knyazev, O.I. Stefanovsky, S.V. Yudintsev, B.S. Nikonov, B.F. Myasoedov, Cold crucible melting and characterization of titanate-zirconate pyrochlore as a potential rare earth/actinide waste form. Ceram. Int. **45**(2019)

80. C.P. Kaushik et al., Barium borosilicate glass—a potential matrix for immobilization of sulfate bearing high-level radioactive liquid waste. J. Nucl. Mater. **358**(2–3), 129–138 (2006)
81. Classification of radioactive waste. IAEA-TECDOC-1744 (Vienna, 2009)
82. Management of radioactive waste. Atomic Energy Regulatory Board, DAE, India. Safety Guide AERB/NPP/SG/O-11 (Mumbai, 2009)
83. Radioactive waste management technology, in Chapter 7: Gaseous Radioactive Wastes (USNRC Technical Training Center), Rev. 0311
84. Air filters for use at nuclear facilities. IAEA TRS No-122 (1970)
85. Treatment and conditioning of the radioactive solid wastes. IAEA TECDOC-655 (Vienna, 1992)
86. R. Vanbrabant, J. Deckers, P. Luycx, M. Detilleux, P. Beguin, 40 years of experience in incineration of radioactive waste in Belgium, IAEA-CSP-6/C (2001)
87. K.C. Pancholi, Suprabha, S. Agarwal, S.K. Solankar, S. Bhandari, S.K. Mishra, S. Ghorui, N.S. Tomar, R.L. Bhardwaj, E. Kandaswamy, M. Martin, A. Sharma, C.P. Kaushik, Plasma pyrolysis and incineration for low level radioactive solid wastes. BARC News Lett. Nov-Dec (2020) 6–10
88. M. Ternovykh, G. Tikhomirov, I. Saldikov, A. Gerasimov, Decay heat power of spent nuclear fuel of power reactors with high burnup at long-term storage, in *EPJ Web of Conferences*, vol. 153 (2017)
89. H.G. Zhao, H. Shao, H. Kunz, J. Wang, R. Su, Y.M. Liu, Numerical analysis of thermal process in the near field around vertical disposal of high-level radioactive waste. J. Rock Mech. Geotech. Eng. **6**, 55–60 (2014)
90. P.D. Ozarde, K.K. Haldar, S. Sarkar, Interim storage of vitrified high-level radioactive waste. Indian Nucl. Soc. News **5**, 29–32 (2008)
91. K. Deepa, A.K. Jakhate, D. Mehta, N.S. Tomar, C.P. Kaushik, K.M. Singh, Estimation of heat generation in vitrified waste product and shield thickness for transportation of vitrified waste product using Monte Carlo techniques. Indian J. Pure Appl. Phys. **50**, 867–886 (2012)
92. Interim storage of radioactive waste packages. Technical Reports Series No. 390 (International Atomic Energy, Vienna, 1998), pp. 54–55
93. https://radioactivity.eu.com/radioactive_waste/is_vitrified_waste
94. https://www.power-technology.com/analysis/featureuk-nuclear-waste-where-its-generated-contained-transported-and-stored (2020)
95. B. Faybishenko, J. Birkholzer, D. Sassani, P. Swift, Geological challenges in radioactive waste isolation: Fifth worldwide review (2016)
96. R.K. Narayan, R.K. Bajpai, Deep geological repositories for vitrified high level long lived wastes, in *Bulletin of Indian Association of Nuclear Chemists and Allied Scientists*, vol. 6 (2007), pp. 224–245
97. D. Savage, The Scientific and Regulatory Basis for the Geological Disposal of Radioactive Waste (Wiley, 1995), pp. 1–64
98. Scientific and technical basis for the near surface disposal of low and intermediate level waste. Technical Reports Series No. 412 (2002)
99. Near surface disposal facilities for radioactive waste. IAEA Safety Standards Series No. SSG-29 (2014)
100. Siting of near surface disposal facilities. IAEA Safety Series No. 111-G-3.1 (1994)
101. Monitoring and surveillance of radioactive waste disposal facilities. Specific Safety Guides No. SSG-31 (2014)
102. J.H. Coobs, J.R. Gissel, History of disposal of radioactive wastes into the ground at Oak Ridge National Laboratory. No. ORNL/TM-10269 (Oak Ridge National Lab., TN USA, 1986)
103. E.R. Vance, B.D. Begg, D.J. Gregg, Geological repository systems for safe disposal of spent nuclear fuels and radioactive waste, in Chapter 10—Immobilization of High-Level Radioactive Waste and Used Nuclear Fuel for Safe Disposal in Geological Repository Systems, 2nd edn. (WP, 2017), pp. 269–295
104. Geological disposal facilities for radioactive waste. Safety Requirements No. WS-R-4 (Vienna, 2006)

105. Geological disposal facilities for radioactive waste. Specific Safety Guide No. SSG-14 (Vienna, 2011)
106. Planning and design considerations for geological repository programmes of radioactive waste. IAEA-TECDOC-1755. (IAEA, Vienna, 2014)
107. R.P. Rechard, Historical background on performance assessment for the waste isolation pilot plant. Reliab. Eng. Syst. Saf. **69**, 5–46 (2000)
108. J. Birkholzer, J. Houseworth, C. Tsang, Geologic disposal of high-level radioactive waste: status, key issues, and trends. Annu. Rev. Environ. Resour. **37**, 79–106 (2012)
109. P. Sellin, O.X. Leupin, the use of clay as an engineered barrier in radioactive waste management—a review. Clays Clay Miner. **61**, 477–498 (2013)
110. D.G. Bennett, G. Sallfors, SSM's external experts' reviews of SKB's safety assessment SR-PSU—engineered barriers, engineering geology and chemical inventory. Initial review phase. Report number: 2016:12. ISSN: 2000-0456
111. M. Buser, K. Verfassen, Repositories for low and intermediate level radioactive waste in Sweden and Finland: a travel report (2019)

Further Reading

1. Handling and processing of radioactive waste from nuclear applications. Technical Report Series No. 402 (IAEA, 2001)
2. Modular design of processing and storage facilities for small volumes of low and intermediate level radioactive waste including disused sealed sources. IAEA Nuclear Energy Series, NW-T-1.4 (IAEA, 2014)
3. Management of radioactive waste from the mining and milling of ores. Safety Guide, WS-G-1.2 (IAEA, 2010)
4. Combined methods for liquid radioactive waste treatment. IAEA TECDOC No. 1336 (IAEA, 2003)
5. Application of ion exchange processes for treatment of radioactive waste and management of spent ion exchangers. Technical Report Series No. 408 (IAEA, 2002)
6. Predisposal management of radioactive waste from nuclear power plants and research reactors. IAEA Safety Standards Series No. SSG-40 (IAEA, 2016)
7. Predisposal management of radioactive waste from nuclear fuel cycle facilities. IAEA Safety Standards Series No. SSG-41 (IAEA, 2016)
8. Application of membrane technologies for liquid radioactive waste processing. Technical Report Series No. 431 (IAEA, 2004)
9. Chemical precipitation processes for treatment of aqueous radioactive waste. Technical Report Series No. 337 (IAEA, 1992)
10. Predisposal management of low and intermediate level radioactive waste. Document No. AERB/NRF/SG/RW-2 (AERB, 2007)
11. Management of radioactive wastes arising during operation of PHWR based NPPS. Document No. AERB/NPP/SG/O-11 (AERB, 2004)
12. Liquid and solid radwaste management in pressurized heavy water reactor based nuclear power plants. Document No. AERB/SG/D-13 (AERB, 2002)
13. *Hazardous and Radioactive Waste Treatment Technologies Handbook*, ed. by C.H. Oh (CRC Press, 2001)
14. *Radioactive Waste Engineering and Management*, ed. by S. Nagasaki, S. Nakayama (Springer, 2015)
15. *Handbook of Advanced Radioactive Waste Conditioning Technologies*, ed. by M. Ojovan (Elsevier, 2011)
16. M.I. Ojovan, W.E. Lee, S.N. Kalmykov, *An Introduction to Nuclear Waste Immobilization*
17. International Atomic Energy Agency, implications of partitioning & transmutation in radioactive waste. Technical Report Series No. 435 (2004)

18. International Atomic Energy Agency, Assessment of partitioning processes for transmutation of actinides. IAEA TECDOC No. 1648 (2010)

19. *RED-IMPACT: Impact of Partitioning, Transmutation and Waste Reduction Technologies on the Final Nuclear Waste Disposal* (2008)

20. *Advanced Separation Techniques for Nuclear Fuel Reprocessing and Radioactive Waste Treatment* (Woodhead Publishing Series in Energy, 2011)

21. *Reprocessing and Recycling of Spent Nuclear Fuel* (Woodhead Publishing Series in Energy, 2015)

22. Ion Exch. Solvent Extr. Ser. Adv. **19** (2009)

23. T. Fanghänel et al., Transuranium elements in the nuclear fuel cycle, in *Handbook of Nuclear Engineering* (2010), pp. 2935–2998

24. B. Bonin, The scientific basis of nuclear waste management, in *Handbook of Nuclear Engineering* (2010), pp. 3253–3419

25. J. VeliscekCarolan, Separation of actinides from spent nuclear fuel: a review. J. Hazard. Mater. **318**, 266 (2016)

26. R.F. Taylor, Chemical engineering problems of radioactive waste fixation by vitrification. Chem. Eng. Sci. **40**(4) (1985)

27. J.A.C. Marples, The preparation, properties, and disposal of vitrified high level waste from nuclear fuel reprocessing. Glass Technol. **29**(6) (1988)

Chapter 9
Nuclear Material Accounting and Control

B. S. Tomar and P. N. Raju

9.1 Introduction to Nuclear Safeguards

The term "nuclear safeguards" refers to all measures established in a state to prevent the diversion of nuclear material from peaceful uses and to enable the timely detection of diversion of any material to the production of nuclear explosive devices [1]. At the international level, the task of ensuring the nuclear safeguards is facilitated by the International Atomic Energy Agency (IAEA), which works with the member states, particularly those which have signed the nuclear non-proliferation treaty (NPT) of 1971. In such cases, the IAEA takes as its responsibility to verify that all nuclear materials in these countries are safeguarded, through periodic inspections and verification programs as a part of comprehensive safeguards agreement (CSA) [2]. Other member states, which have not signed the NPT, are also expected to ensure non-proliferation through other safeguards agreements, such as, item specific safeguards agreement or voluntary offer agreements (VOA). At the state level, the safeguards are implemented by the state system of accounting and control (SSAC), which provides the information about the nuclear materials to the IAEA. Nuclear safeguards have three components, namely.

(a) Nuclear material accounting and control (NUMAC)
(b) Containment and surveillance (C&S)
(c) Physical protection system (PPS)

Nuclear material is a material which can be used for producing nuclear energy either directly through its fission or indirectly by converting it into a fissionable

B. S. Tomar (✉)
Homi Bhabha National Institute, Training School Complex, Anushaktinagar, Mumbai 400094, India
e-mail: tomarbs@hbni.ac.in

B. S. Tomar · P. N. Raju
Formerly Radiochemistry and Isotope Group, Bhabha Atomic Research Centre, Mumbai 400085, India

material. Owing to their potential for threat perception of diversion for non-peaceful purposes, nuclear materials have to be accounted for in any stage of its processing and storage. Further, there has to be complete control over the nuclear material to ensure continuity of knowledge and thereby enhance the ability to deter and detect unauthorized removal of nuclear material. The nuclear material is expected to be always under surveillance of the safeguards personnel. This requires an elaborate physical protection system. The combination of nuclear material accounting and control (NUMAC), containment and surveillance (C&S) and the physical protection system (PPS) constitutes a nuclear safeguards system which every country is expected to have in place so that all nuclear material worldwide is in possession of the authorized agencies only.

Nuclear material accounting refers to activities carried out to establish the quantities of nuclear material present within defined areas and the changes in those quantities within defined periods. For this purpose, the nuclear material has to be accounted for in every stage of its processing in the nuclear fuel cycle, which involves, mining and milling, conversion into nuclear grade form, fuel fabrication, loading into the nuclear reactor for power production or research and radioisotope production, spent fuel storage or its reprocessing and waste management. Depending upon the type of facility, the accounting and control procedures are put in place. The facility is divided into different material balance areas (MBA) for the purpose of NUMAC so that the nuclear material can be accounted and controlled.

The practice of nuclear material accounting as implemented by the facility operator and the SSAC, to satisfy the requirements in the safeguards agreement between the IAEA and the state and as implemented by the IAEA, to independently verify the correctness of the nuclear material accounting information in the facility records and the reports provided by the SSAC to the IAEA is known as nuclear material accountancy [1].

In this chapter, we will focus on the first component, that is, nuclear material accounting and control (NUMAC) as the other two components, namely, C&S and PPS, come under nuclear security and are beyond the scope of this book. A brief description of the nuclear materials is followed by the different aspects of nuclear material accounting and control (NUMAC) practices, which the nuclear facilities should follow as a part of the safeguards system. After explaining the various elements of NUMAC, the accounting procedures are described in detail. This is followed by different methods of physical inventory verification, which includes destructive as well as non-destructive methods depending upon the nature of the sample. The statistical analysis of the NUMAC data is explained, which helps in drawing the inferences about whether any diversion has taken place or not. Finally, the best practices for a reliable accounting and control system of nuclear materials are given for the practicing NUMAC personnel.

9.2   Nuclear Materials

With regard to the safeguards, the nuclear materials refer to special fissionable materials, such as, ^{235}U, ^{239}Pu and ^{233}U as well as source materials, such as, natural or depleted Uranium and Thorium, which can be converted into special fissionable materials.

The most important nuclear material is the nuclear fuel, which consists of compounds of Uranium, Plutonium or Thorium. Some of the isotopes of Uranium (^{235}U, ^{233}U) and Plutonium (^{239}Pu, ^{241}Pu) undergo fission with thermal neutrons and hence are called as fissile isotopes. They are also called as special nuclear materials (SNM) due to their use in nuclear weapons. Some of the isotopes of Thorium (^{232}Th) and Uranium (^{238}U) are converted into above mentioned fissile isotopes, namely, ^{233}U and ^{239}Pu, respectively, upon neutron capture followed by beta decay and hence are called as fertile materials. Nuclear fuels containing high contents of ^{233}U, ^{235}U and ^{239}Pu are susceptible to attaining self-sustaining fission chain reaction, owing to the fission of fissile isotopes due to neutrons generated by spontaneous fission of neighboring even-even isotopes or that generated due to interaction of alpha particles emitted during alpha decay. Hence, these materials have to be handled with utmost care.

The minimum amount of a fissile material, required to achieve a self-sustaining fission chain reaction under stated condition, is known as critical mass. The critical mass for 100% ^{235}U bare sphere is approximately 47 kg (however, in presence of neutron reflecting material, this mass can be reduced significantly) [3], while that for Uranium containing 20% ^{235}U is approximately 400 kg. Accordingly, the Uranium containing more than or equal to 20% ^{235}U is termed as high enriched Uranium (HEU), while that containing ^{235}U below 20% is termed as low enriched Uranium (LEU). Plutonium-based fuels containing more than 90% ^{239}Pu are used in nuclear weapons and are produced in research reactors and hence are also called as research reactor grade Plutonium, while that containing less than 90% ^{239}Pu are produced in power reactors and hence are called power reactor grade Plutonium. The critical mass for research reactor grade Plutonium is 10 kg. These quantities form the basis for arriving at the significant quantities of nuclear materials in their accounting and control exercise as discussed below.

Natural Uranium (as UO_2) is the main fuel element which is used in nuclear power production as fuel in pressurized heavy water reactor (PHWR), while in pressurized water (PWR) or boiling water (BWR) reactors, LEU is used as fuel. Fast breeder reactors (FBR) employ U–Pu mixed oxides or carbides as fuel. The ratio of Uranium to Plutonium may vary depending upon the type of reactor. Thorium-based fuels are expected to be launched in the future in the form of advanced heavy water reactor (AHWR) which may contain mixed oxides of Th-^{233}U and Th-Pu.

Since heavy water is used as moderator in PHWRs, and upon neutron capture reaction with ^2D, gives ^3T, the moderator over a long period of time gets enriched with ^3T, which is a nuclear material of interest for thermonuclear weapons. Hence, heavy water is another important nuclear material from the point of view of nuclear material accounting.

9.2.1 Significant Quantity (SQ)

The quantity of a nuclear material, for which the possibility of manufacturing a nuclear explosive device cannot be excluded, is called significant quantity (SQ). It is different from critical mass defined above and which depends upon the shape and surrounding neutron reflecting materials. The significant quantity essentially reflects the order of magnitude calculated to obtain a critical mass of the fissile material. SQ for different nuclear materials is given in Table 9.1. In case of natural Uranium and Thorium, the SQ is determined by the total quantity which needs to be processed to generate 1SQ of SNM.

9.3 Nuclear Material Accounting and Control (NUMAC)

Implementation of safeguards in a nuclear facility requires credible evidence to show,

(a) The continuous existence (static or dynamic) of the material in safe custody
(b) Detection of any undesirable loss (or theft) of material within certain acceptable statistical limits, and
(c) Identification of other safety, security, economic and efficiency aspects.

NUMAC consists of maintaining records of the physical inventory of nuclear material in every material balance area (MBA), and its movement from one MBA to another. Periodic verification of the inventory of nuclear material in every MBA is carried out by suitable verification method depending upon the type of facility, bulk or item handling. Attempts are made to minimize the uncertainty in measurements, particularly in case of bulk handling facility. The period over which the verification

Nuclear material	Timeliness	SQ (kg)
^{235}U as LEU	1 year	75
^{235}U as HEU	4 weeks	25
Natural Uranium	1 year	10,000
Depleted Uranium	1 year	20,000
Thorium	1 year	20,000
Plutonium (reactor grade)	4 weeks	8
Plutonium (weapon grade)	1 week	4
^{233}U	4 weeks	8

Table 9.1 Time line for reporting of Material Balance Report and Shipper Receiver Report for different nuclear materials [4]

has to be carried out is called as material balance period (MBP). Every facility has to identify the key measurement points (KMP) where the samples will be drawn from the process stream or storage tanks. The quantity of nuclear material present in the facility in the beginning of a MBP is called the beginning inventory, while that present at the end of MBP is called the ending inventory. The difference between the beginning inventory and the ending inventory after taking into account the receipts and shipments at any MBA is called as the material unaccounted for (MUF). In the case of item handling facility, the MUF should ideally be zero and any deviation from zero indicates a gross error in counting or diversion. In the case of bulk handling facility, there is always a chance of mismatch between book inventory and physical inventory owing to several factors, such as, uncertainty in measurement of nuclear material, inadequate sampling and hold up in process lines. In such cases, the standard error of MUF becomes an important quantity to monitor, which will be discussed in detail in the proceeding sections.

9.3.1 Types of Facilities

The state system of accounting and control (SSAC) of special nuclear material in every country envisages every facility to have an in charge of NUMAC. The operator of the facility will submit periodic reports to the NUMAC in charge, who, in turn, is responsible for filing the accounting reports timely to the SSAC for onward transfer to IAEA in case of safeguarded facilities or to maintain its own database so as to ensure the domestic safeguards are in place.

9.3.1.1 Item Handling Facilities

These are the facilities in which the nuclear material is in a form that can be counted item wise. For example, a nuclear reactor contains fuel rods clad in a suitable cladding material and hence each fuel rod or bundle can be identified from an indentation number. Similarly, the spent fuel storage bay containing the fuel rods can be monitored in terms of each fuel rod. Such facilities come under the category of item handling facilities. In such facilities, accounting of nuclear material can be done nondestructively, by placing suitable detectors/ sensors. Many countries do not follow the closed fuel cycle and hence use the interim storage facilities to store the spent fuel which ultimately may be buried in a deep geological repository. Such spent fuel storage facilities also come under the item handling facilities and require surveillance for a long period of time not only from the safeguards point of view but also for the safety and security purposes.

9.3.1.2 Bulk Handling Facilities

A bulk handling facility may involve a large complex, housing different facilities for handling nuclear material. For example, a fuel conversion facility receives, crude uranium product in the form of sodium di-uranate (SDU), ammonium di-uranate (ADU) or magnesium di-uranate (MDU). The crude product may contain Uranium to the extent of 60–70% depending upon the purity and moisture content. It is treated in the purification plant to be converted into nuclear grade Uranium as ADU, which is subsequently converted into Uranium oxide (UO_2) and sent to fuel fabrication plant. The oxide is heated to increase its density and packed into fuel pins, which are assembled in specific configuration to make a fuel bundle. The fuel bundles are then shipped to the reactor for producing electricity. In all these process stages, Uranium has to be accounted for. Owing to the complex nature of operations and facilities, the entire plant is divided into different areas from the point of view of accounting.

Fuel reprocessing plant is another bulk handling facility wherein the spent fuel, cooled for a sufficient length of time, is reprocessed to recover Uranium and Plutonium, leaving minor actinides (MA) and long-lived fission products in the high level liquid waste (HLW). The separation process involves series of solvent extraction of Uranium and Plutonium into organic phase (30% Tributyl Phosphate in dodecane) followed by partitioning of U and Pu and lastly purification of the recovered U, Pu products.

Enrichment plants converting natural Uranium into LEU or HEU are another class of bulk handling facility. The enrichment process may be based on gaseous diffusion or gas centrifuge. In the gaseous diffusion process, the UF_6 gas is made to pass through porous membranes at high pressure. Dependence of the rate of diffusion on mass of the molecule results in isotopic separation of gas molecules after a large number of barriers. In the gas centrifuge process, the UF_6 molecules are made to rotate in a series of cylindrical vessels, wherein the centrifugal force results in concentration of the heavier molecule toward the periphery and smaller molecule toward the center of the cylinder. After a series of such centrifuges, the fraction enriched in lighter isotope is extracted, while that enriched in heavier isotope is fed into the cascade for further enrichment.

9.3.2 Material Balance Area (MBA)

An MBA is a geographical area and a part of the nuclear facility which receives the nuclear materials from another MBA or facility, processes it and ships the product to another MBA or facility. It is one of the basic units of the facility for NUMAC purposes and is expected to have its own records of beginning inventory, all receipts and shipping data as well as the ending inventory. The facility operator will generate materials balance report (MBR) containing above information. Small facilities may be operated as one MBA, while large bulk handling facilities such as fuel fabrication facilities or enrichment plants are divided into more than one MBA. Depending upon the quantity of nuclear material handled in the MBA, the frequency of generating the MBR is decided.

Thus, while designing the facility, it is imperative to include the features so that the flow of nuclear material and its accountancy can be monitored uninterruptedly. In the case of safeguarded facility, the design of the plant has to be approved by IAEA for accounting purposes.

9.3.3   Key Measurement Points (KMP)

The locations in a facility at which the samples are drawn for the purpose of NUMAC are called KMPs. Sampling may be required at the input stage (receiving station) and output stage (final product before packing and shipping) of the plant to determine the beginning inventory (BI) and ending inventory (EI), respectively, for the MBA. Thus, every MBA will have at least two KMPs. In most cases, the sampling process for NUMAC purposes can also be used for quality assurance purposes as well.

9.3.4   Database Management

The information about the inventory of the SNM in different stages of the facility, receipts and shipments and that lying in waste form is stored in database management system. The database contains all the information about the quantity and type of material, including its physical and chemical form, enrichment with respect to the fissile isotope and the location. The database is utilized to obtain the periodic reports on any inventory changes known as inventory change report (ICR), report on the balance of the material in the facility, known as material balance report (MBR) and description of the form and location of the material, known as inventory distribution report (IDR). In addition, all records of receipts and shipments by the facility are maintained and reported periodically in the form of shipper receiver reports (SRR).

9.3.5   Reporting System

The database management system mentioned above is used to prepare the two reports, namely, ICR (or MBR) and SRR for every MBA to be submitted by the operator to the NUMAC in charge who, in turn, communicates the same to the SSAC for the country. The SSAC transmits these reports to the IAEA. Country-specific agreements, which exist between the agency and the member states, determine the type of safeguards measures to be followed by the country.

9.3.6 Material Balance Period (MBP)

The time period in which an MBR has to be generated for an MBA is defined as the material balance period. It is dependent upon the quantity of nuclear material handled in the MBA. For example, if an MBA handles 1SQ of nuclear material in one year, it has to generate one MBR in one year. If more than 1SQ is handled in one year, the MBP will be proportionately reduced, with minimum time period of one month for generating one MBR. Table 9.1 gives the time line for submitting of MBR and SRR for different nuclear materials. Essentially, it is an indicator of the time required to separate one significant quantity of fissile materials for the purpose of making a nuclear device and hence any diversion must be detected within the time period.

9.3.7 Material Unaccounted for (MUF)

The various activities in a plant can be described as, receipt of the nuclear material, its processing to obtain the finished product which may be shipped to another facility. In the process, some nuclear material is generated as waste, which though measured but is not suitable for further use and is called as measured discard (MD). At the end of an MBP, the total quantity of the SNM present in the MBA is measured to determine the physical inventory (PI). This process is called physical inventory taking (PIT). Physical inventory (PI) is the sum of all the measured or derived estimates of batch quantities of nuclear material on hand at a given time within an MBA, obtained in accordance with specified procedures. The PI is compared with the book inventory in the database management system to check for the correctness of the accounting system. However, due to the uncertainty in measurements, the physical inventory may not exactly match with the book inventory. For this purpose, a quantity called, material unaccounted for (MUF) is defined as,

$$MUF = BPI + R - S - MD - EPI \qquad (9.1)$$

where BPI is beginning physical inventory, R is total receipts, S is total shipments, MD is measured discards and EPI is ending physical inventory.

The difference between the book inventory and the actual physical inventory is also called as defect, which can be classified into the following three categories. (i) Gross defect refers to an item or a batch that has been falsified to the maximum extent possible so that all or most of the declared material is missing. (ii) Partial defect refers to an item or a batch that has been falsified to such an extent that some fraction of the declared amount of material is actually present. (iii) Bias defect refers to an item or a batch that has been slightly falsified so that only a small fraction of the declared amount of material is missing.

Ideally, MUF should be zero if there is no loss of material during the MBP. In the case of item handling facility, the MUF should be zero and any deviation from zero is indicative of accounting problem or diversion. However, in the case of bulk handing facility, it is practically not possible to maintain zero MUF due to the uncertainty in measurement of nuclear material at various key measurement points as well as hold up in process streams. However, attempts should be made to keep MUF at lowest practicable level. In practice, the resulting MUF should be so small that it is within the uncertainty of the eventual material balance established for each facility.

The uncertainties in measurement of nuclear materials assume their significance from the point of view of minimizing the MUF. At the end of an MBP, physical inventory taking (PIT) is undertaken to verify if the recorded data in the MBR are correct. For this purpose, the ending inventory is monitored for its correctness. The ending inventory of an MBP becomes the beginning inventory of the next MBP. Thus, for a bulk handling facility, the quantities in MUF Eq. (9.1) are measured quantities and hence have their own uncertainties. The sources of these uncertainties are (i) measurement uncertainties, (ii) nature of processes and (iii) operator's measurement uncertainties associated with BPI, R, S, MD and EPI. Since each quantity has its standard error or standard deviation (σ), the MUF has a standard error (deviation) denoted as σ_{MUF}, which is calculated using Eq. 9.2.

$$\sigma^2_{MUF} = \sigma^2_{BPI} + \sigma^2_R + \sigma^2_S + \sigma^2_{MD} + \sigma^2_{EPI} \tag{9.2}$$

The verification of the physical inventory needs to take note of not only MUF but σ_{MUF}, as a zero MUF associated with large σ_{MUF} can also be indicative of slow diversion of nuclear material from the facility. This is particularly important from the point of view of insider threats. Assuming the errors in measurements of nuclear materials as random in nature, it can be shown that the different MUF values will fall on a Gaussian and hence σ_{MUF} can be used as a measure of the accounting error during PIT. It can be inferred that the $3\sigma_{MUF}$ being more than 1SQ over the time period of one MBP can be taken as a signature of diversion of nuclear material with a confidence level of 99.7% (see Example 1). This forms the index of the performance of the accounting system. So we have to understand the system (of accounting and of the fuel cycle) before arriving at inferences.

9.4 Statistical Aspects in Nuclear Material Accounting

Nuclear material accounting involves different accounting methods depending on the type of nuclear facility. For example, in a reactor, only item counting is done, whereas in case of conversion/fabrication/reprocessing facilities, accounting is based on measurements. This is where statistics comes into picture as measurements are always associated with uncertainties. The MUF data over a long period of time are to be analyzed statistically so that the operator/inspector will have an idea of the performance of the nuclear material accounting system. Different methods of statistical analysis are given below.

9.4.1 Classification of Data

The outcome of a nuclear material accounting exercise in an item handling facility will be a random variable which is discrete, e.g., number of fuel pellets in a pin, number of fuel pins in a bundle, etc. The data are arranged in ascending (or descending) order and classified into number of classes called class intervals. The ends of the class interval are called class limits (lower and upper). The number of items which falls in any class interval is called class frequency. A frequency table is prepared from the data. This can also be presented in the form of a histogram (Fig. 9.1a). If the mid points of the class intervals corresponding to the frequencies are joined, the resulting figure is called the frequency polygon (Fig. 9.1b). As the number of observations increases infinitely, the polygon is called frequency curve.

Mean and variance

The parameters from the random samples involve determination of mean (x) and variance (s^2),

Mean is given by

$$\bar{x} = \sum_{i=1}^{n} \frac{x_i}{n} \tag{9.3}$$

and standard deviation,

$$s^2 = \sum_{i=1}^{n} \frac{(x_i - \bar{x})^2}{n - 1} \tag{9.4}$$

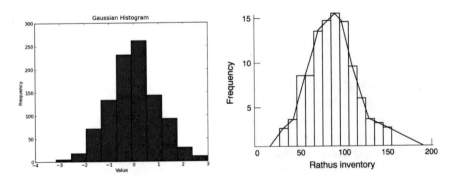

Fig. 9.1 a Frequency histogram. **b** Frequency polygon

where $n-1$ is degrees of freedom. Relative standard deviation, R.S.D $= (s/\bar{x})*100\%$, is a better estimate.

In case of frequency distribution, Mean,

$$\bar{x} = \frac{\sum_{i=1}^{k} f_i x_i}{\sum_{i=1}^{k} f_i} \tag{9.5}$$

$\bar{x}$ and s^2 obtained from the sample are the estimates of mean (μ) and variance (σ^2), respectively, of the entire population. It is important to know how good is the estimate. For this purpose, confidence intervals are defined. The parameter value can be said to lie within the confidence interval within the specified degree of confidence.

The probability that the random variable will lie in a particular interval x_1 to x_2 can be expressed using the population density function, such as,

(a) **Binomial distribution**

$$f(x) = \frac{n!}{x!(n-x)!} p^x (1-p)^{n-x} (x = 0, 1, 2, \ldots n) \tag{9.6}$$

where p is the probability that any given observation is a success and it is constant. N is the total number of observations. The distribution function $F(x)$ in the case of discrete random variables is obtained from density function $f(x)$ by the summation,

$$F(x) = \sum_{(i=1)}^{k} f(x), \text{where } x_1 < x_2 < \cdots x_k \tag{9.7}$$

Binomial distribution is used when the inspector wants to decide the sample size to detect the defects in a large population, e.g., the number of drums (n) to be drawn for testing the proportion (p) of defective drums with an error of e (%), within a certain confidence interval.

The confidence interval for population proportion ($P^\wedge$) is given by,

$$P^\wedge = p \pm z\sqrt{(pq/n)} \tag{9.8}$$

where $p =$ sample proportion, $q = 1-p$, $z = 1, 2$ and 3 for confidence interval of 68.3%, 95.4% and 99.7%, respectively.

Then with the given precision rate, the acceptable error "e" can be expressed as, $e = z\sqrt{(pq/n)}$, $e^2 = z^2 pq/n$, which gives,

$$n = \frac{z^2 p \cdot q}{e^2} \tag{9.9}$$

This formula gives the sample size in case of infinite population. For finite population,

$$n = \frac{z^2 pq N}{e^2(N-1) + z^2 pq} \tag{9.10}$$

(b) Poisson distribution

In the binomial distribution, if probability of occurrence of an event is infinitely small and n is so large that the factor np is a finite number, then the distribution is called Poisson distribution. The probability of n successes is given by

$$\Pr(n\ successes) = e^{-\mu}\mu^n/n! \tag{9.11}$$

where e is the base of the natural logarithms ($e = 2.7183$).
  In Poisson distribution, Mean = Variance = μ.
  Ex: 1. Emission of particles by radioactive element due to decay.
  Ex: 2. Receiving telephone calls during a time period.

(c) Normal density distribution

When the measured random variable takes continuous values, the data follow normal density function given as,

$$f(x) = \frac{1}{\sqrt{2\pi\sigma^2}} e^{-\frac{(x-\mu)^2}{2\sigma^2}}, \quad -\infty < x < \infty \tag{9.12}$$

The distribution function $F(x)$ is given by,

$$F(x) = \frac{1}{\sqrt{2\pi\sigma^2}} \int_{-\infty}^{x} e^{-\frac{(x-\mu)^2}{2\sigma^2}} dx \tag{9.13}$$

The integral cannot be evaluated in the closed form and requires numerical integration.
  Substituting, $z = (x-\mu)/\sigma$ in Eq. (9.13) gives,

$$F(z) = \frac{1}{\sqrt{2\pi}} \int_{-\infty}^{z} e^{-\frac{z^2}{2}} dz \tag{9.14}$$

The mean and variance of z are equal to 0 and 1, respectively. $F(z)$ values are available in tabular form for positive values of z. Figure 9.2 shows the normal curve. Area under normal curve is 1.

Example: Assuming that the random variable x has $\mu = 3$ and $\sigma = 4$ units. Find the probability that a given value of z is less than 4 units, that is, find $\Pr(x < 4)$.

Using the transformation $z = (x-3)/4$,
$\Pr(x < 4) = \Pr((x-3)/4 < (4-3)/4) = \Pr(z < 0.25) = 0.5987$ from the table for F.

Fig. 9.2 Normal density distribution

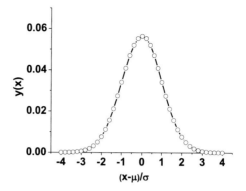

9.4.2 Sampling and Statistical Inference

Statistical inference consists of making conclusions of a population on the basis of a given random sample. The sample must be a representative of the population. We make certain observations on the sample, and based on those results, we make meaningful inferences. Random sampling involves selection of items from a population such that every item in the population has the same probability of being selected. For selection of UO_2 drums, random number tables can be used and for taking representative samples from the drums, we may have to make use of proper mixing, blending, etc., for homogenization of the material. Still, if the material is heterogeneous, samples from different positions of the drum can be taken and mixed properly (stratification).

9.4.3 Hypothesis Testing

While proceeding with statistical treatment of the data, a statement is made which is called null hypothesis and is denoted by H_0.

$$H_0 : \overline{x} = \mu \tag{9.15}$$

where $\overline{x}$ = sample mean, and μ = population mean. After calculations, we make a decision to accept or reject H_0. Here, four possibilities exist:

1. H_0 is actually true, but it is rejected.
2. H_0 is actually false, but it is accepted.
3. H_0 is actually true, it is accepted.
4. H_0 is actually false, it is rejected.

With 3 and 4, there is no problem. If 1 happens, a type I error is committed. If 2 happens, a type II error is committed.

Type I error is designated by α, and type II error by β. Usually, we take $\alpha = 0.05$, that is, we are $(1-\alpha)$ $100\% = 95\%$ confident that the observed value lies within the interval of expected value. This interval is called confidence interval. The region of rejection is called critical region.

Suppose we have values $x_1, x_2, ... x_n$ and $\overline{x}$ is the mean. According to the central limit theorem, means $(\overline{x})$ of samples drawn from a population always follow normal distribution, with mean equal to μ, irrespective of the distribution of x_i.

$$E(\overline{x}) = \mu \qquad (9.16)$$

Further, we assume that co-variance among the values does not exist.
Variance of

$$\overline{x}, \sigma_{\overline{x}}^2 = \frac{\sigma^2}{n} \text{(Infinite population)} \qquad (9.17)$$

$$= \frac{N-n}{N-1} \frac{\sigma^2}{n} \quad \text{(Finite population)} \qquad (9.18)$$

$(N-n)/(N-1)$ is called the finite population correction factor.

9.4.4   Test for Outliers

9.4.4.1   Grubb's Test

During the measurements on the samples drawn from the bulk, it is possible that one or more values may not belong to the set. With a view to test for these outliers, Grubb's test or T_n test may be used. It is assumed that the values follow normal distribution. The values are arranged in ascending order.

$$x_1 < x_2 < ...... < x_n$$
$$Tn = (x_n - \overline{x})/s \qquad (9.19)$$

where $\overline{x}$ and s are the mean and standard deviation. Suppose x_n is the suspect value. Then, Null hypothesis H_0 is that x_n belongs to the set.

If T_n < critical value (from table), at given significance level, the hypothesis is accepted.

If the suspect value is x_1, then

$$T_1 = (\overline{x} - x_1)/s \qquad (9.20)$$

9.4.4.2 Dixon's Q Test

$$Q = (x_2 - x_1)/(x_n - x_1) \tag{9.21}$$

$$or \quad Q = (x_n - x_{n-1})/(x_n - x_1) \tag{9.22}$$

where x_1 and x_n are suspected outliers in Eqs. 9.21 and 9.22, respectively. The denominator is the range of the values. The observed Q value is compared with the critical (from the Q tables) value at a certain confidence level (usually 95%) and a decision is made.

9.4.5 Confidence Limits/Intervals

For μ with known σ (Fig. 9.2):

$$(\text{z statistic})z = \frac{(\bar{x} - \mu)\sqrt{n}}{\sigma} \tag{9.23}$$

$$\mu = \bar{x} \pm z_{1-\alpha/2} \; \sigma/\sqrt{n} \tag{9.24}$$

For μ with unknown σ:

$$\mu = \bar{x} \pm t_{1-\alpha/2} \; s/\sqrt{n} \quad \text{(Student's } t)$$

Taking $\alpha = 0.05$ means that 95% confidence level is being used. Let us compare sample mean with population mean or true/theoretical value, as in NM accounting, where observed MUF is compared with expected MUF. From the above,

$$t = (\mu - \bar{x})\sqrt{n}/s \tag{9.25}$$

For a given value of α, and degree of freedom, observed 't' value is compared with the tabulated value to conclude, if the sample mean is a true representative of population mean.

Comparing two sets of values

Let us consider two sets of data with sample sizes n_1 and n, means $\bar{x}_1$ and $\bar{x}_2$ and standard deviations s_1 and s_2. Here, the pooled standard deviation (s_p) is calculated as below:

$$s_p = \sqrt{[\{(n_1 - 1)s_1^2 + (n_2 - 1)s_2^2\}/(n_1 + n_2 - 2)]} \tag{9.26}$$

The t value is given by,

$$t = (\bar{x}_1 - \bar{x}_2)/s_p\sqrt{(1/n_1 + 1/n_2)} \tag{9.27}$$

Examples are:

a. Shipper receiver difference comparison. (Steps to reduce S-R difference)
b. Difference between operator's statement and corresponding audit team's statement.
c. Difference between two analytical methods.
d. Difference between two analysts, etc.

Chi square test for normalcy

This test is used to find whether there is any difference between actual and expected frequencies.

$$\chi^2 = \sum [(f - f_1)^2/f_1] \tag{9.28}$$

where f is observed frequency.

F test for comparing two sample variances

We have to test whether one dataset is more reliable than the other. Let A and B be the two methods by which the two sets of data are produced. We should determine whether method A should be replaced by method B. If the variance of B is greater than that of A, then it will not be replaced. Number m and n of observations are made on a single well blended sample of material by the two methods A and B, respectively, and the variances are calculated for each set. Let the variances be s_A^2 and s_B^2. The F statistic is given by

$$F = S_A^2/S_B^2 \tag{9.29}$$

with $m-1$ and $n-1$ degree of freedom.

Comparing observed F value with the value in F-table, if the observed value is less than the tabulated value, we can say method B is better.

9.4.6 Statistical Treatment of Uncertainties Associated with MUF

Since all the quantities in RHS of MUF Eq. (9.1) are measured quantities and they are associated with uncertainties, the uncertainties associated with MUF can be calculated using Eq. 9.2. As the uncertainties are random in nature, they follow normal distribution, we can say that at 95% confidence level,

$$\text{Obs. MUF} = \text{True MUF} \pm 1.96\sigma_{\text{MUF}} \tag{9.30}$$

In an ideal case, we expect MUF to be zero, that is,

$$\text{Obs. MUF} = \text{Zero} \pm 1.96\sigma_{\text{MUF}} \approx \pm 2\sigma_{\text{MUF}} \tag{9.31}$$

This is called Limit of Error on MUF (LEMUF). The significance of LEMUF is that if observed MUF in a particular MBP is less than LEMUF, then the nonzero MUF could arise due to uncertainties in measurement. However, if MUF in a particular MBP is greater than LEMUF, then MUF should be declared as significant and needs investigation.

Expected measurement uncertainties associated with closing of material balance are given below.

Uranium enrichment 0.002.
Uranium fabrication 0.003.
Plutonium fabrication 0.005.
Uranium reprocessing 0.008.
Plutonium reprocessing 0.010.
Scrap storage 0.04.
Waste storage 0.25.

In a large reprocessing plant, large MUF requires investigation to find the reasons for it. Under diversion-free and theft-free conditions, MUF is only due to measurement errors, then measurement control program needs to be implemented.

9.5 Physical Inventory Taking (PIT)

The facility is expected to undertake physical inventory taking (PIT) at the end of MBP. It is at this point of time that the SSAC, the regulating body comes into action to verify, not only the records at the facility but also the physical inventory. The inspectors from the regulatory body are expected to visit the facility and verify the physical inventory recorded by the facility operator in the book. Thus, the physical inventory verification (PIV) by the regulator can be taken along with PIT by the operator.

The measurement methods and techniques must be selected in such a way as to ensure that the measurement accuracy conforms to requirements defined by the SSAC and is based on internationally accepted values, procedures and standards. Each nuclear facility must establish and maintain documented procedures to control, calibrate and maintain measuring and test equipment and methods to ensure the credibility of the facility material balance statements. The different experimental techniques employed for physical inventory taking and the internationally accepted target uncertainties are given in Table 9.2.

Some of the techniques used for PIT are described below.

Table 9.2 Experimental techniques for physical inventory taking and their internationally accepted target measurement uncertainties [5]

Technique	Analyzed for	Type of material	Uncertainty random (%)	Uncertainty systematic (%)
Biamperometry/ Davis Gray	U, Pu	Oxides	0.2	0.2
K-edge densitometry	Th, U, Pu	U–Pu, U-Th, Pu-Th	0.2	0.2
K-X-ray fluorescence	Pu	Pu materials	0.2	0.2
Wavelength dispersive X-ray fluorescence	U, Pu	Pure U, Pu oxides, MOX	0.3	0.3
Isotope dilution mass spectrometry	U, Pu	Dissolver solution, U, Pu solutions, HLW	0.1	0.1
Pu(VI) spectrophotometry	Pu	Pu, U–Pu	0.2	0.2
α-spectrometry	Am, Cm, Np	HLW, dissolver solution	5.0	5.0
Thermal ionization mass spectrometry	Pu and U isotopes	All U, Pu materials, dissolver solution	0.05	0.05
High resolution γ-ray spectrometry (HPGe)	Pu isotopes, Am, Np	Pure U and Pu materials	0.05–2.0	0.05–2.0
γ-ray spectrometry (NaI(Tl))	U235	LEU	0.2–0.5	0.2–0.5

9.5.1 Destructive Methods of PIT

9.5.1.1 Electro-analytical Techniques for Precise Determination of Uranium

(a) **Davis and Gray method**: The method was developed by Davies and Gray [6] and is based on (i) reduction of U(VI) to U(IV) by excess of Fe(II) in presence of strong phosphoric acid and sulphamic acid, followed by (ii) selective oxidation of remaining Fe(II) to Fe(III) by Mo(VI) as catalyst, and finally (iii) dilution of the titrant solution with water to reduce phosphoric acid concentration and potentiometric titration of U(IV) with dichromate solution of known concentration. The method has been found to determine Uranium concentration with high precision ((<0.2%) and accuracy. Further, there is no interference from other metal ions as well as nitric acid. The method is now accepted internationally for accounting of Uranium for safeguards purposes. Further details are given in Chap. 4.

(b) **Biamperometry**: A convenient and rapid method for determination of Uranium by biamperometric end point detection has been developed by Mary Xavier et al. [7]. The method is based on reduction of U(VI) by Ti(III) and destruction of excess Ti(III) by nitric acid. Fe(III) is added to oxidize U(IV) to U(VI) till the amperometric end point, and finally, the Fe(II) formed during the above redox titration is titrated with standard dichromate solution to amperometric end point. The concentration of Fe(II) is equivalent to that of U(VI) in the sample solution. The method does not have any interference from Pu and Fe present in the sample. More details are given in Chap. 4.

9.5.1.2 Electro-analytical Techniques for Precise Determination of Pu

Drummond and Grant developed a potentiometric method for determination of Plutonium in fuel samples [8]. The method is based on oxidation of all Plutonium to Pu(VI) using AgO as oxidant in nitric acid medium, destruction of the excess oxidant by sulfamic acid, reduction of Pu(VI) with excess Fe(II) and potentiometric titration of excess Fe(II) with standard dichromate solution. The analysis can be done in a single vessel under cold condition and is complete in about 20 min. Further details are given in Chap. 4.

9.5.1.3 Mass Spectrometry

(a) Determination of isotopic composition of U, Pu: Uranium-based fuels contain Uranium isotopes of ^{234}U, ^{235}U and ^{238}U, while Plutonium bearing fuels contain Plutonium isotopes of ^{238}Pu, ^{239}Pu, ^{240}Pu, ^{241}Pu and ^{242}Pu. Accurate and precise determination of isotopic composition of Uranium and Plutonium is of paramount importance from the point of an accurate estimate of fissile material inventory in any material balance area. The most commonly employed technique for determination of isotopic composition has been thermal ionization mass spectrometry (TIMS), in which the element of interest (viz., U, Pu) is chemically separated and is deposited as a solid matrix on a filament (Re or Ta) which is heated electrically to evaporate and ionize the atoms of the element of interest. The ionized atoms are analyzed using a sector focused magnetic analyzer and detected by Faraday cup (single or multiple). The advantage of TIMS is the high precision and accuracy as low as 0.01%. Chemical separation is required to eliminate isobaric interference, e.g., between ^{238}U and ^{238}Pu. Further, the method requires very small (micrograms to nanograms) quantity of the sample. The technique can also be used for determining the concentration of the element, in isotope dilution mass spectrometry (IDMS) mode using one of the rare isotopes as spike (viz., ^{242}Pu for Pu and ^{236}U for U). The principle of IDMS is based on the dilution in the isotopic ratio of the different isotopes with respect to the spike which is proportional to the concentration of the element. Table 9.3 gives the list of isotopes of Uranium and Plutonium along with their half-life and decay characteristics [9].

Table 9.3 Nuclear data of U and Pu isotopes [9]

Isotope	Half-life	Decay	Eγ (keV)	Iγ(%)
^{234}U	2.455×10^5 y	α	53.20	0.123
^{235}U	7.04×10^8 y	α	185.715	57.0
			143.76	10.96
			58.570 D (^{231}Th)	0.462
^{236}U	2.342×10^7 y	α		
^{238}U	4.468×10^9 y	α	63.29 D (^{234}Th)	3.7
^{238}Pu	87.7 y	α	152.72	9.29×10^{-04}
^{239}Pu	24,110 y	α	129.296	6.31×10^{-03}
			144.201	2.83×10^{-04}
			161.450	1.230×10^{-04}
			203.550	5.69×10^{-04}
			332.845	4.94×10^{-04}
			345.013	5.56×10^{-04}
			413.713	1.466×10^{-03}
^{240}Pu	6561 y	α	160.308	4.02×10^{-04}
^{241}Pu	14.329 y	β-	148.567	1.860×10^{-04}
			208.005 D (^{237}U)	21.2
			164.61 D(^{237}U)	1.86
			332.35 D(^{237}U)	1.200
			267.35 D(^{237}U)	0.712
^{242}Pu	3.75×10^5 y	α		

D stands for daughter isotope in secular equilibrium

IDMS coupled with alpha spectrometry has been used to determine the half-life of several actinide isotopes. In nuclear forensics, isotopic composition measurements provide valuable information about the age of the fuel sample, e.g., buildup of ^{241}Am due to beta decay of ^{241}Pu in an aged Pu sample is a measure of the time elapsed after the Pu sample was purified.

9.5.2 Non-destructive Methods for Physical Inventory Taking

From the point of view of accounting of nuclear materials, the precision and accuracy of inventory should be better than 0.5%, which is achieved using destructive methods such as, electro-analytical and /or mass spectrometry techniques as described above. However, in many situations, particularly if the nuclear material is in a finished form or in a waste form which cannot be dissolved for the assay by above mentioned techniques, it is required to assay the fuel material by non-destructive assay (NDA) method. NDA methods are based on the measurement of nuclear radiations emitted by the isotopes, either spontaneously (passive methods) or upon bombardment with suitable projectiles to induce a nuclear reaction, viz., fission (active method).

9.5.2.1 Passive Methods for Non-destructive Assay (NDA) [10]

Passive methods of NDA rely upon the measurement of radiations, such as, gamma rays or neutrons emitted during the radioactive decay of the isotopes of U, Pu. Alternatively, the heat produced by the decay radiations in a medium can also be used to deduce the quantity of the nuclear fuel in a sample.

(a) **Gamma ray spectrometric determination of U, Pu**: Fortunately, most of the isotopes of U and Pu emit gamma rays during their primary mode of decay (alpha, beta). Table 9.3 gives the gamma rays and their abundances for the different isotopes of U and Pu. For a sample of infinitesimally small mass, the gamma ray spectrometry can be used to determine the amount of a radioisotope in a sample. The gamma ray spectra are measured using high purity Germanium (HPGe) detector coupled to a multichannel analyzer, which records the gamma ray spectrum. The peak area (PA) of a gamma line of a radioisotope is related to the amount of the isotope (w in g) by the following equation.

$$PA = \frac{wN_0}{A} x \frac{ln2}{T_{\frac{1}{2}}} I_\gamma \varepsilon_\gamma t \tag{9.32}$$

where N_0 is the Avogadro's number, A is the mass number of the radioisotope, $T_{1/2}$ is the half-life of the radioisotope, I_γ is the branching intensity (abundance) of the gamma ray, ε_γ is the detection efficiency of the gamma ray and t is the counting time. The units of $T_{1/2}$ and t are same. The data on half-life and gamma ray branching intensity are available in the literature, while the detection efficiency is determined by counting a standard source emitting gamma rays of different energies for efficiency calibration of the detector system assuming that there is no attenuation of gamma rays in the sample. However, in the case of a sample of finite mass, the attenuation of the gamma rays in the sample is dependent upon their energy and hence the efficiency calibration obtained with standard source cannot be used for the sample. In view of these limitations, gamma ray spectrometric measurements are generally used for determination of isotopic composition of the elements (U, Pu) as in this case, the relative efficiency calibration is sufficient and which can be generated using the γ-ray spectrum of the sample itself. This method is widely used in determining the isotopic composition of U (^{234}U, ^{235}U and ^{238}U) as well as Pu (^{238}Pu, ^{239}Pu, ^{240}Pu and ^{241}Pu). The fraction of ^{242}Pu (which does not emit any gamma rays) is deduced using isotopic correlations obtained from mass spectrometric measurements.

^{235}U fraction in a Uranium sample, known as enrichment factor, is commonly determined by gamma spectrometric measurements using a HPGe detector. In natural Uranium, the ^{235}U abundance is 0.72%, while in the fuel for boiling water or pressurized water reactors, ^{235}U enrichment is enhanced to 1–3% so as to maintain the neutron economy with light water as a moderator and coolant. Measurement of ^{235}U in these type of low enriched Uranium (LEU) fuels is one of the most important tasks carried out by the safeguards inspectors while verifying the physical inventory of the fuels at any facility under safeguards. Low energy gamma ray spectra of ^{234}U (53 keV), ^{235}U (38 keV) and ^{238}U (63 keV) in a LEU sample can be used for isotopic composition of Uranium using low energy photon spectrometry (LEPS) employing a small volume HPGe detector.

In the case of Pu bearing samples, the determination of isotopic composition of Pu is a prerequisite for determination of total Pu content by techniques based on neutron counting or calorimetry as described below. The sealed sample is counted in a HPGe-based gamma ray spectrometer for a time period sufficiently high to acquire statistically good counts for the weakest of the gamma lines of interest. For this purpose, a low energy gamma ray spectrometer having good energy resolution (typically 550 eV at 121.8 keV) is used. The relative efficiency calibration is carried out using the multiple gamma lines of ^{239}Pu and ^{241}Pu, which in turn are normalized to generate common relative efficiency calibration curve. This relative efficiency calibration curve is used to determine the ratio of amounts of ^{238}Pu, ^{240}Pu and ^{241}Pu with respect to ^{239}Pu. Typical gamma spectrum of NBS Pu standard measured using a low energy photon spectrometer is given in Fig. 9.3.

The ratio of ^{242}Pu with respect to ^{239}Pu is obtained using the isotopic correlations deduced from mass spectrometric measurements on samples of similar reactor origins. The precision and accuracy of isotopic compositions determined using gamma ray spectrometry are in the range of 3–4% [11].

(b) Neutron assay for determination of Pu

Even-even isotopes of Pu, namely, ^{238}Pu, ^{240}Pu and ^{242}Pu undergo spontaneous fission which is accompanied by emission of 2–3 prompt fission neutrons. Measurement of these neutrons by suitable neutron detectors is employed to obtain the total Pu content in a Pu bearing sample in conjunction with the data on isotopic composition of Pu measured by gamma ray spectrometry or mass spectrometry. Detection of neutrons is performed using a high efficiency (^{3}He) gas-filled neutron coincidence counter. The high efficiency is achieved by putting a series of ^{3}He gas-filled proportional counters to form a well in which the sample is placed thereby achieving near 4π geometry and hence high counting efficiency. A typical neutron well coincidence counter (NWCC) has about 24 ^{3}He gas counters arranged vertically to form a cylinder of height 50 cm and diameter 30 cm. A high density polyethylene (HDPE) sheet surrounds the inner wall of the counter to moderate the neutrons before they reach the counters so as to increase the detection probability. As the Pu isotopes undergo alpha decay, which in turn may undergo (α, n) reactions with the low Z elements of the fuel (e.g., PuO_2, PuF_4, etc.), thereby generating accidental (unwanted) neutrons. With a view to minimize the interference from these accidental neutrons, the fission neutrons are

Fig. 9.3 Gamma spectrum
of NBS Pu standard

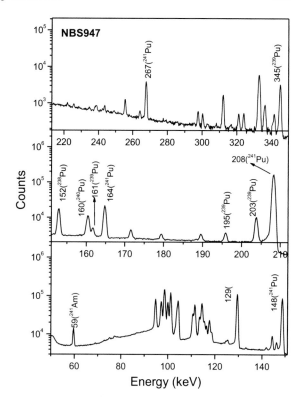

counted in coincidence mode using a shift register logic system, which counts the
neutrons in coincidence with each neutron within a time gate of width equivalent to
the die-away time of the counting system. The die-away time (τ) is the time required
for the neutron to leak out of the system by capture or escape. Typical die-away
time of the NWCC is about 64 μs. Thus, any neutrons counted during the window
(τ) register the real + accidental (R + A) neutrons. After a long gate of about 10
times the τ, another gate equivalent to τ is opened to register any accidental neutrons
(A). By subtracting the accidental (A) neutrons from the total neutrons (R + A),
the real coincidence neutrons can be obtained which upon correcting for the neutron
detection efficiency provide the effective ^{240}Pu content and which after correcting
for the isotopic composition gives the total Pu content in the sample. The method
provides Pu content with an accuracy and precision of about 4–5%. A photograph
of the neutron well coincidence counter is given in Fig. 9.4.

NWCC measures only singles and double coincidence neutrons and thereby can
resolve fission neutrons from accidental neutrons. However, in case of higher Pu
bearing samples, self multiplication of neutrons in the sample cannot be corrected in
NWCC. Neutron multiplicity counter having higher detection efficiency can correct
for self multiplication and hence provides accurate data of Pu content in such samples
[12].

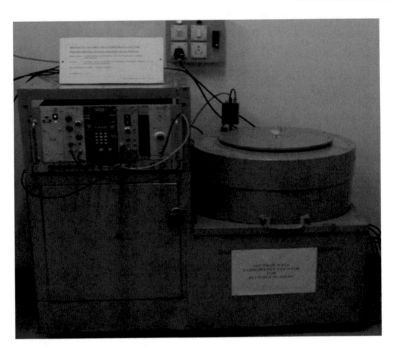

Fig. 9.4 Photograph of neutron well coincidence counter

(C) Calorimetry

The radioactive decay of Pu isotopes generates heat which can be monitored in a calorimeter to obtain the total number of decays in a particular interval of time and hence the total Pu content provided the isotopic composition of Pu is known. The rate of heat energy released (P in MeV/s) during the decay of a radioisotope can be calculated using the following equation,

$$P(MeV/s) = QN\lambda \tag{9.33}$$

where Q is the energy of emitted radiation in MeV, N is the number of atoms in the sample and λ is the decay constant of the radioisotope. Taking 1 MeV/s = 1.602×10^{-13} J/s, N atoms = wxN_0/A (m = mass of the isotope in g, A = atomic mass of the radioisotope, N_0 = Avogadro number and λ = $\ln2/T_{1/2}$) one obtains,

$$P(Watt) = 2119.3 \, QmAT_{1/2} \tag{9.34}$$

where $T_{1/2}$ is half-life of the radioisotope in years.

The specific power of the different isotopes of Pu is given in Table 9.4.

The total power of the sample (W) can be expressed as in terms of the sum of the products of mass of individual isotopes (m_i) and their specific powers (P_i)

Table 9.4 Specific power of different isotopes of Plutonium [9]

Radioisotope	Alpha decay energy (MeV)	Specific power (mW/g)
^{238}Pu	5.592	567.57 ± 0.26
^{239}Pu	5.243	1.9288 ± 0.0003
^{240}Pu	5.255	7.0824 ± 0.002
^{241}Pu	0.0055 (β)	3.412 ± 0.002
^{242}Pu	4.985	0.1159 ± 0.0003
^{241}Am	5.640	114.2 ± 0.42

$$W = \sum m_i P_i \qquad (9.35)$$

Defining the mass fraction of each isotope as $R_i = m_i/M$,

$$W = M \sum R_i P_i \qquad (9.36)$$

The quantity in the summation is the effective specific power, P_{eff} (Watts/g). Thus,

$$M = W/P_{\text{eff}} \qquad (9.37)$$

Measurement of total power of the sample using a calorimeter and deducing the effective specific power from measured isotopic composition and known nuclear data on decay energies and half-life provides the total Pu content (M) in the sample.

Heat flow type isothermal calorimeters are preferred over adiabatic calorimeters for NDA of Pu. The system consists of two chambers, namely, sample chamber and reference chamber thermally isolated from each other. The rate of heat flow between the two chambers is given by,

$$dQ/dt = k\left(T_{\text{sample}} - T_{\text{reference}}\right) \qquad (9.38)$$

where Q is the heat energy, k = thermal conductivity of the medium between the two chambers, T_{sample} and $T_{\text{reference}}$ are the temperature of the sample and reference chambers.

The temperature difference between the sample chamber and the reference chamber is measured by a Wheat Stone bridge circuit. When the sample is placed in the sample chamber, the temperature rises until the heat loss through the medium between the sample and reference chamber becomes equal to the heat generated by the sample. As the equilibrium is established, the temperature difference between the sample and reference chamber is measured and which is proportional to the heat generated by the sample (Eq. 9.38), which in turn gives the Pu content in the sample.

Calorimetric method of NDA of Pu has several advantages, namely, (i) the entire sample can be assayed, (ii) it does not require any standard as the measurements can be traced to SI units (electric current and potential), (iii) the measurements are independent of the composition of the matrix and distribution of Pu in the sample container, (iv) the precision and accuracy of the method is comparable to destructive methods (electro-analytical methods, mass spectrometry), provided the isotopic composition is measured with the same precision and accuracy. The only disadvantage of calorimetry is the long time (typically one day) taken for each assay, owing to the time required for the equilibrium to be established between the sample and the standard.

9.5.2.2 Active Methods for NDA [13]

As mentioned above, the NDA of Uranium-based fuels by passive methods is not very successful owing to attenuation of the gamma rays in the sample. Further, the fission neutron yields in LEU-based Uranium fuels are negligible owing to long spontaneous fission half-lives of ^{234}U and ^{238}U. However, bombardment of Uranium bearing fuels with neutrons can induce fission in fissile isotopes, (^{235}U) and thereby generating prompt and delayed neutrons which can be taken as a measure of ^{235}U content in the sample.

(a) **Active neutron interrogation followed by neutron counting for NDA of ^{235}U**

Fission of ^{235}U with thermal neutrons gives ~2.5 neutrons most of which are emitted promptly, while a small fraction (1 in 500) is emitted following beta decay of some fission products, viz. ^{137}I, ^{87}Br, etc. Active well coincidence counter (AWCC), an analog of NWCC for measurement of SF neutrons in case of Pu is used to determine ^{235}U content in enriched Uranium-based fuels. A neutron source, viz., ^{241}Am, Li is used to induce fission in the fissile isotope and the prompt fission neutrons are counted using the neutron well coincidence counter. In a typical AWCC system, the sample is bombarded with neutrons from two Am, Li sources placed above and below the sample, while the prompt fission neutrons are detected in coincidence mode using a ring of ^{3}He detectors. The system can be used in fast mode, wherein the ^{3}He detectors are covered with Cadmium sheets to remove thermal neutrons, or without any Cadmium sheets around ^{3}He detectors in thermal mode. The fast mode is suitable for higher content of high enriched Uranium, while thermal mode can be used for small quantity of HEU or LEU.

(b) **Active interrogation delayed neutron counting (AIDNEC) for NDA of ^{235}U**

Irradiation of Uranium sample with a neutron source to induce fission followed by counting of delayed neutrons using ^{3}He counters provides better precision and accuracy than AWCC system primarily due to reduced background.

Historically Scientists at Los Alamos National Laboratory in 1960s developed the ^{252}Cf neutron source-based system for inducing fission in ^{235}U and counting the delayed neutrons, while the ^{252}Cf source was placed back in the shielded storage position. Due to the shuffling of the ^{252}Cf source between the storage position and to near the sample, the equipment was called "Shuffler". 1 mg ^{252}Cf source gives 2.34×10^9 neutrons/s. Typical shuffler contains few mg of ^{252}Cf and can be used for about 10 years after which the neutrons yields become too low to generate delayed neutrons significantly above the background counts. The number of delayed neutrons detected is directly proportional to the fissile content in the sample, and a calibration graph with standards is required to obtain the fissile content of routine samples. Typical accuracy of shufflers has been found to be less than 1% for large quantities of Uranium samples. Shufflers were widely used by many laboratories in US for assay of ^{235}U in HEU bearing scraps for accounting purposes. It was also used for assay of fissile content of 200 L waste drums. The disadvantage of the Shuffler is the continuously decreasing neutron yield and hence frequent calibrations.

Plasma focus device is a neutron source which gives a pulse (of few nanoseconds) of 2.45 MeV neutrons following D + D or 14.6 MeV neutrons following D + T fusion. Typical neutron intensity generated in these PF sources is about 10^9 neutrons per pulse. Bombardment of the ^{235}U bearing Uranium sample with thermalized neutrons from a PF source can be used to induce fission in ^{235}U and the delayed neutrons are measured by ^{3}He neutron detector bank. The prompt neutrons vanish in a short time of few micro seconds and the delayed neutrons can be measured after a few milliseconds up to about 100 s to encompass all the six delayed neutron groups having half-lives (seconds) of 0.23, 0.61, 2.3, 6.22, 22.72 and 55.72. The neutron background in such a system is negligible. Typical detection limit for ^{235}U in such AIDNEC system is 18 mg [14]. Such a system has been developed at Bhabha Atomic Research Centre laboratories. With the scarcity of ^{252}Cf sources, PF sources have a potential role in NDA of ^{235}U bearing fuels for NUMAC purposes.

9.5.2.3 Hold Up Monitoring

In a bulk handling facility, such as, fuel fabrication, enrichment, reprocessing plant, significant quantities of nuclear material may be accumulated, over long period of operation, in equipment, transfer lines or ventilation system. Measurement of such held up nuclear materials is important not only from the point of view of accounting and control, but also for radiological safety, criticality as well as waste management. Hold up measurements are carried out non-destructively using passive gamma ray spectrometry and/or neutron counting [10]. Determination of absolute amounts from hold up measurement requires simulation of the detection efficiency for the finite source dimensions and self-attenuation effects.

9.5.2.4 Real-Time NUMAC

Real-time accounting and control of nuclear materials is an important aspect of nuclear safeguards, which also aids in ensuring security of nuclear material. This is achieved by following in process physical inventory taking. For example, K-edge densitometry technique helps in monitoring the concentration of U, Pu in process streams of reprocessing plant while the plant is in operation [15]. A more refined version of this in the form of hybrid K-edge K-XRF densitometer has been developed at Karlsruhe, Germany, for simultaneous determination of concentrations of Uranium and Plutonium in dissolver solution, which provides capabilities for in situ nuclear material accounting [16].

9.6 Measurement Control Program [17]

Monitoring and control of the quality of the measurements of special nuclear materials is an important element of nuclear material accounting and control. This is carried out using a measurement control program. The program involves (i) evaluation of measurement methods, wherein the existing data are checked for the precision and accuracy of measurements, which in turn helps in monitoring the quality. If required, additional data may be collected. The evaluation process also involves calibration of the measuring instruments as well as inter-laboratory comparison experiments, and the statistical techniques used for drawing inferences about the MUF. (ii) Administrative control over the selection or design of facilities, equipment and measurement methods, (iii) control over selection, training and qualification of personnel performing the measurement of SNM and (iv) periodic audit of the facility including records of the material balance. Control of the measurement quality is needed to ensure that a loss, theft or diversion of SNM will not be masked either by bias or large random errors in the measurement data used to calculate MUF.

9.7 Containment and Surveillance Techniques

Safeguards approach for a facility is based on Nuclear Material Accountancy as the fundamental measure of safeguards complimented by Physical Protection and Containment/Surveillance Techniques. According to the International Atomic Energy Agency, Containment is referred to as structural features of a facility, containers or equipment which are used to establish the physical integrity of the area or items including data and equipment. In other words, the proper containment gives assurance about the continuous existence of the nuclear material at a given place. For example, walls of a fuel storage room, spent fuel storage pool, transport flasks and storage containers. This containment integrity is achieved by using doors, locks, vessel lids and seals etc.

Surveillance is the term used to describe the collection of information by (inspectors) authorities using instruments to monitor any unauthorized movements of nuclear material or interference with containment or tampering with seals or data.

Containment techniques involve:

(a) Using tamper indicating locks for storage areas/rooms.
(b) Using tamper resistant seals for the containers.

(c) Using CCTV cameras at the proper places such as fresh fuel storage, spent fuel storage to record the unauthorized or unexplained removal of material.

Seals: Seal is a tamper indicating device used to join movable segments of a containment in a manner such that the access to its contents without opening the seal or breaking of the containment is difficult, e.g., metal cap seals, ultrasonic and electric seals with fiber optic loop, etc. Tamper indicating paper tape seals are used for short-term use.

Normally, C/S measures are applied:

(1) During flow and inventory verification to ensure that each item is verified without duplication and that the integrity of the samples is preserved.
(2) To confirm that there has been no change to the inventory previously verified. They reduce the need for re-measurement.
(3) To isolate nuclear material that has not been verified until it can be measured.

9.8 Analysis of Environmental Samples

Measurement of environmental samples in and around a nuclear facility is an important task, especially when the inspectors want to find out if there is any undeclared activity going on in and around the facility. This exercise has its genesis in the past as some of the member states which signed the NPT, and hence abide by INFCIRC153, but continued the production of SNM clandestinely, though the declared facilities were filing the MBRs as per the requirements from IAEA. The additional protocol INFCIRC 540 calls for voluntarily offering the facilities (other than those declared) for inspection. Measurement of environmental samples for detecting any undeclared activity requires highly improved techniques having very low limits of detection for SNM. For instance, high resolution inductively coupled plasma mass spectrometry (HR-ICP-MS) is used to determine the isotopic composition of U, Pu in swipe samples at ultra-trace (ppt) levels. Determination of ultra-trace (μBq) levels of alpha emitters requires high efficiency surface barrier silicon detectors or solid state nuclear track detectors (SSNTD). High resolution gamma ray spectrometry using high volume shielded HPGe detectors can also provide the information about very low levels of gamma emitters, e.g., ^{235}U and Pu. The samples could be swipe samples or that of soil, water, etc. The same facilities can also be used for nuclear forensic investigations of the interdicted materials.

9.9 Summary

A robust safeguards system at state level helps in giving confidence to the international community of the commitment of the country to the peaceful uses of atomic energy. The state system of accounting and control SSAC has a very important role to play in implementing the safeguards both at national as well as at international level. In this endeavor the accounting and control of nuclear materials is the most important aspect. Trained NUMAC personnel well laid out material balance areas with notified key measurement points coupled with facilities for measurement of inventories of SNM help in establishing the robustness of the NUMAC system. The quality infrastructure for verifying the correctness of the measured quantities with availability of standard reference materials with traceability to primary standards and SI units will generate confidence in verifying the inventories and hence in ensuring the effectiveness of the safeguards system.

Illustrative examples using statistics for drawing safeguards conclusions

Example 1: (Construction of Two-Sided Confidence Interval on the Mean) Five samples of UO_2 powder are measured for percent uranium and the results are the following:

$x_1 = 87.538, x_2 = 87.682, x_3 = 87.553, x_4 = 87.682, x_5 = 87.548$.

Sample Mean $\bar{a} = 87.601$ and Std. Dev. $s = 0.074$, $t = 2.776$ for 95% confidence interval and degree of freedom $= 4$.

The 95% confidence interval for the mean is given by

$$\mu = \bar{a} \pm ts/\sqrt{(n-1)}$$
$$= 87.601 \pm (2.776)(0.074)/2 = 87.601 \pm 0.103$$
$$= 87.498 \text{ to } 87.704$$

Example 2: (Test for Outliers) Suppose $x_1 = 87.627, x_2 = 87.649, x_3 = 87.642, x_4 = 87.571, x_5 = 87.637$.

Let us test the value $x_1 = 87.571$ if it is an outlier.

Here, Sample Mean $\bar{a} = 87.625$ and Std. Dev. $s = 0.031$

$$T_1 = (\bar{a} - x_1)/s = (87.625 - 87.571)/0.031 = 1.729$$

Since T_1 is less than 1.75, the critical value, we conclude that x_1 is not an outlier.

Example 3: (Physical Inventory Verification in Item Handling Facility) During the physical inventory verification of a large inventory of items, 20 items are selected at random and checked for any defect. If there are ≤ 1 defects among the 20 selected, inventory is accepted as having been verified. What is the probability, $F(1)$, of accepting the inventory if p the probability that any item is a defect is (a) 0.05, (b) 0.1?

We find the probability that $x = 0$ and $x = 1$ and add the two.

(a) $p = 0.05, n = 20$

$$\text{For } x = 0, f(0) = \frac{20!}{0!(20)!} 0.05^0 (0.95)^{20} = 0.358$$

$$\text{For } x = 1, \ f(1) = \frac{20!}{1!(19)!} 0.05^1 (0.95)^{19} = 0.377$$

Thus, $F(1) = f(0) + f(1) = 0.358 + 0.377 = 0.736$

Thus, the probability $F(1)$ that the inventory is accepted with ≤ 1 defect among the 20 selected is 0.736, given that $p = 0.05$.

(b) $p = 0.1, n = 20$

$$\text{For } x = 0, \ f(0) = \frac{20!}{0!(20)!} 0.1^0 (0.9)^{20} = 0.212$$

$$\text{For } x = 1, \ f(1) = \frac{20!}{1!(19)!} 0.1^1 (0.9)^{19} = 0.270$$

Thus, $F(1) = f(0) + f(1) = 0.212 + 0.270 = 0.482$

Thus, the probability $F(1)$ that the inventory is accepted with ≤ 1 defect among the 20 selected is 0.481, given that $p = 0.1$.

Example 4: (Physical Inventory Verification in Bulk Handling Facility) In a fuel fabrication plant, an MBA receives uranium powder, transports uranium pellets for fabrication to fuel assemblies and pellet storage and transports uranium powder scrap to waste storage. U powder received by MBA is weighed to measure the receipts (R), U pellets are counted and average mass per pellet is used to determine total U mass (S) shipped out of MBA and U powder shipped to scrap is weighed to measure the measured discard (MD). It is assumed that MBA is cleaned after the shipment, meaning BPI and EPI are zero. $49{,}873{,}000 \pm 393{,}500$ gU is received with enrichment 3.0% (± 0.02). 3,985,000 pellets are shipped from the MBA with average mass of each pellet = 12.21 ± 0.20 gU/pellet, $551{,}000 \pm 3700$ g U is shipped from MBA as scrap. Determine the MUF and σ_{MUF} and find out if they are within the acceptable limits.

$$\text{MUF} = \text{BPI} + \text{R} - \text{S} - \text{MD} - \text{EPI}$$

$$\text{BPI} = 0, \ \text{EPI} = 0$$

$$R = 49873000 \text{ gU} * 0.03 \text{ gU235/gU}$$
$$= 1496190 \pm 9975 \text{ g U235}$$

$$S = 3985000 \text{ pellets} * 12.21 \text{ gU/pellet} * 0.03 \text{ gU235/gU}$$
$$= 1459706 \pm 23910 \text{ g U235}$$

$$\text{MD} = 551000 \text{gU} * 0.03 \text{ gU235/gU} = 16530 \text{ gU235}$$

$$\text{MUF} = 0 + (1496190 - 1459706 - 16530) - 0 = 19955 \text{ g U235}$$

Error analysis on MUF

$$3\sigma_{\text{MUF}} = \text{SQRT} (9975 * 9975 + 23910 * 23910 + 111 * 111) = 25908 \text{ gU235}$$

$$\text{MUF} < 3\sigma_{\text{MUF}}, \text{ but } 3\sigma_{\text{MUF}} > 1\text{SQ}(25 \text{ kg})$$

Thus, though the MUF is less than 1SQ, σ_{MUF} is more than 1SQ, hence, the results do not rule out the possibility of diversion over a long time period.

Questions

(1) What do you understand by the term "nuclear safeguards"? What are the different components of nuclear safeguards?

(2) Explain the reasons behind the need for establishing a nuclear safeguards system at national as well as at international level.

(3) Describe the steps involved in accounting and control of nuclear materials as a part of nuclear safeguards system.

(4) Explain the term "significant quantity" (SQ) in nuclear material accounting. List the 1SQ values for different nuclear materials.

(5) What is the difference between the item handling facilities and bulk handling facilities with regard to accounting and control of nuclear materials?

(6) Give an example of bulk handling facility and explain how the accounting of nuclear materials is carried out in such a facility.

(7) Explain the rationale behind the different time lines for submission of material balance report and shipper receiver report for special nuclear materials.

(8) Explain the term "material unaccounted for". What is its significance with regard to nuclear material accounting and control?

(9) Explain how the variance of MUF can be used as an indicator of slow diversion of nuclear material from the bulk handling facility.

(10) Briefly describe the experimental methods which provide the lowest uncertainties during physical inventory taking of Uranium and Plutonium.

(11) Describe a neutron based non-destructive assay method used for physical inventory taking of Plutonium based nuclear materials.

(12) What is the different between passive and active methods used for physical inventory taking of nuclear materials? Give an example of each.

References

1. *Nuclear Material Accounting Handbook*, IAEA *Services Series* 15 (IAEA, 2008)
2. *The Structure and Content of Agreements between the IAEA and States required in Connection with the Treaty on the Non-Proliferation of Nuclear Weapons* (INFCIRC/153 (Corrected), IAEA, Vienna, 1972)
3. P. Mohanakrishnan, O.P. Singh, K. Umashankari, *Physics of Nuclear Reactors* (Academic Press, 2021), p. 74
4. B. Goddard, A. Solodov, V. Fedchenko, Non Proliferation Rev. **23**(5–6), 677–689 (2016). https://doi.org/10.1080/10736700.2017.1339934
5. *International Target Values 2010 for Measurement Uncertainties in Safeguarding Nuclear Materials, Vienna* (STR-368, 2010)
6. W. Davies, W. Gray, Talanta **11**, 1203–1211 (1964)
7. M. Xavier, P.R. Nair, K.V. Lohitakshan, S.G. Marathe, H.C. Jain, J. Radioanalyt, Nucl. Chem. **148**, 251–256 (1991)
8. J.l. Drummond, R.A. Grant, Talanta **13**, 477–488 (1966)
9. Chart of Nuclides—Brookhaven National Laboratory. https://www.nndc.bnl.gov/nudat2/chartNuc.jsp
10. D. Reilly, N. Ensslin, H. Smith Jr., *Passive Non Destructive Assay of Nuclear Materials* (Los Alamos National Laboratory Report, LA-UR-90-732, 1991)
11. S. Muralidhar, R. Tripathi, B.S. Tomar, G.K. Gubbi, S.P. Dange, S. Majumdar, S.B. Manohar, Nondestructive assay of the fissile content in FBTR fuel pins. Nucl. Inst. Meth. A **511**, 437–443 (2003)
12. N. Ensslin, M.S. Krick, D.G. Langner, M.M. Pickrell, T.D. Reilly, J.E. Stewart, *PANDA* (Addendum, Chapter 6, 2007), p. 163
13. *Active Nondestructive Assay of Nuclear Materials, Principles and Applications*, NUREG/CR-0602 (U.S. Nuclear Regulatory Commission, 1981)
14. B.S. Tomar, T.C. Kaushik, S. Andola, Ramniranjan, R.K. Rout, A. Kumar, D.B. Paranjape, P. Kumar, K.L. Ramakumar, S.C. Gupta, R.K. Sinha, Non-destructive assay of fissile materials through active neutron interrogation technique using pulsed neutron (plasma focus) device. Nucl. Instr. Meth. A **703**, 11–15 (2013)
15. M.L. Brooks, P.A. Russo, J.K. Sprinkle, Jr, *A Compact L-edge Densitometer for Uranium Concentration Assay* (Los Alamos National Laboratory report LA-10306-MS, 1985)
16. H. Ottmar, H. Eberle, *The Hybrid K—Edge/K—XRF Densitometer: Principles—Design—Performance* (Kernforschungszentrum Karlsruhe GmbH, Karlsruhe Report KFK 4590, 1991)
17. R.J. Brours, F.P. Roberts, J.A. Merril, W.B.Brown, *A Measurement Control Program for Nuclear material Accounting* (USREG. Report PNL-3021, 1980

Further Reading

IAEA Safeguards, Serving Nuclear Non Proliferation (IAEA, 2015), p. 6
IAEA Safeguards Glossary (International Nuclear Verification Series No. 3, 2001), p. 13
D.D. Sood, *Nuclear Materials*, 2nd edn. (Indian Association of Nuclear Chemists and Allied Scientists, 2016)
G. Janssens-Maenhout (ed.), *Nuclear Safeguards and Non-Proliferation, Syllabus of ESARDA Course* (European Safeguards Research and Development Association, IRC, 2008)
J.E. Doyle (ed.), *Nuclear Safeguards, Security and Non-Proliferation, Achieving Security with Technology and Policy* (Elsevier Inc., 2008)
Passive Non Destructive Assay of Nuclear materials (PANDA 2007 Addendum, Douglas Reilly, Los Alamos National Laboratory 2007)

J.L. Jaech, *Statistical Methods in Nuclear, Material Control* (US Atomic Energy Commission, 1973)

J.D. Hinchen, *Practical Statistics for Chemical Research* (Egmont UK, 1969)

Use of Nuclear Material Accounting and Control for Nuclear Security Purposes at Facilities (IAEA Nuclear Security Series No. 25-G. Implementation Guide, IAEA Vienna, 2015)

Chapter 10
Transport and Storage of Nuclear Materials

A. N. Nandakumar, Manju Saini, and G. K. Panda

10.1 Safety Standards for Transport of Nuclear Material

Every year a very large number of packages containing radioactive material are transported globally in connection with the nuclear fuel cycle. Movement of nuclear materials through public domain needs to be carried out with proper care because of the radiological hazards associated with the materials. For this reason, transport of nuclear material is governed by national and international regulations. This chapter discusses the safety standards specified in the regulations for the safe transport of radioactive material published by the International Atomic Energy Agency which are applicable to all modes of transport [1]. Nuclear materials are the radioactive materials that are used for nuclear power production. Nuclear materials are a sub-set of radioactive materials. The two terms are used interchangeably in this chapter.

There are nine classes of dangerous goods. Radioactive material is identified as class 7 dangerous good. Radioactive material constitutes a small fraction of all dangerous goods transported as can be seen from Table 10.1 [2].

Nuclear materials are transported by all modes of transport, namely, road, rail, inland waterway, sea and air. The various national [e.g. 3] and mode-specific international regulations are essentially based on or derived from the IAEA regulations. The term "Safety Standards" used in this chapter refers to the IAEA regulations. In any scientific discussion on transport and storage of radioactive material, reference to the IAEA Safety Standards is inevitable.

A. N. Nandakumar (✉)
Formerly Radiological Safety Division, Atomic Energy Regulatory Board, Mumbai, India
e-mail: a.n.nandakumar@gmail.com

A. N. Nandakumar · M. Saini · G. K. Panda
Radiological Safety Division, Atomic Energy Regulatory Board, Mumbai, India
e-mail: manjusaini@aerb.gov.in

G. K. Panda
e-mail: gkpanda@aerb.gov.in

Table 10.1 Radioactive material as a fraction of all dangerous goods transported worldwide

Mode of transport	Estimated fraction of all dangerous goods that are radioactive (%)
Road	<2
Rail	<2
Air	<10
Sea or inland waterways	<1

Radioactive material can be classified on the basis of the potential hazard associated with the material. This classification helps in optimally ensuring protection during transport and storage. In some nuclear reactors, isotopes may be produced for use in medical and industrial facilities. These isotopes are transported to "source" preparation facilities. Depending on the nature of use for which they are intended, some of these sources may need to be encapsulated and may qualify as special form radioactive material, which is safer than a dispersible radioactive material of comparable activity from contamination point of view. Certain radioactive materials have additional hazardous properties (e.g. corrosiveness, pyrophoricity) warranting special additional protective measures. This chapter addresses the salient aspects of the requirements for the safe transport and storage of nuclear materials in connection with the nuclear fuel cycle.

10.2 Radioactive Material Transported in Nuclear Fuel Cycle

Transport of nuclear material in connection with nuclear fuel cycle activities may be broadly described as follows:

A. excepted radioactive material (e.g. small quantities of very low activity, reference sources used in the Health Physics units, empty package having contained radioactive material)
B. uranium/thorium ore from the mine to the mill to produce ore concentrate
C. uranium concentrate from the mill to the conversion facility (e.g. uranium hexafluoride)
D. uranium from the conversion facility to the enrichment plant
E. enriched uranium from the enrichment plant to the fuel fabrication facility
F. fabricated fuel assemblies from the fuel fabrication facility to the nuclear reactor
G. start-up neutron source and isotopes produced in the reactor for supply to users
H. spent fuel from the nuclear reactor to the reprocessing facility
I. recovered fuel from the reprocessing facility to the fuel fabrication facility
J. nuclear waste from the nuclear reactor/reprocessing plant to the waste repository
K. contaminated objects arising during operation or following an accident or during decommissioning

Fig. 10.1 Transport of
nuclear material in nuclear
fuel cycle

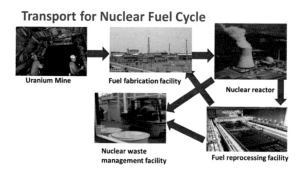

Examples of transport of nuclear materials in nuclear fuel cycle are shown in Fig. 10.1.

10.3 Transport of Radioactive Material Relevant to the Front End of the Nuclear Fuel Cycle

Transport of radioactive material in the front end is essentially ore concentrates, UF_6, and fresh uranium fuel [4]. The radioactivity contained in uranium ore is distributed over a large mass to such an extent that its specific activity, that is, activity per unit mass, is always low. Hence, it does not pose any significant radiological health hazard. Such materials are classified as Low Specific Activity (LSA) materials. However, when fabricated fuel assembly is transported, one has to take into consideration the radioactivity and, if enriched, the fissile nature of the material to ensure radiological and criticality safety during transport and storage. Uranium hexafluoride (UF_6), which in addition to being a radioactive material, is also a corrosive substance, would need special attention. If it is enriched in ^{235}U, criticality safety also has to be assured during transport and storage of UF_6.

10.4 Transport of Radioactive Material Relevant to the Back End of the Nuclear Fuel Cycle

Back-end materials are essentially solid products. The solid nature of the products—spent fuel, MOX fuel, and vitrified high-level waste—is an important safety factor [5]. The materials are characterised by long-term stability and low solubility in water and would stay contained in a solid form after an accident. Spent fuel and MOX fuel are both made of hard ceramic pellets that are contained in zirconium alloy tubes or SS (fuel rods). The difference lies in the content; spent fuel contains uranium (96%), plutonium (1%) and fission products (3%) and is highly radioactive, while MOX fuel is made of uranium and plutonium oxides. In the case of vitrified high-level waste, the vitrification process allows the fission products to be incorporated into a molten glass which is then poured into a stainless-steel canister, where it solidifies.

Spent fuel would contain a large quantity of mixed fission products and would be highly radioactive. The nuclear waste may have low or medium or high activity. Sometimes the radioactive material, contained in the waste, could have low to medium specific activity (that is, activity per unit mass).

10.5   Radiation Protection

Transport of nuclear material through public domain is governed by the principle that safety should be ensured, and people, property and the environment are protected from harmful effects of ionizing radiation during the transport of nuclear material under all conditions of transport, viz., routine, normal and accident conditions [1].

Routine conditions refer to incident-free conditions of transport. Normal conditions pertain to minor mishaps such as improper handling, over-stacking and exposure to sudden rain, which a consignment is likely to encounter during transport and storage in transit. Accident conditions are events that may result in mechanical and thermal impacts that may occur during transport, and those that may cause increase in pressure on the external surface/containment system of the package as can happen if a package falls in a waterbody during transport.

If the estimated annual dose to transport workers (i.e. workers routinely engaged in transport of radioactive material) does not exceed 1 mSv, there is no need for radiation monitoring. If the estimated annual dose to transport workers is likely to exceed 1 mSv but not exceed 6 mSv, workplace monitoring should be carried out. If the transport workers are likely to receive an estimated annual dose in excess of 6 mSv, they have to be provided with individual monitors. The individual dose limit for occupational exposure, namely, 20 mSv/year should never be exceeded.

The safety of transport is achieved on the basis of three levels of safety, namely, inherent safety, passive safety and active safety. These safety levels are characterized by the nature and quantity of radioactive material transported, selection of appropriate and optimal packages and implementing administrative control over transport of radioactive material.

10.6   Inherent Safety

Inherent safety consists in limiting the potential radiological risk by limiting the activity/quantity of the nuclear material in a package of a given design on the basis of the nature of the material. Nuclear materials are transported in different physical and chemical forms. The design of the package in which a nuclear material is to be transported and the control measures to be implemented during transport and storage would be determined by the potential hazard associated with the nuclear material.

Table 10.2 A_1/A_2 and exemption values for some commonly transported radionuclides

Radionuclide	A_1 (TBq)	A_2 (TBq)	Limit for exempt activity concentration (Bq/g)	Activity limit for an exempt consignment (Bq)
^{3}H	4×10^1	4×10^1	1×10^6	1×10^9
^{60}Co	4×10^{-1}	4×10^{-1}	1×10^1	1×10^5
^{131}I	3×10^0	7×10^{-1}	1×10^2	1×10^6
^{90}Sr	3×10^{-1}	3×10^{-1}	1×10^2	1×10^4
^{137}Cs	2×10^0	6×10^{-1}	1×10^1	1×10^4
^{235}U	Unlimited	Unlimited	1×10^1	1×10^4
^{239}Pu	1×10^1	1×10^{-3}	1×10^0	1×10^4

Radioactivity is present in all materials. Many of these materials do not warrant any safety procedures for transport. As can be seen from Table 10.2 [1], only if a consignment of plutonium-239 has an activity concentration in excess of 1 Bq/g **and** contains an activity in excess of 10^4 Bq, the requirements for safe transport would apply to the consignment. If either the total activity or the activity concentration of a radioactive material in a consignment is less than the basic values specified for the radionuclide (columns 4 and 5 of Table 10.2), transport of such material is **exempt** from the safety requirements. Considerations for exemption from the requirements are specified in the Safety Standards. The significance of A_1 and A_2 values of radioactive materials listed in Table 10.2 is explained in the following paragraphs.

10.7   The Significance of A_1/A_2 Values

The potential radiological hazard associated with the transport of nuclear material depends on the total activity contained within the package, the physical form of the radioactive material, the external radiation levels and the dose per unit intake (e.g. by ingestion or inhalation or through an open wound) of the radionuclide. Radioactive material in an indispersible form or sealed in a strong metallic capsule that is **specially designed** to offer protection against dispersion during accident conditions of transport is designated as special form radioactive material.

Inherent safety can be achieved through limiting the activity of the nuclear material transported in a package. For this purpose, two quantities, namely, A_1 and A_2 values are specified for each radionuclide (Table 10.2). The A_1/A_2 values have been derived on the basis of a dosimetric model for limiting external exposure and internal exposure considering different pathways (ingestion, inhalation, skin absorption, submersion) which could be received by a person in an accident [6]. In an accident, the external exposure, due to an unshielded special form radioactive material of activity A_1, would not exceed 50 mSv. The value of A_2 is so determined that the internal exposure due to an accident involving a Type A package containing a radionuclide of activity A would not result in a committed equivalent dose in excess of 50 mSv. Type A package is discussed in detail in Sect. 10.9.2.5. The limit on the activity of a radioactive material that may be permitted in a package is specified in multiples or sub-multiples of A_1/A_2 values.

For mixtures of radionuclides, the A_1/A_2 and the exemption values may be determined as follows:

$$X_m = [1/\{\Sigma_i f(i) / X(i)\}]$$

where

f(i) is the fraction of activity or activity concentration of radionuclide i in the mixture.

X(i) is the appropriate value of A_1 or A_2 or the activity concentration limit for exempt material or the activity limit for an exempt consignment as appropriate for radionuclide i.

X_m is the derived value of A_1 or A_2 or the activity concentration limit for exempt material or the activity limit for an exempt consignment in the case of a mixture.

In the case of radionuclides that are not listed in the Safety Standards, the A_1/A_2 values have to be calculated using the methods specified in the Safety Standards.

10.8   Classification of Radioactive Materials

Radioactive materials are classified based on their potential hazard. Any radioactive material would belong to one of the following classes:

(a)   Special form radioactive material
(b)   Low dispersible radioactive material (LDRM)
(c)   Low specific activity material, LSA I/II/III
(d)   Surface contaminated object, SCO-I/II/III
(e)   Fissile material
(f)   Uranium hexafluoride (UF_6)

Each of these classes is discussed in the following paragraphs.

10.8.1   Special Form Radioactive Material

A radioactive material is said to be in special form if it is either an indispersible solid radioactive material or a sealed capsule containing radioactive material which is designed to withstand the regulatory tests. It should not break or shatter or melt or disperse when subjected to the tests specified in the Safety Standards.

The start-up neutron source may be expected to be a special form radioactive material.

10.8.2   Low Dispersible Radioactive Material

A low dispersible radioactive material (LDRM) is designed to withstand very severe air accidents and therefore offers a very high degree of protection. An LDRM should be designed to withstand the regulatory tests. It should not break or shatter or melt or disperse when subjected to the tests specified in the Safety Standards.

10.8.3   Inherently Safe Low Specific Activity Material

Certain nuclear materials have very low specific activities. Therefore, the quantity of intake of such material that would cause a significant hazard is too large. Uranium ore that is transported from the mine to the mill, uranium concentrate from the mill to the conversion facility, uranium from the conversion facility to the enrichment plant and certain low level nuclear waste fall in this category. Such radioactive materials are described as Low Specific Activity (LSA) materials. External shielding materials surrounding the LSA material should not be considered in estimating the average specific activity. A solid compact binding agent, such as concrete and bitumen that is mixed with the LSA material, is not an external shielding material, though the binding agent may attenuate radiation to some extent. An illustration is provided in Fig. 10.2 [6].

LSA materials are brought under 3 groups, namely, LSA-I, LSA-II and LSA-III. Examples of radioactive materials that qualify as LSA-I include uranium and thorium ores and concentrates of such ores and other ores containing naturally occurring radionuclides and natural uranium, depleted uranium, natural thorium or their compounds or mixtures, that are unirradiated and in solid or liquid form. These may include fissile materials that are excepted from the criticality safety requirements by virtue of conditions such as quantity, concentration, composition and enrichment.

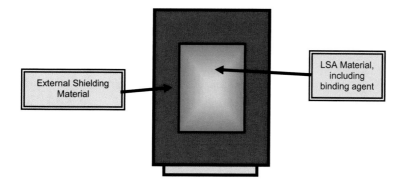

Fig. 10.2 LSA material including binding agent within shielding

Fissile material may be considered as LSA-I only if it is excepted from considerations of criticality hazard.

Examples of radioactive materials that qualify as LSA-II include tritiated water containing only ^{3}H with a concentration not exceeding 0.8 TBq/L and other materials in which the activity is distributed throughout and the average specific activity does not exceed 10^{-4} A_2/g for solids and gases and 10^{-5} A_2/g for liquids.

Nuclear reactor process wastes which are not solidified, such as lower activity resins and filter sludges and other similar materials from reactor operations, similar materials from other fuel cycle operations and activated equipment from the decommissioning of nuclear plants qualify as LSA-II material. In addition, LSA-II could include many items of activated equipment from the decommissioning of nuclear plants. Waste from nuclear power plants such as evaporator bottoms, mechanical filters, filter media and absorbed liquids in rubble exceeding LSA-I limits, activated metals, organic liquids, e.g. liquid scintillation fluids and paints from contaminated surfaces are examples of LSA-II. Activated metals or radioactive material, which is solidified or absorbed on non-radioactive material and wastes generated during decontamination activities, which exceed LSA-I limits, may also be classified as LSA-II.

A Low Specific Activity-III (LSA-III) material has to be a solid (e.g. consolidated wastes, activated materials), excluding powders, in which the radioactive material is distributed throughout a solid or a collection of solid objects or is essentially uniformly distributed in a solid compact binding agent (such as concrete, bitumen and ceramic), and the estimated average specific activity of the solid, excluding any shielding material, does not exceed 2×10^{-3} A_2/g.

Generally, activated metals could be LSA-III. Their surfaces could be contaminated. Solids that absorb radioactivity such as towels, rags or tape may be classified as LSA-I or II or III based on the criteria specified above.

10.8.4 Objects Not Radioactive but Contaminated on Their Surfaces

A surface contaminated object (SCO) is a solid object that is not itself radioactive but has radioactive material distributed on its surface. Objects which are classified as SCOs include those parts of nuclear reactors or other equipment that have come into contact with primary or secondary coolant or process waste, resulting in contamination of their surface with mixed fission products. SCO is brought under three groups, namely SCO-I, SCO-II and SCO-III. The limits on contamination for SCO-I and SCO-II are specified in Table 10.3.

SCO-I: A solid object with contamination levels that do not exceed the limits.

SCO-II: A solid object with contamination levels that are higher than either of the applicable limits for SCO-I without exceeding the corresponding limits for SCO-II.

Table 10.3 Contamination levels on surfaces* of SCOs

Surface	Fixed/non-fixed contamination	Nature of contamination	Activity (Bq/cm^2)	
			SCO-I	SCO-II
Accessible	Non-fixed	β, γ and low toxicity α#	4	4×10^2
		Other α	0.4	4×10^1
	Fixed	β, γ and low toxicity α	4×10^4	8×10^5
		Other α	4×10^3	8×10^4
Inaccessible	Fixed + non-fixed	β, γ and low toxicity α	4×10^4	8×10^5
		Other α	4×10^3	8×10^4

Low toxicity alpha emitters are as follows: natural uranium, depleted uranium, natural thorium, ^{235}U, ^{238}U, ^{232}Th, ^{228}Th and ^{230}Th when contained in ores or in physical and chemical concentrates or alpha emitters with a half-life of less than 10 days
* The contamination levels have to be averaged over an area of 300 cm^2

An SCO-III is a large solid object which, because of its size, cannot be transported in any type of package described in this chapter.

10.8.5 Distinguishing Between LSA or SCO

LSA material and SCOs allow the use of less expensive industrial packagings because of the inherent safety associated with LSA materials and SCOs. Difficulties may be experienced in practice while categorizing a material to be transported as LSA or SCO and further as LSA-I, II or III or SCO-I or II. In ambiguous situations, one adopts a conservative approach in classifying such material. Experience would help in making a proper classification.

10.8.6 Fissile Material and Criticality Safety During Transport and Storage

Certain special requirements for ensuring criticality safety have to be met for transport of fissile material. Fissile material is a material that contains any of the fissile nuclides. Fissile nuclides are ^{233}U, ^{235}U, ^{239}Pu and ^{241}Pu. The following are individually and in combination excluded from the definition of fissile material, provided they are not transported together with any fissile material (that is, materials that warrant criticality safety measures):

(a) Natural uranium or depleted uranium that is unirradiated;

(b) Natural uranium or depleted uranium that has been irradiated in thermal reactors only;

(c) Material with fissile nuclides less than a total of 0.25 g.

If LSA material or SCO is fissile or contains fissile material (i.e. requiring criticality safety measures), criticality safety requirements should also be satisfied in addition to radiation safety requirements.

10.8.7 Fissile-Excepted Material

Not all fissile materials pose criticality hazard. Certain materials containing fissile nuclides may not pose criticality hazards in view of their composition, mass or concentration. Such materials are referred to as fissile-excepted because they do not need to meet the criticality safety requirements that normally apply to fissile material. The criteria for identifying fissile-excepted materials may be found in the Safety Standards [1].

10.8.8 Uranium Hexafluoride (UF₆)

After conversion of uranium ore concentrate into UF_6, it is transported to the enrichment facility. The mass of UF_6 contained in a package should not be different from that allowed by the package design. The filling of UF_6 should be controlled such that the ullage (i.e. unfilled space above the liquid level for accommodating evaporation) should not be less than 5% at the maximum temperature of the package. When presented for transport, UF_6 should be in solid form, and the internal pressure should not be above the atmospheric pressure.

The chemical toxicity of UF_6 is significant. When UF_6 is exposed to moist air, it reacts with the water in the air to produce uranyl fluoride (UO_2F_2) and hydrogen fluoride (HF) both of which are highly corrosive and toxic. Special safety provisions need to be implemented in the transport of UF_6. If the UF_6 includes enriched uranium, the design of the package should satisfy the specific additional requirements applicable to packages for transport of fissile material.

10.9 Passive Safety

Passive safety is provided by the packaging. Package is packaging plus the radioactive content. Under normal conditions of transport, external exposure may occur due to leakage radiation outside the shielding. Shielding is provided in the packaging to reduce the dose rate due to leakage radiation to levels within the regulatory limits.

An accident may challenge the integrity of the shielding and/or the containment of the packaging thereby potentially resulting in external exposure and/or internal exposure. The effectiveness of a packaging has to be demonstrated in terms of its ability to withstand all conditions of transport.

In order to limit external and internal exposures, the activity of radioactive material permitted in different types of packages is limited. The activity limits of the nuclear material in packages are specified in units of multiples or sub-multiples of A_1/A_2. The required level of safety could be achieved by designing the package to meet the standards discussed in Sect. 10.9.2 and withstand the challenges posed by transport that are simulated by regulatory tests described in Sect. 10.9.3.

10.9.1 Unpackaged Transport

Certain LSA material and SCO, that is LSA-I and SCO-I may be transported, unpackaged, under certain conditions. This is because of the low potential radiological hazard associated with these nuclear materials. Under routine conditions of transport, the radioactive contents should not escape from the conveyance, and if any shielding is available, it should remain intact during transport. Where it is suspected that non-fixed contamination exists on inaccessible surfaces in excess of the values specified for SCO-I, it should be ensured that the radioactive material is not released into the conveyance.

The conveyance should be under exclusive use for certain nuclear materials specified in the Safety Standards. (**Exclusive use** means that a single consignor has the sole use of a conveyance or of a large freight container, and all initial, intermediate and final loading and unloading and shipment are carried out in accordance with the directions of the consignor or consignee. Clearly, a higher level of administrative control is achievable under exclusive use.) If a fissile-excepted material has to be transported in an unpackaged condition, the total mass of fissile nuclides in the consignment should not be greater than 45 g in a conveyance, and it should be transported under exclusive use. However, when transporting only SCO-I on which the contamination on the accessible and the inaccessible surfaces is not greater than $4 \, Bq/cm^2$ for beta, gamma and low toxicity alpha emitters and $0.4 \, Bq/cm^2$ for other alpha emitters is transported, it need not be under exclusive use.

Since SCO-III is transported unpackaged, the following conditions must be met:

(a) Transport should be under exclusive use by road, rail, inland waterway or sea.
(b) Stacking should not be permitted.
(c) All activities associated with the shipment, including radiation protection, emergency response and any special precautions or special administrative or operational controls that are to be employed during transport should be described in a transport plan. Upon subjecting the package specimen to the tests for demonstrating ability to withstand normal conditions of transport, there should be no release of activity, and the increase in dose rate should not exceed 20%.

(d) The object and shielding, if any, are secured to the conveyance such that it can be easily and safely transported.

(e) The shipment should be subject to "multilateral" approval. [That is, prior to transporting nuclear material, the consignor should obtain an approval certificate from the competent authority of each country through or into which the radioactive material would be transported. Hence, it is referred to as multilateral approval. Here the term, "through or into" specifically excludes countries over which a consignment is carried by air, with no scheduled stops in those countries. If there are scheduled stops in those countries, approval would be required from the competent authorities of those countries also.]

10.9.2 Transport in Packages

Generally, all nuclear materials have to be transported in appropriate packagings. The protection offered by the package should be commensurate with the hazard associated with the nuclear material for which the packaging is intended. This leads to the concept of a graded approach to packaging requirements. The more hazardous the material, the studier should the packaging be. For this purpose, different types of packages are deployed for transport of nuclear material, namely,

(a) Excepted package
(b) Industrial package Type IP-1, IP-2 and IP-3
(c) Type A
(d) Type B(U)/(M)
(e) Type C

On certain occasions, one may use freight containers which are specially designed to facilitate the transport of goods by one or more modes of transport without intermediate reloading, and secured and readily handled. There are specific regulatory provisions which should be complied with, if freight containers are used.

There are certain basic **general design requirements** which should be satisfied by all types of packages. In addition to the general requirements, **specific design requirements** should be met for each type of package. Appropriate **additional design requirements** should be satisfied for certain nuclear materials such as fissile material and uranium hexafluoride (UF_6).

The important aspects of design are containment of the radioactive contents, control of external dose rate (shielding), prevention of criticality (in the case of transport of fissile material), prevention of damage caused by heat (for nuclear material of large activity), appropriate exterior of the package (smooth surface, minimal protrusions, plain surface to prevent collection and retention of water and facilities for tie-down and safe lifting) and ability to withstand routine/normal/accident conditions of transport. These aspects are discussed in some detail below.

10.9.2.1   General Design Requirements for All Packages

It should be possible to handle the package easily and safely and to secure it properly in or on the conveyance during transport. The external surfaces should be smooth to facilitate decontamination and free from protruding features. The package should be capable of withstanding the effects of any acceleration, vibration or vibration resonance that may arise under routine conditions of transport without any deterioration in its integrity. The materials of the packaging and its components should be physically and chemically compatible with each other and with the radioactive contents even under irradiation. All valves through which the radioactive contents could escape should be protected against unauthorized operation.

The dose rate at any point on the external surface of the package should not exceed the regulatory limits. For packages to be transported by air, the package should be capable of withstanding a wide range of temperature (-40 to $+55\ °C$) and the specified pressure differential and retain its containment integrity. The criteria for making an optimal choice of package based on the graded approach are discussed below.

10.9.2.2   Excepted Package

Radioactive materials of very low activity (within the specified activity limits) are termed as excepted. The package that is ideal for transporting excepted radioactive material is described as excepted package. It should satisfy all the general design requirements describe above.

Since an excepted package is of simple design, the activity that can be permitted in such a package is limited. The limit would depend on the physical state (i.e. solid, liquid, gas, special form, other form) of the radioactive material. For solid/gaseous form, the activity limit is typically $10^{-3}\ A_1$ for special form and $10^{-3}\ A_2$ for other form. For liquid form, the limit is $10^{-4}\ A_2$. For tritium gas, the limit is $2 \times 10^{-2}\ A_2$.

The package should be capable of retaining its radioactive contents under routine conditions of transport. The package should bear the mark "RADIOACTIVE" on an internal surface in such a manner that a warning of the presence of radioactive material is visible on opening the package. If it is impractical to mark on internal surface, the mark may be outside of the package.

One may ask whether UF_6 can be transported in an excepted package. The answer is yes, provided that (a) the activity of the content does not exceed the limit for an excepted package, (b) the mass of UF_6 in the package is less than 0.1 kg, and it is non-fissile or fissile-excepted, (c) the mass of UF_6 is so restricted that the ullage would not be less than 5% at the maximum temperature of the package, (d) UF_6 is in solid form, and (e) the internal pressure of the package is not above atmospheric pressure when presented for transport.

10.9.2.3 Industrial Package Types IP-1/IP-2/IP-3

Industrial Packages Types IP-1/IP-2/IP-3 are ideal for transporting LSA materials and SCOs. The optimal package for each of these groups of low-hazard radioactive material is specified in Table 10.4.

There is a limit on the quantity of LSA material or SCO in a single package. That is the dose rate at 3 m from the nuclear material (LSA/SCO) **without any shielding** must not exceed 10 mSv/h.

Industrial Package Type IP-1

An Industrial Package Type IP-1 should be designed to satisfy all the general design requirements and certain additional specific regulatory requirements. The quantity of nuclear material in an IP is restricted as specified above. A typical Industrial Package Type IP-1 is shown in Fig. 10.3.

Industrial Package Type IP-2

A package to be qualified as Type IP-2 should be designed to meet the requirements for Industrial Package Type IP-1, and, in addition, if it were subjected to the free drop and stacking tests specified in the Safety Standards, there should be no dispersal of the radioactive contents, and the dose rate at the external surface of the package should not increase by more than 20%.

Industrial Package Type IP-3

A package to be qualified as Type IP-3 should be designed to meet the requirements for Industrial Package Type IP-2 and the requirements relating to the dimensions

Table 10.4 Optimal packages for LSA materials and SCO-I/II

IP-1	IP-2	IP-3
LSA-I solid	LSA-I liquid	LSA-II liquid/gas
LSA-I liquid Exclusive use	LSA-II solid	LSA-III
SCO-I	LSA-II liquid/gas Exclusive use	
	LSA-III exclusive use	
	SCO-II	

Fig. 10.3 Industrial package Type IP-1

of the package, a tamper-proof arrangement, features to ensure safe tie-down to the conveyance, ability to withstand the temperature and pressure ranges specified for its components and the containment system and ability to satisfy the test requirements specified in the Safety Standards.

10.9.2.4 Packaging for Uranium Hexafluoride

Uranium hexafluoride can give off a toxic vapour. It is transported in steel cylinders with a capacity of approximately 12 tonnes. These packages are designed to satisfy the requirements of the free drop test applicable to Type A packages, a pressure test which they must withstand without leakage and the thermal test prescribed for Type B(U)/(M) packages.

The package should be designed to meet the requirements that pertain to the radioactive and fissile properties of the material. Further, uranium hexafluoride in quantities of 0.1 kg or more should also be packaged and transported in accordance with the provisions of ISO 7195 and meet the requirements of the Safety Standards. However, if the packages are designed (a) to meet standards other than ISO 7195, (b) to meet a test pressure of less than 2.76 MPa and/or (c) to contain 9000 kg or more of UF_6, then multilateral approval of the design would be required.

A sub-atmospheric cold pressure test should be used to demonstrate suitability of the cylinder for transport of UF_6. The operating procedure for the package should specify the maximum sub-atmospheric pressure allowed, which will be acceptable for shipment, and the method and results of this measurement should be included in appropriate documentation.

Generally, cylinders are filled with UF_6 at pressures above atmospheric pressure under gaseous or liquid conditions. Until the UF_6 is cooled and solidified, a failure of the containment system in either the cylinder or the associated plant fill system could result in a dangerous release of UF_6. However, since the triple point of UF_6 is 64 °C at normal atmospheric pressure of 1.013×10^5 Pa, if the UF_6 is presented for transport in a thermally steady state, solid condition, it is unlikely that during normal conditions of transport it will exceed the triple point temperature.

Satisfying the requirement that the UF_6 be in solid form with an internal cylinder pressure less than atmospheric pressure for transport would ensure that the containment boundary of the package is functioning properly and maximize the structural capability of the package and the safety margin while handling the package.

10.9.2.5 Type A Package

If the activity of the nuclear material is greater than permitted in an excepted package, one should examine whether its specific activity/surface contamination would qualify the material as LSA or SCO so that the material could be transported in an IP-1/2/3, as appropriate. If the nuclear material does not qualify as LSA/SCO, and its activity is not greater than A_1 or A_2, depending upon whether it is special form or other form

Fig. 10.4 Type A package

radioactive material, it should be transported in a Type A package. Figure 10.4 is an example of a Type A package.

A Type A package is designed to satisfy the requirements specified for an IP-3 package. The package should satisfy the tests demonstrating the ability to withstand normal conditions of transport. In the case of a Type A package designed to contain a radioactive liquid or gas (except for tritium gas or noble gas), the following conditions should be met:

(i)   There should be no loss or dispersal of the radioactive contents if the package is subjected to the free drop test from a height of 9 m and the penetration test with a drop of the bar through a height of 1.7 m, and

(ii) The package should be provided with sufficient absorbent material that can absorb twice the volume of the liquid contents. The absorbent material must be placed such that in the event of leakage, the liquid would directly contact the absorbent material. Alternatively, the containment system should be composed of inner (primary) and outer (secondary) containment components to ensure that even if the primary inner components leak, the contents would be retained within the secondary containment components.

10.9.2.6 Type B(U)/(M) Package

If the nuclear material to be transported does not qualify as LSA/SCO and its activity exceeds A_1 or A_2, it has to be transported in a Type B(U)/(M) package or Type C package, as appropriate.

Examples of radioactive materials that would require to be transported in a Type B(U)/(M) package are neutron start-up sources, spent fuel and certain fissile materials. Since the activities of the nuclear material permitted in a Type B(U)/(M) package are relatively high, a design approval certificate should be obtained from the competent authority of the country of origin of the design.

Packages that are designed to satisfy all the requirements of the Safety Standards would, in general, be universally acceptable. In such cases, the competent authority would issue a Type B(U) approval certificate, where "U" stands for "Unilateral".

There may be certain packages that would not meet some specific performance standards for Type B(U) package. In such cases, the competent authority would issue a Type B(M) approval certificate, where "M" stands for "Multilateral". That is, multilateral approval certification would be required from the competent authorities of the countries through or into which the package would be transported.

The activity of nuclear material that is transported in a Type B(U)/(M) package should not exceed the limit specified in the design approval certificate. However, Type B(U) and Type B(M) packages, **if transported by air**, should not contain activities greater than the following:

(i) For special form radioactive material—3000 A_1 or 10^5 A_2, whichever is the lower;

(ii) For all other radioactive material—3000 A_2.

If the radioactive content of a Type B(U)/(M) package, transported by air, is a low dispersible radioactive material, its activity should not exceed the limit authorized for the **package design** as specified in the certificate of approval.

Type B(U) packages should be designed to meet the requirements for Type A packages. Heat may be generated in the package because of the large activity of the radioactive contents. The design should ensure that the heat, in addition to insolation (solar heat input), does not affect the shielding and containment integrity of the package nor cause alteration of the arrangement, the geometrical form or the physical state of the radioactive contents. The temperature on the outer surface of the package, during normal conditions of transport, should be within the limits specified in the Safety Standards.

Any thermal protection included in the design of the package for the purpose of satisfying the requirements of the thermal test specified in the Safety Standards should remain effective if the package is subjected to the regulatory tests. The package should satisfy the requirements for the drop tests, thermal test and water immersion test. (These tests are discussed in Sect. 10.9.3.)

If a package that meets the design requirements of a Type B(U) package, except that the requirements relating to ambient temperature, temperature on the external surface of the package, insolation and the performance of the containment system are met only to a limited extent, it could qualify as a Type B(M) package. In some instances, intermittent venting of Type B(M) packages may be permitted during transport. A picture of a Type B(U)/(M) package can be seen in Fig. 10.5.

10.9.2.7 Type C Package

If the activity of a nuclear material which is intended to be transported by air exceeds the limits permitted in a Type B(U)/(M) package for transport by air, then Type C package is used. The design of a Type C package should satisfy the design requirements specified for a Type B(U) package. The package should satisfy the test requirements, namely, Drop I and Drop III, puncture-tearing test, enhanced thermal test, impact test and water immersion test (Sect. 10.9.3). If the package was at the

Fig. 10.5 Type B(U)/(M)
package

maximum normal operating pressure, its containment system is required to retain its integrity after burial in an environment defined by a thermal conductivity of 0.33 W/(m°K) and a temperature of 38 °C in the steady state. Initial conditions for the assessment should assume that any thermal insulation of the package remains intact. Compliance with this requirement may be demonstrated through conservative calculations or validated computer codes.

10.9.2.8 Packages Containing Fissile Material

Special additional safety provisions need to be met in the design of a package for transport of fissile material. The nuclear materials that could qualify as fissile material that would need to be transported in connection with the nuclear fuel cycle operations include the following:

- Uranium hexafluoride enriched in ^{235}U
- Fresh reactor fuel for a power reactor or a research reactor using enriched uranium fuel
- Irradiated reactor fuel from a power reactor or a research reactor
- Nuclear waste from any of the facilities handling fissile materials

The optimal type of package is selected depending on the activity and the nature of the nuclear material. Fissile material should be transported in such a way that subcriticality is maintained during routine, normal and accident conditions of transport. Therefore, certain specific **additional** safety requirements for packages containing fissile materials have to be met. For a package designed for fissile material, a quantity called Criticality Safety Index (CSI) will have to be determined as described in this chapter. This index is used for the purpose of restricting the number of packages containing fissile material that can be accumulated in a conveyance or storage area so as to ensure that they remain subcritical with adequate safety margin.

Factors such as geometry of the fissile material, presence of reflectors, moderators and neutron absorbers have a bearing on criticality safety during transport

Fig. 10.6 Relation between the critical mass and the shape of the fissile material

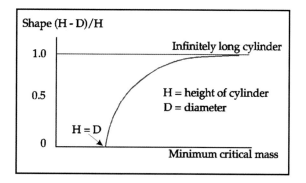

and storage. Critical mass depends on the number of neutrons escaping and, therefore, the shape of the system. A sphere is the most efficient configuration for criticality. A cylinder requires more mass for criticality due to its higher surface area. As a cylinder becomes longer, more neutrons escape, and the critical mass of fissile material increases (Fig. 10.6).

Hydrogenous substances including water are very good neutron reflectors. Water should be prevented from entering a package containing fissile material during transport and storage. There may be materials amongst other cargo that may act as unintended reflectors. They are packaged merchandise, bulk timber, oil and foodstuffs and structures of concrete and brick.

Neutrons from fission generally have greater kinetic energy than is ideal to cause further fissions. However, collisions with surrounding nuclei could result in the loss of kinetic energy. Collisions with light nuclei (e.g. hydrogen) are very effective at slowing neutrons (Fig. 10.7). The package design should minimize the effects of reflection and moderation during transport and storage.

Most nuclides surrounding the fissile material can capture neutrons without causing fission. Elements such as cadmium, boron, gadolinium and lithium are very strong neutron absorbers. Absorbers can be used within the confinement system of the

Fig. 10.7 Critical mass versus hydrogen content in a solution [H = number of hydrogen atoms in the solution and X = number of fissile atoms in the solution]

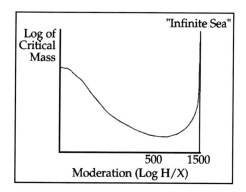

package to control neutron population. Confinement system refers to the assembly of fissile material and packaging components designed to preserve criticality safety.

Exceptions from specific fissile-related requirements

Apart from packages containing fissile-excepted material, there are certain **packages containing fissile materials** which need not meet the fissile material-specific requirements. Conditions in which certain fissile materials do not need to meet the specific provisions for criticality safety are addressed in the Safety Standards.

10.9.2.9   Specific Additional Requirements for Packages Containing Fissile Material

Criticality safety assessments of an individual package in isolation and of arrays of package under normal conditions of transport and under accident conditions of transport should be performed. In designing a package for containing fissile material, criticality safety in the following contingencies should be considered:

(a) Leakage of water into or out of packages;
(b) Loss of efficiency of built-in neutron absorbers or moderators;
(c) Rearrangement of the contents either within the package or as a result of release from the package;
(d) Reduction of spaces within or between packages;
(e) Packages becoming immersed in water or buried in snow;
(f) Temperature changes.

Sometimes the chemical or physical form, isotopic composition, mass or concentration, moderation ratio or density or geometric configuration may not be known. In such cases, the assessments should be performed, assuming that each parameter that is not known has the value that gives the maximum neutron multiplication consistent with the known conditions and parameters in these assessments.

For irradiated nuclear fuel, assessments for individual package and array of packages should be based on an isotopic composition that would demonstrably provide either the maximum neutron multiplication during the irradiation history or a conservative estimate of the neutron multiplication for the package assessments. Prior to shipment, a measurement should be performed to confirm the conservatism of the isotopic composition.

The package, after being subjected to the tests for demonstrating ability to withstand normal conditions of transport specified in the Safety Standards, should preserve the minimum overall outside dimensions of the package to at least 10 cm and prevent the entry of a 10 cm cube.

The package should be designed for an ambient temperature range of −40 to +38 °C unless the competent authority specifies otherwise in the certificate of approval for the package design.

10.9.2.10   Criticality Safety Assessment of an Individual Package in Isolation

Assessment should be made to confirm that an individual package in isolation remains subcritical. In the assessment, it should be assumed that water can leak into or out of all void spaces of the package, including those within the containment system. Absence of leakage may be assumed in respect of those void spaces if the design incorporates special features to prevent such leakage of water into or out of certain void spaces, even as a result of error. In the assessment, it should be assumed that the confinement system is closely reflected by at least 20 cm of water or any greater reflection that may be provided by the surrounding material of the packaging. The purpose of the assessment is to confirm that the package remains subcritical under conditions that result in the maximum neutron multiplication consistent with incident-free conditions of transport and in the tests specified below for assessment of package arrays under normal conditions of transport and accident conditions of transport.

Criticality Safety Assessment for Packages to Be Transported by Air

Packages should be assessed for their ability to remain subcritical under conditions that result in maximum neutron multiplication, under routine conditions and after the tests for demonstrating the ability of the package to withstand, normal conditions and accident conditions of transport, assuming reflection by at least 20 cm of water but no water in-leakage. In making the assessment, one may take advantage of any special design features intended to prevent leakage of water into or out of void spaces, even as a result of error. These design features should be effective when the package is subjected to the tests prescribed for Type C package followed by the water leakage test requiring the test specimen to be immersed under a head of water of at least 0.9 m for a period of at least 8 h in the attitude for which maximum leakage is expected.

Assessment of Package Arrays Under Normal Conditions of Transport

A number N should be derived, such that five times N packages should be subcritical for the arrangement and package conditions that provide the maximum neutron multiplication consistent with the following:

(a) There should not be anything between the packages, and the package arrangement should be reflected on all sides by at least 20 cm of water.
(b) The state of the packages should be their assessed or demonstrated condition if they had been subjected to the tests demonstrating ability of the package to withstand normal conditions of transport.

Assessment of Package Arrays Under Accident Conditions of Transport

For the purpose of the assessment, the package should be subjected to tests demonstrating ability of the package to withstand normal conditions of transport followed by whichever of the following is the more limiting:

(i)   The Drop II test and either Drop I or Drop III test, as appropriate, followed by the thermal test and completed by the water leakage test specified for packages containing fissile material or

(ii) The water immersion test.

However, when it can be demonstrated that the confinement system remains within the packaging following the tests for assessment of an array of packages under accident conditions of transport, close reflection of the package by at least 20 cm of water may be assumed.

Where any part of the fissile material escapes from the containment system following the tests specified for assessment of package arrays under accident conditions of transport, it should be assumed that fissile material escapes from each package in the array and that all the escaped fissile material should be arranged in the configuration and moderation that results in the maximum neutron multiplication with close reflection by at least 20 cm of water.

Now, the maximum number of packages following the test simulating accident conditions of transport that would remain subcritical should be determined. Half of that number should be designated as N. Thus, again a number N has been derived, such that $2 \times N$ packages would remain subcritical following the accident conditions of transport, for the arrangement and package conditions that provide the maximum neutron multiplication consistent with hydrogenous moderation between the packages and the package arrangement reflected on all sides by at least 20 cm of water.

Determination of Criticality Safety Index for Packages

The CSI for packages containing fissile material should be obtained by dividing the number 50 by the **smaller of the two values of N** derived as described above (i.e. CSI = 50/N). The value of the CSI may be zero, provided that an unlimited number of packages are subcritical (i.e. N is effectively equal to infinity both under the normal and the accident conditions of transport).

In summary, the nature of the content (radioactive, fissile and chemical) dictates the design requirements of the package.

Radioactive nature of contents: The A_1 or A_2 activity values and the physical form of the nuclear material define package type.

Fissile nature of contents: For transport and storage of fissile material, criticality safety should be assured for individual package in isolation and array of packages under normal and accident conditions of transport by sufficient margin.

Chemical nature of contents: Special characteristics such as chemical hazards associated with nuclear material, for example uranium hexafluoride, must also be taken into account.

10.9.3 Test Requirements for Packages

Packages are required to satisfy tests simulating normal conditions and accident conditions of transport. Following the graded approach, test requirements are classified as (a) tests simulating normal conditions of transport (b) tests simulating accident conditions of transport and (c) special additional tests for Type C packages. Packages containing UF_6 and for fissile material should, over and above, satisfy special additional test requirements.

10.9.3.1 Tests for Demonstrating Ability to Withstand Normal Conditions of Transport

Packages designed for containing nuclear material with low potential radiological hazard, e.g. small activities, such as Industrial Package Type IP-2/IP-3, and Type A package should maintain their integrity during normal transport conditions and are required to withstand a series of tests simulating normal conditions of transport, e.g. water spray test, free drop test, stacking test and penetration test. These tests simulate normal conditions such as the package getting exposed to rain, falling during handling, experiencing a compression load due to stacking and getting pierced by a sharp object.

10.9.3.2 Tests for Demonstrating Ability to Withstand Accident Conditions of Transport

Type B(U)/(M) and Type C packages are required to withstand stringent accident conditions of transport. In accidents, a package could experience mechanical impact caused by a fall through a height or an impact with a vehicle, thermal impact caused by a fire accident and external pressure caused by immersion in water [7]. The regulatory tests simulate the accident conditions as described below.

Mechanical Test

Drop I test: The test specimen is dropped through 9 m onto a flat **unyielding** surface.
Drop II test: The test specimen is dropped through 1 m onto a steel bar.
Drop III test: A 500-kg steel plate is dropped through 9 m on the test specimen.

The test specimen should be subjected to Drop II test and either Drop I or Drop III test as specified in the Safety Standards. The 9-m drop test simulates a severe accident because an unyielding surface which is an ideal surface that transfers all the kinetic energy of the impact to the package itself so that the package sustains the maximum damage that the impact can cause. In real-life situations, a package could impact such objects as concrete roads, bridge abutments and piers. All these objects would yield to some extent, and therefore, a fraction of the impact energy would be absorbed by the target. The 9-m drop test onto an unyielding surface is therefore relevant to impacts onto real-life objects as a result of a high-speed crash, e.g. in the case of a spent fuel transport cask, the test is equivalent to a crash onto a concrete slab at 250 km per hour.

Thermal Tests

In the thermal test for Type B(U)/(M) package, the test specimen is subjected to a fully engulfing fire of 800 °C for 30 min. In the **enhanced** fire test for Type C package, the test specimen is subjected to a fully engulfing fire of 800 °C for 60 min.

Fire accident is an important contingent to be considered in the transport of nuclear fuel cycle materials since it increases the potential for release of radioactive materials to the environment and for this reason; Type B(U)/B(M)/C packages should be able to withstand fires without significant release of activity. Analytical and experimental studies have shown that the conditions generated in this regulatory test are more severe than in a realistic fire accident in a storage area or during transport.

Water Immersion Tests

The water immersion test is designed to ensure safety in the event of accidents at sea. For packagings for spent fuel, fresh fuel and vitrified high-level radioactive waste, the test specimen is subjected to an immersion test equivalent to a water depth of 15 m for 8 h without loss of shielding or significant release of radioactivity. In addition, for Type B(U)/B(M)/C packagings designed for containing very high activity, the test specimen is subjected to the **enhanced** water immersion test equivalent to a water depth of 200 m for 1 h, and the containment system must not rupture. It should be demonstrated that as a result of the tests simulating accident conditions of transport, the dose rate at 1 m from the surface of the package would not exceed 10 mSv/h with the maximum radioactive contents that the package is designed to contain and that the accumulated loss of radioactive contents in a period of one week would not exceed $10A_2$ for krypton-85 and A_2 for all other radionuclides.

10.9.3.3 Special Additional Tests for Type C Packages

The tests for a Type C package for demonstrating ability to withstand severe air accidents [8] are described here.

Puncture-Tearing Test

A solid steel probe of mass 250 kg is dropped through 3 m on the specimen for a package of mass 250 kg or less. If the package weighs more, it is dropped on to the probe. The probe is a cylinder of diameter 30 cm with the striking end forming a frustrum of a cone with a height of 30 cm and diameter at the striking end being 2.5 cm.

Impact Test

The specimen should be subjected to an impact on a target at a velocity of not less than 90 m/s, at such an orientation as to suffer maximum damage. The target surface should be normal to the specimen path.

10.9.3.4 Special Additional Tests for Packages for UF_6 and for Fissile Material

For packaging designed to contain 0.1 kg or more of UF_6, the specimen is tested hydraulically at an internal pressure of at least 1.38 MPa (When the test pressure is less than 2.76 MPa, the design would require multilateral approval).

For a packaging designed for fissile material, the test specimen should be subjected to the tests for demonstrating ability to withstand normal conditions of transport followed by either (i) Drop II test and Drop I or Drop III test, as appropriate, then the fire test and completed by a water leakage test immersing the specimen under a head of water of at least 0.9 m for a period of not less than 8 h and in the attitude for which maximum leakage is expected or (ii) the water immersion test, whichever is more limiting. The tests form part of the assessment of an array of packages intended for fissile contents for criticality safety under accident conditions.

10.9.4 Graded Approach

The requirements specified in the Safety Standards are commensurate with the radiological risk associated with the radioactive material to be transported. Accordingly, a graded approach is adopted in specifying the requirements. As can be seen in Fig. 10.8, selection of the optimal package would depend on the nature of the radioactive content.

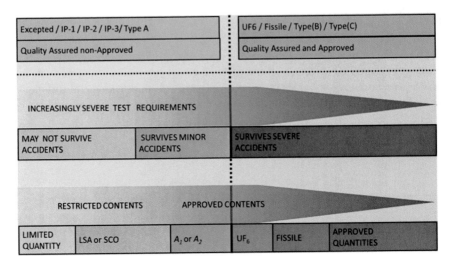

Fig. 10.8 Graded approach for selecting the optimal package based on the nature of the content

10.10   Active Safety

Upon taking advantage of the inherent safety offered by the nuclear material and implementing passive safety through selection of the appropriate packaging for optimal protection, active safety is brought into play through operational control during transport and administrative control. Many control measures are implemented during transport and storage of nuclear material.

10.10.1   Preparation of the Package

The control measures include (a) preparation of the package, ensuring that the total activity contained in the package does not exceed the regulatory limit and (b) marking of the package (Fig. 10.9). The packages communicate with the outer world through the labels affixed on their exterior. Three categories of labels are identified for packages. The category of a package depends on the radiation level on the external surface of the package and its transport index.

10.10.2   Determination of the Transport Index of the Package

The TI of a package is the number obtained by multiplying by 100 the maximum dose rate at 1 m from the external surface of the package expressed in mSv/h. The meaning of TI is explained in Fig. 10.10.

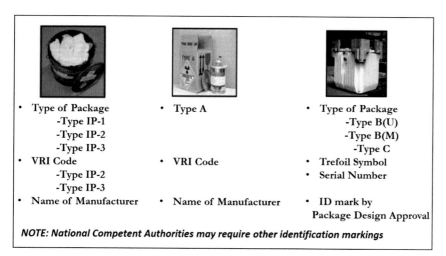

- **Type of Package**
  -Type IP-1
  -Type IP-2
  -Type IP-3
- **VRI Code**
  -Type IP-2
  -Type IP-3
- **Name of Manufacturer**

- **Type A**

- **VRI Code**

- **Name of Manufacturer**

- **Type of Package**
  -Type B(U)
  -Type B(M)
  -Type C
- **Trefoil Symbol**
- **Serial Number**

- **ID mark by**
  Package Design Approval

NOTE: National Competent Authorities may require other identification markings

Fig. 10.9   Some examples of markings on packages

Fig. 10.10 Transport index of a package

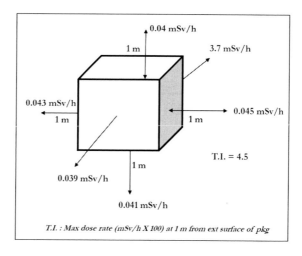

T.I. : Max dose rate (mSv/h X 100) at 1 m from ext surface of pkg

Uranium and thorium ores and concentrates are transported and stored in large quantities. The TI for a load of these materials **may be taken as**, 0.4 mSv/h for ores and physical concentrates of U and Th, 0.3 mSv/h for chemical concentrates of Th and 0.02 mSv/h for chemical concentrates of U, other than UF_6. For tanks, freight containers and unpackaged LSA-I, SCO-I and SCO-III, the value as stated above should be multiplied by the appropriate factor from Table 10.5.

Table 10.5 Multiplication factors for tanks, freight containers and unpackaged LSA-I, SCO-I and SCO-III

Size of load[a]	Multiplication factor
Size of load $\leq 1\ m^2$	1
$1\ m^2 <$ size of load $\leq 5\ m^2$	2
$5\ m^2 <$ size of load $\leq 20\ m^2$	3
$20\ m^2 <$ size of load	10

a Largest cross-sectional area of the load being measured

Table 10.6 Categories of packages

Conditions		Category
TI	Maximum dose rate at any point On the external surface of the package	
0	Not more than 0.005 mSv/h	I-WHITE
More than 0 but not more than 1	More than 0.005 mSv/h but not more than 0.5 mSv/h	II-YELLOW
More than 1 but not more than 10	More than 0.5 mSv/h but not more than 2 mSv/h	III-YELLOW
More than 10	More than 2 mSv/h but not more than 10 mSv/h	III-YELLOW

10.10.3 Determination of the Category of the Package

Three categories of packages are identified. The category of a package is determined on the basis of the maximum dose rate at the external surface of the package and the transport index (T.I.) of the package. The three categories of a package are designated as Category I-WHITE, Category II-YELLOW and Category III-YELLOW as shown in Table 10.6.

10.10.4 Labelling

Labels showing appropriate category, as explained below, should be affixed on two opposite sides of the package, and the required information should be inscribed on the labels. The three category labels are shown in Fig. 10.11.

Each package, containing fissile material, should bear labels shown in Fig. 10.12.

The vehicle carrying the nuclear material should be placarded (Fig. 10.13) as required by the national/international regulations.

The dimensions for the trefoil symbol and the labels should be as specified in the Safety Standards.

Fig. 10.11 Labels of Category I-WHITE, Category II-YELLOW and Category III-YELLOW

Fig. 10.12 Labels for package containing fissile material

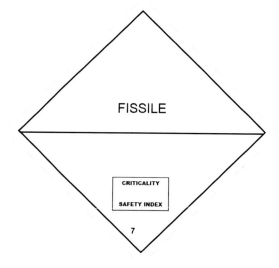

Fig. 10.13 Vehicle placard

10.10.5 Transport Documents

Transport documents constitute an important element of active safety and include a declaration by the consignor (sender of the package), stating that the relevant regulatory requirements have been duly complied with, instructions to the carrier as to the actions, if any, to be taken during carriage including emergency response actions. The requirements in this regard, including notification of the competent authorities of the relevant states, are specified in the Safety Standards.

10.10.6 Loading and Accumulation of Packages During Transport and Storage in Transit

Accumulation of packages should be controlled for limiting the dose that could be received by transport workers and public. The total number of packages aboard a single conveyance should be limited (Fig. 10.14) so that the sum of the TIs aboard the conveyance does not exceed the values given in Table 10.7. (Consignments of LSA-I material are excepted from this limit.)

The sum of the T.I.s in the large freight containers in the hold, compartment or defined deck area of a sea-going vessel should not exceed 200. The sum of the CSIs in a freight container and aboard a conveyance should not exceed the values shown in Table 10.7.

Any package having a TI greater than 10, or any consignment having a CSI greater than 50, should be transported only under exclusive use.

Though LSA materials and SCOs pose limited radiological hazard, the loading and accumulation of IPs containing these materials in conveyances should be controlled as shown in Table 10.8.

For SCO-III, the limits in Table 10.8 may be exceeded provided that the transport plan contains precautions which are to be employed during transport to obtain an

Fig. 10.14 Loading of packages in vehicles

Table 10.7 TI and CSI limits for conveyances [*]

Type of conveyance	Limit on sum of TI-not under exclusive use	Limit on sum of CSI	
		Not under exclusive use	Exclusive use
Vehicle (road/rail vehicle)	50	50	100
Passenger aircraft	50	50	Not applicable
Cargo aircraft	200	50	100
Inland waterway craft	50	50	100
Hold, compartment or defined deck area of a sea-going vessel, packages, overpacks, freight containers	50	50	100
Total vessel	200	200	200

[*] Please also see Sect. 10.10.9

Table 10.8 Conveyance activity limits for LSA materials and SCOs in IPs

Material	Activity limit (other than inland waterway craft)	Activity limit in inland waterway craft
LSA-I	No limit	No limit
LSA-II/LSA-III Non-combustible solids	No limit	$100 A_2$
LSA-II/ LSA-III Combustible solids and all liquids and gases	$100 A_2$	$10 A_2$
SCO-I and SCO-II	$100 A_2$	$10 A_2$

overall level of safety at least equivalent to that which would be provided if the limits had been applied. Transport of SCO-III should be under exclusive use by road, rail, inland waterway or sea. Stacking should not be permitted. The shipment should be subject to multilateral approval.

10.10.7 Segregation of Packages During Transport and Storage

Category II-YELLOW or III-YELLOW packages should not be carried in compartments occupied by passengers. If packages have to be carried in a passenger compartment, they have to be accompanied by specially authorized couriers, and it is necessary to exclusively reserve the compartment for such packages.

Packages containing radioactive material and unpackaged radioactive material have to be segregated during transport and during storage in transit. Distance of segregation has to be calculated depending on the nature of occupancy in the immediate vicinity.

(a) Dose limit for workers in regularly occupied working areas is 5 mSv in a year;
(b) Dose limit for members of the public in areas where the public has regular access is 1 mSv in a year;
(c) Dose limit for undeveloped photographic film is 0.1 mSv per consignment;
(d) For other dangerous goods (e.g. explosives), the relevant regulations should be additionally complied with for determining the segregation distance.

10.10.8   Stowage of Package

Consignments should be securely stowed on the conveyance. A package may be carried or stored amongst packaged general cargo. The average surface heat flux should not exceed 15 W/m^2. The immediate surrounding cargo should not be in sacks or bags. Any special stowage provisions specified by the competent authority must be satisfied.

10.10.9   Additional Requirements Relating to Transport and Storage in Transit of Fissile Material

Any group of packages and freight containers containing fissile material stored in transit in any one storage area should be so limited that the sum of the CSIs in the group does not exceed 50. Each group should be stored so as to maintain a spacing of at least 6 m from other such groups.

Where the sum of the CSIs on board a conveyance or in a freight container exceeds 50, as permitted in Table 10.7, storage should be such as to maintain a spacing of at least 6 m from other groups of packages, overpacks or freight containers containing fissile material or other conveyances carrying radioactive material.

10.10.10   Special Additional Requirements for Transport by Air

A single package of non-combustible solid LSA-II or LSA-III material, if carried by air, should not contain an activity greater than 3000 A$_2$. Type B(M) packages and consignments under exclusive use should not be transported on passenger aircraft. Vented Type B(M) packages, packages that require external cooling by an ancillary cooling system, packages subject to operational controls during transport and packages containing liquid pyrophoric radioactive materials should not be transported by air.

Packages having a surface dose rate greater than 2 mSv/h may be transported by air only under special arrangement.

10.10.11   Special Arrangement

It may happen that a shipment has to be made, but it may be impractical to comply with some of the regulatory requirements. On such occasions, the consignor has to make arrangements for provision of additional safety features such as improvised fittings and administrative control that would ensure at least the same level of safety that would have been provided if all the regulatory requirements had been duly met. This is called special arrangement. Competent authority approval has to be obtained for shipment under special arrangement.

10.10.12   Approvals

Competent authority approval has to be obtained for design of (a) special form radioactive material, (b) low dispersible radioactive material and (c) packaging. Competent authority approval is required for certain shipments. Some approvals have to be unilateral and some multilateral. For the requirements for approval, one should refer to the Safety Standards and the different national and international regulations.

10.11   Regulatory Limits for a Package at a Glance

The regulatory limits pertaining to the dose rate, the non-fixed contamination level and the temperature at the external surface of the package which need to be complied with are given here.

The **dose rate at the external surface of a package** should not exceed the following limits:

Whether exclusive use	At the external surface of the package	T.I
Not under exclusive use	2 mSv/h	10
Exclusive use	1000 mSv/h	(0.1 mSv/h at 2 m from the external surface of the package)

The dose rate at the external surface of an excepted package should not exceed 5 µSv/h

For shipments under exclusive use by road or rail, the dose rate at any point on the package may exceed 2 mSv/h but should not exceed 10 mSv/h provided the dose rates at the outer surface of the vehicle and at 2 m from the external surface of the vehicle do not exceed 2 mSv/h and 0.1 mSv/h, respectively.

The **non-fixed contamination on the external surface of a package**, under routine conditions of transport, should not exceed the following limits:

(a) 4 Bq/cm^2 for beta and gamma emitters and low toxicity alpha emitters
(b) 0.4 Bq/cm^2 for all other alpha emitters.

The **temperature of the accessible surface of the package** assessed assuming an ambient temperature of 38 °C and absence of solar heat input should not exceed the following limits:

Mode of transport	Exclusive use (°C)	Not under exclusive use (°C)
Air	50	50
Other than air	85	50

10.12   Emergency Response

Management of emergency involving radioactive shipments is a comprehensive topic in itself. The consignor has the responsibility of providing emergency instructions [e.g. TREMCARD] to the carrier. Since an emergency can occur anywhere during a shipment, certain response measures have to be implemented by the crew at the site of emergency before the arrival of the first responders. The TREMCARD should include the particulars of the radioactive material, the nature of the potential hazard, the precautionary measures to be implemented before the first responders arrive such as cordoning a specified distance and the contact details of the consignor and the national emergency response authority. The consignor would be expected to provide support to the experts managing the emergency by way of sending trained personnel with tools to the site of emergency. The personnel of all the stake holders (e.g. consignor's and the carrier's personnel and the emergency response personnel) need to be trained in handling an emergency involving transport of radioactive material [9].

10.13   Security of Transport of Radioactive Material

Security of transport of radioactive material is important since a breach of security such as theft of a package or diversion of a shipment could have safety implications. Depending on the nature of shipment, three levels of security are identified, namely, prudent management, basic level and advanced level [10–12]. The consignor has to determine the appropriate level of security on the basis of the consequences of a breach of security and threat perception on the route of shipment and at the relevant period and accordingly develop a security plan that would deter, delay, detect security events and respond to them promptly. The security plan should also make provisions, where required, to mitigate the consequences of a security event. Any security measure should not undermine the safety requirements. Co-operation amongst the consignor, law enforcing authorities and the national and local emergency response authorities should ensure that any security event is handled safely and effectively.

10.14   Summary

The safety of transport and storage of nuclear material is ensured by a judicious combination of the inherent safety of the radioactive material, the standards of the design of the material and of the design of the packages adopting a graded approach and implementing control during transport and storage and adhering to the required administrative measures. Inherently safe nuclear materials such as uranium ore and concentrates can be transported in Industrial Packages. Nuclear materials of small activities may be transported in Type A packages. Spent fuel and certain fissile materials should be transported in Type B(U)/(M) packages. Fissile materials and UF_6 warrant additional safety measures. In addition, there are specific safety provisions for transport of large quantities of nuclear material (e.g. spent fuel) by dedicated ships, details of which can be found in the relevant literature [e.g. 13].

Exercise Problems

Q (1)   Mark "X" in the applicable cell(s) to indicate the monitoring required.

Assessed dose received by transport worker, D	Monitoring required		
	None	Area monitoring	Individual monitoring
$D \leq 1$ mSv/a			
1 mSv/a $< D \leq 6$ mSv/a			
$D > 6$ mSv/a			

Q (2) It is estimated that an object weighing 200 g contains ^{137}Cs of total activity 0.9 kBq. Will the safety requirements apply to the transport of this object through public domain or will it be exempt from the safety requirements?

Q (3) Identify the following nuclear material as LSA-I or II or III or none of these:

(a) Radioactive dust of specific activity 10^{-3} A_2/g
(b) Liquid nuclear waste of specific activity 6×10^{-6} A_2/g
(c) Unshielded solid nuclear waste of specific activity 10^{-3} A_2/g

Q (4) Swipe samples over an area of 300 cm^2 were taken on three surface contaminated objects. The measured activities were as given below. Determine the SCO group of the objects.

(a) Non-fixed ^{90}Sr contamination on accessible as well as inaccessible surfaces is 1 kBq; fixed contamination ^{90}Sr on accessible surface is 2 kBq;
       fixed plus non-fixed contamination ^{90}Sr on inaccessible surface is 10^5 Bq.

(b) Non-fixed ^{239}Pu contamination on accessible surface is 60 Bq;
       fixed ^{239}Pu contamination on inaccessible surface is 400 kBq
       fixed plus non-fixed ^{239}Pu contamination on inaccessible surface is 10^8 Bq.

(c) Non-fixed ^{90}Sr and ^{137}Cs contamination on accessible surfaces is 5 KBq;
       only fixed ^{137}Cs contamination on inaccessible surfaces is 15 KBq;
       only non-fixed ^{239}Pu contamination on inaccessible surfaces is 30 kBq.

Q (5.1) What is the difference between a fissile-excepted material and a non-fissile material?

Q (5.2) Write an algorithm to determine whether a consignment containing less than 1% of ^{235}U by mass and some plutonium is fissile-excepted or not [1].

Q (6) Write an algorithm to determine whether a consignment containing a maximum enrichment of 5% of ^{235}U by mass and some plutonium and ^{233}U is fissile-excepted or not [1].

Q (7) Identify the Industrial Package Type (IP-1/2/3) that would be the optimal choice for the transport of the following materials:

(a) radioactive dust of specific activity 10^{-4} A_2/g by exclusive use,
(b) liquid nuclear waste of specific activity 6×10^{-6} A_2/g by exclusive use,
(c) unshielded solid nuclear waste of specific activity 10^{-3} A_2/g by non-exclusive use.

Q (8) Can the following radioactive materials that are not in special form be transported in a Type A package?

(a) A consignment of ^{131}I of activity 1 TBq,

(b) A consignment of ^{239}Pu of activity 2 GBq.

Q (9) In what type of package can a sample of ^{239}Pu of mass 15 g in powder form be transported by air? [The specific activity of ^{239}Pu is 2.3×10^9 Bq/g].

Q (10) The criticality safety assessment of a package shows that the maximum number of packages that remain subcritical under normal conditions of transport is 45 and that the maximum number of packages that would remain subcritical under accident conditions of transport is 9. What is the CSI of the package?

Q (11) If the CSI of a package of a certain design is determined to be 5, what is the maximum number of packages of that design can be expected remain subcritical under (a) normal conditions of transport (b) accident conditions of transport.

Q (12) A cask for which multilateral approval certificates have been issued, containing spent fuel with a burn up of 9000 MWd/t from a CANDU type reactor is transported to a waste repository. What will be the proper shipping name and the UN Number of that shipment? [1]

Q (13) Indicate the nature of approval (unilateral/multilateral) required in each case [1].

(a) Design approval: IP-1/2/3, Type A, Type B(U)/(M)/C: fissile content
(b) Type C package (non-fissile/fissile-excepted)
(c) Low dispersible radioactive material (LDRM)
(d) Type B(U) package containing LDRM
(e) Special form radioactive material
(f) Shipment approval: Special arrangements

References

1. International Atomic Energy Agency, Regulations for the safe transport of radioactive material, specific safety requirements, SSR 6, Vienna (2018)
2. International Atomic Energy Agency, Safe transport of radioactive material, fourth edition, training course series no. 1, IAEA-TRC-01/04, Vienna (2006)
3. International Civil Aviation Organization, Technical instructions for the safe transport of dangerous goods by air 2021–2022, ICAO, Montreal (2021)
4. World Nuclear Transport Institute, Nuclear fuel cycle transport-nuclear fuel cycle transport—front end materials, Fact Sheet, England
5. World Nuclear Transport Institute, Nuclear fuel cycle transport-nuclear fuel cycle transport—back end materials, Fact Sheet, England
6. International Atomic Energy Agency, Advisory material for the IAEA regulations for the safe transport of radioactive material (2012 Edition), Specific safety guide SSG 26, Vienna (2014)
7. World Nuclear Transport Institute, Nuclear fuel cycle transport—the IAEA regulations and their relevance to severe accidents, Information Paper, England
8. International Atomic Energy Agency, Assessment of air transport accident conditions in the context of test requirements for type C packages containing radioactive material, final report

of a coordinated research project on accident severity during the air transport of radioactive material, (1998–2006), IAEA-TECDOC-1965, Vienna (2021)

9. International Atomic Energy Agency preparedness and response for a nuclear or radiological emergency involving the transport of radioactive material, IAEA safety standards series no. SSG-65, Vienna (2022)

10. International Atomic Energy Agency security in the transport of radioactive material, implementing guide, IAEA nuclear security series no. 9, Vienna (2008)

11. International Atomic Energy Agency security nuclear material in transport, implementing guide, IAEA nuclear security series no. 26-G, Vienna (2015)

12. Atomic Energy Regulatory Board security of radioactive material during transport, AERB safety guide no. AERB/NRF-TS/SG-10, Mumbai (2008)

13. International Maritime Organization, International code for safe carriage of packaged irradiated nuclear fuel, plutonium and high-level radioactive wastes on board ships (INF), London (2000)

Further Reading

1. R. Gibson (ed.), *The Safe Transport of Radioactive Materials* (Pergamon Press, 2013)

2. Atomic Energy Regulatory Board, Safe transport of radioactive material, AERB safety code no. AERB/NRF-TS/SC-1 (Rev.1) (2016)

3. US Nuclear Regulatory Commission, Part 71—Packaging and transportation of radioactive material, 10 CFR Part 71 (2004)

4. Australian Radiation Protection and Nuclear Safety Agency, Safe transport of radioactive material, safety guide, radiation protection series no. 2.1, Melbourne (2008)

Chapter 11
Radiation Protection

A. N. Nandakumar

11.1 Part I Basic Principles of Radiation Protection

11.1.1 Introduction

Radiation protection is a vast subject in itself. This chapter briefly identifies some, not all, of the important considerations for radiation protection in nuclear facilities. In respect of safety requirements for specific nuclear facilities, the reader is encouraged to refer to appropriate further reading material [1]. Working with ionizing radiations can result in biological effects. The safety standards recommended for protection and safety should be adhered to. The safety standards implemented for radiation safety are generally based on the recommendations of the International Commission on Radiological Protection (ICRP).

The biological hazard to individuals resulting from exposure to radiation depends on a number of factors such as the magnitude of radiation exposure received, period of exposure, quantity of intake of radioactive material in the case of handling unsealed radioactive material and the radiotoxicity of the radioactive material [2].

For enabling a detailed discussion, one should get familiar with the quantities and units that are relevant to radiation protection.

11.1.2 Quantities and Units

The quantities that are important to radiation protection are activity of a radioactive source, exposure, absorbed dose, equivalent dose, effective dose, committed equivalent dose and committed effective dose [3].

A. N. Nandakumar (✉)
Formerly Radiological Safety Division, Atomic Energy Regulatory Board, Mumbai, India
e-mail: a.n.nandakumar@gmail.com

© The Author(s), under exclusive license to Springer Nature Singapore Pte Ltd. 2023
B. S. Tomar et al. (eds.), *Nuclear Fuel Cycle*,
https://doi.org/10.1007/978-981-99-0949-0_11

11.1.2.1 Activity

Activity of a radioactive material represents the quantity of emission of particulate radiation and/or gamma photons per unit time from the material. Activity is the rate of disintegration of a radionuclide.

The unit of activity is Becquerel (Bq).

1 Bq = 1 Disintegration Per Second.

11.1.2.2 Exposure Unit

In order to be able to detect and measure the radiation exposure, one has to measure the ionization produced by radiation in air. This quantity is called the Exposure Unit. One Exposure Unit is defined as that quantity of X- or gamma radiation that produces in air, ions of either sign, carrying 1 C of charge per kg of air at STP. Because of practical difficulties in designing instruments that can accurately collect all the ions produced by radiation at higher energies, the Exposure Unit is defined only up to 3 meV. The application of the Exposure Unit is restricted to X-radiation and gamma radiation only.

11.1.2.3 Absorbed Dose

Radiation causes ionization and excitation in the medium it traverses. Each ionization event results in loss of energy by the radiation. This energy deposited by the radiation per unit mass of the medium is called the **absorbed dose** and is given by,

$$D = \delta E / \delta m,$$

δE is the energy deposited by radiation and

δm is the mass of the matter where the energy is deposited.

The unit of absorbed dose is Gray (Gy) and is defined as one Joule per kg of the medium (Fig. 11.1). Transfer of energy by ionization is common to all types of radiation. Thus, absorbed dose and the corresponding unit Gray are common to all radiations, viz., alpha, beta, gamma, X-ray, protons and neutrons for both external exposure and internal exposure of all energies as against the unit of exposure which applies only to gamma and X-radiations of energy up to 3 MeV.

Damage to biological tissues depends on the type of radiation encountered, the absorbed dose in the tissue and the radiosensitivity of the tissue or organ.

11.1.2.4 Equivalent Dose

Biological effect depends on the energy transferred by the radiation along its path of travel. This is called linear energy transfer or LET (keV/micron). Some radiations

Fig. 11.1 Absorbed dose

Absorbed energy = $\delta E = E_1 - E_2$

Absorbed dose =Absorbed energy per unit mass

Absorbed dose = $\delta E/ \delta M$ J/kg

1 Gy (Gray)= 1 J/kg

Gray is used for alpha, beta, gamma and X radiations

cause more ionization events per unit path length, and they are called high LET radiations. An example of high LET radiation is α radiation. Radiations that cause fewer ionization events per unit path length are called low LET radiations, for example, β, Υ and X-radiations. The biological effect caused by one Gy of α radiation which is a high LET radiation would be more than that caused by one Gy of low LET radiations such as β, Υ and X-radiations. For example, on a standard scale of biological damage, α radiation is 20 times as effective as β or Υ or X-radiation. Hence, a weighting factor is associated with each radiation. It is called the radiation weighting factor, W_R. The W_R of β, Υ and X-radiations is 1. The W_R of α radiation is 20. For neutrons, the W_R value depends on the neutron energy.

From the biological effect point of view, 20 Gy of Υ radiation is *equivalent* to 1 Gy of α radiation.

The unit called **equivalent dose** is introduced to describe the dose received by a person without having to specify the radiation. The unit of equivalent dose is Sievert, Sv.

1 Gy of α dose multiplied by the W_R of $\alpha = 1$ Gy $\times 20 = 20$ Sv.

20 Gy of Υ dose multiplied by the W_R of $\Upsilon = 20$ Gy $\times 1 = 20$ Sv.

Clearly, 1 Sv of α radiation dose is equivalent to 1 Sv of β or Υ or X-radiation dose.

Heavy charged particles and neutrons have a higher LET than beta, gamma and X-rays. The equivalent dose to tissue or organ T, $H_{T,R}$, is defined as:

$$H_{T,R} = w_R \cdot D_{T,R},$$

where $D_{T,R}$ is the absorbed dose delivered by radiation type R averaged over a tissue or organ T and w_R is the radiation weighting factor for radiation type R. When dose is delivered to a tissue, T, by radiations with different values of w_R, the equivalent dose to the tissue, T, is:

$$H_T = \Sigma w_R \cdot D_{T,R}.$$

Table 11.1 Radiation weighting factors

Radiation type and energy	Radiation weighting factor (WR)
Gamma and X-rays	1
Beta particles and electrons	1
Neutrons energy $\varepsilon_n < 10$ keV	5
10 keV $\leq \varepsilon_n \leq 100$ keV	10
100 keV $< \varepsilon_n \leq 2$ MeV	20
2 MeV $< \varepsilon_n \leq 20$ MeV	10
20 MeV $< \varepsilon_n$	5
Alpha particles and fission fragments	20

The unit for the equivalent dose is the same as for absorbed dose, J kg^{-1}, and its special name is Sievert (Sv). The quantity, equivalent dose, applies to external exposure as well as internal exposure. Radiation weighting factors are given in Table 11.1.

For example, consider a tissue T that has received a dose of D_Υ Gy of Υ radiation and a dose of D_α Gy of α radiation. The w_R of Υ radiation is 1, and the w_R of α radiation is 20. Therefore, the total equivalent dose to tissue T $= (D_\Upsilon \times 1 + D_\alpha \times 20)$ Sv.

11.1.2.5 Effective Dose

When different individual organs of a person are exposed to radiation, the effective harm to the exposed person is the summation of the damage to the individual organs. The sensitivity and vulnerability to radiation damage and the repairability of tissues vary from one tissue to another. Accordingly, tissues are associated with weighting factors, viz., tissue weighting factors, W_T. The w_T values are given in Table 11.2.

Table 11.2 Tissue weighting factors

Tissue	Tissue weighting factor, W_T	ΣW_T(*)
Bone marrow (red), colon, lung, stomach, breast, remaining tissues*	0.12	0.72
Gonads	0.08	0.08
Bladder, esophagus, liver, thyroid	0.04	0.16
Bone surface, brain, salivary glands, skin	0.01	0.04
	Total	1.00

(*) **Remaining tissues: Adrenals, extrathoracic region, gall bladder, heart, kidneys, lymphatic nodes, muscle, oral mucosa, pancreas, prostate (σ), small intestine, spleen, thymus, uterus/cervix (φ).**

Note that, the whole-body dose involves all the tissues, and hence, the sum of the W_T of all tissues is 1. If individual tissues receive different doses, the **effective dose** is given by

$$E = \sum w_T H_T.$$

The effective dose is taken to represent the overall harm to the body due to radiation exposure received by individual tissues. The effective dose resulting from exposure to different types of radiation is given by

$$E = \sum w_T H_T = \Sigma w_T \cdot \Sigma w_R \cdot D_{T,R}.$$

The unit for the effective dose is the same as for absorbed dose, J kg^{-1}, and its special name is Sievert (Sv). The quantity, effective dose, applies to external exposure as well as internal exposure.

For example, if a person receives a gamma dose of D_Υ Gy to thyroid and an α dose of D_α Gy to the lungs, the effective dose to the person is given by

$$E = \left[\{D_\Upsilon \times (w_R \text{for } \Upsilon) \times (w_T \text{for thyroid})\} + \{D_\alpha \times (w_R \text{ for } \alpha \times w_T \text{ for lung})\} \right] Sv$$
$$= [\{(D_\Upsilon \times 1 \times 0.04\} + \{D_\alpha \times 20 \times 0.12)\}] Sv.$$

From Table 11.1, the w_R for Υ is 1 and w_R for α is 20.
From Table 11.2, the w_T of thyroid is 0.04 and the w_T of lung is 0.12.
The W_T values have been arrived at for low doses and low dose rates. Hence, the unit Sv is appropriate for radiation protection purposes only, that is, for low doses and low dose rates.

11.1.2.6 Committed Equivalent Dose

Committed dose is defined for internal exposure, i.e., the dose received due to intake of radioactive material. The radioactive material may get deposited in a tissue or organ or the whole body. The activity of the radioactive material keeps reducing according to its physical half-life, t_P, and due to the natural biological elimination process that is characterized by a biological half-life, t_B.

The effective half-life is defined as

$$1/t_{\text{eff}} = (1/t_P + 1/t_b).$$

The effective decay constant can be derived as $\lambda_{\text{eff}} = \lambda_p + \lambda_b,$

where λ_{eff}, λ_p and λ_b are given by

$$\lambda_{eff} = 0.693/t_{eff}, \ \lambda_p = 0.693/t_p \ and \ \lambda_b = 0.693/t_b.$$

Given the dose rate at the time of intake, $D_{T,R}(0)$, for the quantity of radioactive material that has been taken up by the tissue, T, the dose rate at any time, t, after intake is

$$D_{T,R}(t) = D_{T,R}(0) \exp(-\lambda_{eff}t).$$

The total absorbed dose to tissue, T, integrated over a period of τ from the time of intake is

$$D_{T,R} =_\tau \int D_{T,R}(t) \, dt.$$

The **committed *equivalent* dose** to the tissue is the total equivalent dose due to intake of radionuclide over a period τ, from the time of intake, and is given by,

$$H_T(\tau) = w_{R\tau} \int D_{T,R}(t) \, dt.$$

11.1.2.7 Committed Effective Dose

The **committed *effective* dose** is given by

$$E(\tau) = \Sigma w_T \cdot H_T(\tau),$$

where $H_T(\tau)$ is the committed equivalent dose to tissue, T, and w_T is the tissue weighting factor for tissue, T. When τ, the time following intake, is not specified, it is taken as 70 years for children and 50 years for adults.

11.1.2.8 Collective Dose

The term collective dose is the summation of the product of the dose and the number of persons who received the dose, that is, $\Sigma \ N(E) \times E$. The concept of collective dose should be used only for dose optimization purposes and not for making risk estimates.

11.1.3 Biological Effects of Radiation

Biological effect of radiation is broadly classified as *stochastic effects* and *deterministic effects.*

Examples of deterministic effects are change in blood counts, nausea, vomiting and diarrhea (NVD), temporary loss of hair, erythema (skin burn), cataract, etc. Deterministic effects would not occur if the dose received by a person is less than the threshold dose defined for the effect. However, as the dose increases, the *severity* of the deterministic effect may increase.

Examples of stochastic effects are incidence of cancer, leukemia and hereditary effects. No threshold dose is defined for stochastic effects. As the dose increases, the *probability* of occurrence of the effect increases (Fig. 11.2).

Some biological effects occur within a few hours to days after exposure and are described as *immediate effects.* Some effects occur after a few years and are known as *delayed effects.* Biological effects are also classified as *somatic effects* and *hereditary effects.* Somatic effects are those which appear in the person who was exposed to radiation. Hereditary effects are those which appear in the progeny of the person exposed to radiation.

Examples of somatic effects are change in blood picture, NVD, erythema, temporary loss of hair, temporary or permanent sterility and fatality due to failure of the central nervous system. All immediate effects are somatic effects. The converse is not true. That is, there are some somatic effects that manifest a few years after exposure to radiation. Examples of delayed somatic effects are cataract and different kinds of cancer. Leukemia is the most common, with a delay period of 2–5 years. Other common examples of radiogenic cancers are those of colon, lung and stomach.

Hereditary effects occur due to mutations that cause malformation such as mongolism.

The biological effects depend on the nature of exposure to radiation which is classified as *acute exposure* or *chronic exposure.* Acute exposure results in a high dose within a short period. Chronic exposure represents low dose rate over an extended period of time.

An important effect that radiation can have on a cell is that it can break the chromosomes which are found inside the nucleus of the cell. The broken chromosomes

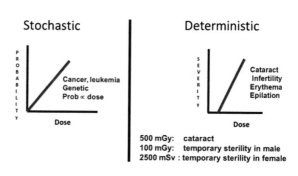

Fig. 11.2 Stochastic and deterministic effects—an illustrative example

often recombine, but if the recombination is not as per the original structure of the chromosome, one sees a misshapen chromosome. This condition is called *chromosome aberration* (CA). The CA can be seen through a microscope and counted! This is called CA test. The number of CA is a measure of the dose that caused the CA. In situations where a person got exposed to radiation without wearing a personal radiation monitor, a CA test can provide an approximate information about the dose received by the person when the dose received is suspected to be above 100 mSv.

11.1.4   External and Internal Exposures

In the case of external exposure, there need not be physical contact of the exposed individual with the radiation source. When the source is removed or when one moves sufficiently far from the source, external exposure ceases or reduces. In the case of external exposure, the radiation dose received by a person from a source is determined by measuring the exposure level in the work place [4].

Internal exposure can occur when a person comes into contact with a radioactive material that is in the form of liquid, powder or gas. Radioactive contamination on any surface and the environment should be avoided as it can cause internal exposure. Internal radiation exposure occurs when radionuclides from environmental contamination enter the body. Radioactive substances can gain entry into the body through three pathways:

- Inhalation—by breathing radioactive gases and aerosols.
- Ingestion—by consuming contaminated water, food.
- Absorption—through the intact skin or through wounds.

Therefore, safety measures to counter internal radiation are designed to either block the portals of entry into the body or to intercept the transmission of radioactivity from the source to the worker. This can be effected at the source, by enclosing and confining it, and at the worker's, position through the use of protective clothing and respiratory protective devices. Additionally, work practices should be designed to minimize contamination and exposure to contaminated environments.

The radiation dose due to internal contamination cannot be directly measured; it can only be calculated. In many instances, the level of contamination is known only after a lengthy investigation, which may include bioassay of samples (e.g., urine, feces, mucous and saliva) collected from persons who may have incurred an intake of radionuclides. Safety is concerned mainly with preventing or minimizing the intake of radionuclides into the body and the deposition of radioactivity on the body.

When radioactivity is deposited in the body, irradiation of the contaminated person continues even after the person leaves the area where the contamination occurred. It must be emphasized, however, that the fact that an internally deposited radioisotope continues to irradiate as long as it is in the body is explicitly considered in calculating the dose from an internally deposited radionuclide. In the context of potential harm, the radiation dose from an internally deposited radionuclide is not different from the same dose absorbed from external radiation. Therefore, dose for dose, the consequences of internal radiation are the same as those from external radiation; a milligray is a milligray and a milliSv is a milliSv, regardless of whether it was delivered from an internally deposited radionuclide or from external radiation [4].

11.1.5 System of Dose Limitation

The International Commission on Radiological Protection (ICRP) recommends a system of radiological protection based on three requirements. They are Justification of practices, Optimization of protection and safety and Dose limitation [2].

Justification of practice: No practice involving exposure to radiation should be adopted unless it produces sufficient benefit to the exposed individuals (e.g., a patient undergoing diagnosis with radiation) or to the society (e.g., production of nuclear power) to offset the detriment caused by the exposure to radiation because of the practice.

Optimization of protection and safety: It is assumed that no dose is safe. The dose to individuals should be kept as low as practicable. Just as detriment caused by radiation such as dose received by workers and public, however small, has a social cost, reduction of dose has an economic cost of protection, viz., expenditure incurred in achieving the required standard of safety. The aim is to keep the dose to individuals as low as reasonably achievable and never in excess of the prescribed dose limits. This is referred to as the ALARA principle.

Dose limitation: Even if an exercise in optimization of protection determines a dose value above the dose limit, it should be ensured that the dose limit is never exceeded.

11.1.6 Exposure Situations

Three types of exposure situations are considered, viz., *planned exposure situations, emergency exposure situations* and *existing exposure situations.*

11.1.6.1 Planned Exposure Situation

A *planned exposure situation* is one that arises from the operation of a source or from an authorized practice such that the exposure is controlled by good design of installations and equipment, safe operating procedures and implementation of dose limits (Table 11.3). The dose limits are specified with a view to minimize stochastic effects. Planned exposure situations include occupational exposure, exposure of pregnant

Table 11.3 Dose limits for planned exposure situations

Type of dose limit	Occupational exposure	Apprentices/students (age: 16–18)	Public exposure
Effective dose	20 mSv per year, averaged over defined periods of 5 consecutive years, with no single year exceeding 50 mSv After a worker declares a pregnancy, the dose to the embryo/fetus should not exceed about 1 mSv during the remainder of the pregnancy	6 mSv in a year	1 mSv in a year In special circumstances, a higher value could be allowed in a single year, provided that the average over 5 years does not exceed 1 mSv per year
Equivalent dose to eye lens	20 mSv per year, averaged over defined periods of 5 consecutive years, with no single year exceeding 50 mSv	20 mSv in a year	15 mSv in a year
Equivalent dose to skin	500 mSv in a year	150 mSv in a year	50 mSv in a year
Equivalent dose to hands and feet	500 mSv in a year	150 mSv in a year	–

radiation workers, exposure received by apprentices and students, public exposure and medical exposure.

Doses received from medical exposures and natural background radiation are excluded from the dose limits.

A worker who becomes pregnant while working in a nuclear facility is required to inform her superiors as soon as possible after the pregnancy has been confirmed. The dose received by the fetus should not exceed 1 mSv during the rest of the pregnancy period because the fetus is more sensitive to radiation than adults [5].

The dose limits apply to chronic exposures, that is, delivery of low dose over an extended period of time, as characterized by occupational exposure. The dose limits specified above do not take into consideration the *medical exposures* incurred by individuals and the exposure from *natural background radiation*.

Medical exposure means exposure incurred by individuals for the purposes of medical or dental diagnosis or treatment, by carers and comforters of such individuals and by volunteers subjected to exposure as part of a program of biomedical research.

Natural background radiation has always been present and is all around us in many natural forms. Many radioisotopes are naturally occurring and originated during the formation of the solar system and through the interaction of cosmic rays (radiation originating from the outer space) with molecules in the atmosphere. Carbon-14 is an example of a radioisotope formed by this interaction. Natural background radiation arises also from terrestrial sources, such as radon gas from ground, building walls and floors and traces of naturally occurring radioactive material in food and drinks. Radioisotopes such as polonium-210, carbon-14 and potassium-40 naturally occur within the human body. Potassium-40 is present in many common foods including red meats, white potatoes, carrots, bananas and Brazil nuts. The dose resulting from natural background radiation is generally very low and varies from place to place but may be relatively high in certain places. Worldwide average of effective dose from background natural radiation is about 2.4 mSv/year.

The dose limits are based on the nominal risk coefficient of 4.2% Sv^{-1} for workers and 5.7% Sv^{-1} for the whole population [6]. *The dose limit for members of the public is less than that for occupational exposure because public includes children who are very sensitive to radiation.*

11.1.6.2 Emergency Exposure Situation

An *emergency exposure situation* is one that arises as a result of an accident, a malicious act or any other unexpected event and requires prompt action in order to avoid or reduce adverse consequences. Exposures can be reduced only by implementing protective actions. Guidance levels recommended for exposure of emergency workers are 100 mSv when actions to restrict large collective dose are undertaken and 500 mSv when actions are required to be taken to prevent severe deterministic effects and/or catastrophic situations. The latter value may be exceeded for life-saving actions provided that the emergency workers are informed of the risk [7].

11.1.6.3 Existing Exposure Situations

Existing exposure situations include exposure to natural background radiation and to residual radioactive material from past practices that were never subjected to regulatory control or from a nuclear or radiation emergency after the termination of an emergency situation.

11.2 Part II Application of the Principles of Radiation Protection in Nuclear Facilities

11.2.1 General Practical Considerations for All Nuclear Facilities

Radiation protection aims at preventing deterministic effects and minimizing stochastic effects by keeping exposures as low as reasonably achievable. This objective is accomplished through (a) safe design of the plant layout so as to minimize the likelihood of contamination of the personnel and shielding of structures, systems and components (SS&C) containing radioactive materials and (b) operational safety including safe management of waste.

11.2.1.1 Classification of Areas and Zones

The nuclear facility is divided into different areas, for prevention and control of external exposure and contamination, as described below.

(i) **Supervised area**: It is a defined area for which occupational exposure conditions are kept under review, even though no specific protection measures or safety provisions are normally needed.

(ii) **Operations area**: It is the area that contains the nuclear facility. It is enclosed by a physical barrier (the operations boundary) to prevent unauthorized access.

(iii) **Controlled area**: It is a defined area where specific protection measures are, or could be, required for controlling normal exposures or preventing the spread of contamination during normal working conditions. The controlled area may sometimes be further classified into four zones, viz., white, green, amber and red zones.

White zone is where radioactive material is not handled in unsealed or sealed form. The possibility of contamination is almost negligible.

Green zone is where no radioactive material or contaminated enclosures are present, but possibilities of contamination cannot be ruled out due to proximity to amber areas. No protective clothes are required.

Amber zone is within an active laboratory and area around process equipment where personnel may be permitted to work without respiratory protective equipment during routine operations. Here, contamination is likely to occur due to the adjoining enclosures containing radioactive materials. Entry to the amber zone is restricted to the workers with adequate individual and area monitoring devices and protective clothing.

Red zone is the area where radioactive material is physically present such as hot cells, the inside of glove boxes and fume hoods and glove box exhaust ducting. This area is accessible only with proper planning and adequate protective clothing and protective equipment. The ventilation in the whole facility is maintained once through such that air moves from white zone to green zone followed by amber zone and finally through red zone and never in the reverse order.

Differential negative pressures are maintained between the zones, and they are separated by airlocks to prevent spread of airborne contamination.

11.2.1.2 Radiation Monitoring

Radiation monitoring is of two types, namely, personnel monitoring (also known as individual monitoring) and workplace monitoring (also known as area monitoring).

Area monitoring is done by measuring the dose rate at various locations of the nuclear facility and measuring radioactive contamination on surfaces of the work place.

Individual Monitoring

Individual monitors are of many types. Film badges are based on measuring the darkening produced by radiation in photosensitive films. The developed films serve as records. The thermoluminescence dosimeter (TLD) is presently the most popular individual monitor. The excitation electrons produced by incident radiation in certain crystals are locally trapped in certain impurities that are deliberately included in the crystals. The trapped electrons are released by heating the crystals under controlled conditions resulting in release of light as the electrons return to their ground states. The luminescence is a measure of the dose recorded in the crystals. The response of TL dosimeters displays linearity over a wide range (6 decades) of dose. This feature makes TLD a good option as an individual monitor. Commonly used TLD phosphors are LiF and $CaSO_4$.

Optically stimulated luminescence dosimeters (OSLDs) use materials that are similar to those used in thermoluminescent dosimetry, that is, they are crystalline solids, such as, carbon-doped aluminum oxide (Al_2O_3:C). Radiation energy deposited in the material promotes electrons from the valence band to the conduction band. These electrons move to traps in the band gap. The number of electrons trapped is proportional to the dose received by the crystal. The trapped electrons are freed by exposing the dosimeter to light. The freed electrons fall to a lower energy level and emit light photons. The intensity of the emitted light is a measure of the dose received.

An electronic personal dosimeter (EPD) is a battery-powered wearable device. It uses either a small GM tube or a semiconductor, usually, silicon. The ionizing radiation releases charges resulting in measurable electric current. It can give a continuous readout of cumulative dose or current dose rate and sounds an alarm when a specified dose rate or cumulative dose is exceeded. EPDs are especially useful in high-dose areas where residence time of the wearer is limited due to dose constraints.

The TLD or an OSLD is a suitable individual monitor that records the integrated dose received by an individual over a period of time, typically one to three months. Where the dose received needs to be known instantly pocket dosimeters (also known as Direct Reading Dosimeters, DRD) may be worn by workers. A DRD is a supplement and not a substitute for a TLD monitor. Where neutron dose is likely to be received, neutron monitoring badge is to be worn by the workers. A fast neutron dosimeter operates on the principle of measuring the tracks of protons released in an emulsion by incident neutrons. A TLD, e.g., LiF(Mn), can measure thermal neutrons.

Fig. 11.3 Schematic diagram of collection of ions in an ion chamber

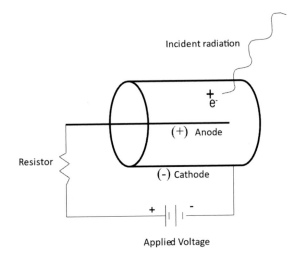

Incident radiation

+
e⁻

(+) Anode

Resistor

(-) Cathode

+   -

Applied Voltage

Workplace Monitoring

Area monitoring is carried out in the active laboratory area and around process equipment during routine operations and also in workplaces where dose rate needs to be measured due to proximity to operations areas.

The radiation monitors operate on the principle of measuring the ionization produced by radiation in a medium (Fig. 11.3). The medium is gas in the case of ionization chamber, Geiger counter and proportional counter-based monitors. The response to the incident radiation is magnified by the secondary ionization caused by the ions produced by the incident radiation in Geiger counters and proportional counters.

The inherent detection efficiency of gas-filled counters is about 100% for alphas or betas that enter the counter. However, their detection efficiency for gamma rays is very low, usually less than 1%. Certain solid scintillating crystals have relatively high detection efficiencies for gamma rays. A scintillation detector is a transducer that converts the energy of the incident radiation into photons. The photons are allowed to pass through a photomultiplier tube housing a number of photodiodes. The photons incident on the photodiodes release electrons which produce pulses that are amplified, sorted by size and counted. Thallium-activated sodium iodide, NaI(Tl), is ideal for detection of gamma, whereas silver-activated zinc sulfide, ZnS(Ag), suitable for alpha. Plastic scintillators such as anthracene can be used for detecting and measuring beta radiation. The intensity of the light emitted by the scintillation detector is proportional to the energy of the incident gamma. Therefore, a scintillation detector can be used as a spectrometer and can identify the radionuclide emitting the radiation.

In a semiconductor, the detector acts as a solid-state ionization chamber. Electrons are produced in the detector by the incident radiation. The collection of these ions leads to an output pulse. The mean ionization energy for most gases is 30–35 eV, while it is 3.5 eV in a semiconductor detector such as silicon. Thus a semiconductor device is a more sensitive detector of ionizing radiation.

Neutrons are detected by the transformation induced by neutrons that result in charged particles or gamma radiation. Boron trifluoride (BF_3) gas-filled proportional counters are the most commonly used thermal neutron detectors. Fast neutrons are first thermalized by a paraffin sphere that houses the BF_3 counter. The α particles released by the $^{10}B(n, \alpha)^7Li$ reaction are detected by a proportional counter that is sensitive to charged particle.

Radiation monitors have to be properly maintained and regularly calibrated.

11.2.1.3 Design and Layout of Nuclear Facilities

The important elements of design and layout of the plant for minimizing external exposure and contamination from all sources are shielding from direct and scattered radiation such that dose rates in operating areas are within the prescribed limits, ventilation and filtration for control of airborne radioactive materials, control of access to the plant, making provisions to minimize movement of radioactive materials and personnel within the plant and inclusion of facilities for decontamination and for storage of radioactive sources and radioactive waste.

11.2.1.4 Protection Against External Exposure

Protection against external exposure is achieved by employing one or more of the three tools, namely, time, distance and shielding.

Time: The total dose received by a person is just the product of the dose rate at the position occupied by the person and the period of exposure. To **reduce the dose** received, one should reduce the period of exposure.

Distance: The radiation dose rate from a source varies inversely as the square of the distance from the source (inverse square law). If the dose rate at a distance of r_1 m from the source is I_1 mSv/h, then the dose rate I_2 mSv/h at a distance r_2 m varies as the inverse of the square of the distance from the source such that $I_2 = I_1 [(r_1)^2/(r_2)^2]$. That is, $I_1 (r_1)^2 = I_2 (r_2)^2$ (Fig. 11.4).

To **reduce the dose rate**, one should increase the distance, that is move away, from the source.

Fig. 11.4 Inverse square law reduction of intensity

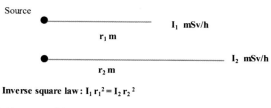

Inverse square law: $I_1 r_1{}^2 = I_2 r_2{}^2$

'r' is measured from source to point of interest

Shielding: A suitable shielding material interposed between the source and the place occupied by a person would **reduce the dose rate** at the place, as briefly discussed below.

Alpha radiation has a short range in materials as it loses energy quickly before traveling a short distance through a medium and can be shielded even by a thin sheet of paper.

An approximate general formula for the range of alpha particles in air is:

$$R_{air} = 0.56\,E \qquad cm \qquad for\ E\ \le\ 4\,MeV$$
$$R_{air} = 1.24\,E\ -\ 2.62\,cm \qquad for\ 4\ <\ E\ <\ 8\,MeV.$$

In other media, the range in the media (R_m, in units of mg/cm^2) is given as:

$$R_m\ =\ 0.56\ A^{1/3}\,R_{air},$$

where A is the mass number of the absorber and R_{air} is in mg/cm^2.

Beta radiations have a limited range in materials. They can be shielded by using a thin sheet of aluminum or plastic or similar low <u>atomic number</u> material. If a high Z material is used for beta shielding, it would give rise to the production of *bremsstrahlung* radiation which is the same as X-radiation. An approximate formula for beta particle range in low Z materials as a function of beta energy is given as:

$$R\ =\ 412E^{1.265-\,0.0954\,lnE}\ \ mg/cm^2 \qquad for\ 0.01\ \le\ E\ <\ 2.5\,MeV$$
$$R\ =\ 530E\ -106 \qquad mg/cm^2 \qquad for\ E\ >\ 2.5\ MeV.$$

The thickness of the shielding required for beta radiation has to be more than the range of beta particle in the shielding material. Depending on the bremmstrahlung production, additional shielding for X rays may be required.

Gamma radiation is an indirectly ionizing radiation. Since it does not carry an electric charge, gamma radiation has infinite range in matter. The shielding for gamma radiation is achieved by reducing the intensity of radiation (Fig. 11.5).

Fig. 11.5 Attenuation of radiation

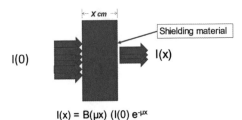

$$I(x) = B(\mu x)\ (I(0)\ e^{-\mu x}$$
μ: Linear absorption coefficient , cm $^{-1}$
The value of μ depends on shielding material and energy of radiation
HVL = 0.693/μ, cm
B(μx) is the build up factor

As gamma radiation passes through the shielding material, only a fraction of the incident radiation is absorbed by the material and the remaining fraction is directly transmitted with the same energy. The radiation that is neither directly transmitted nor absorbed by the material is scattered in all directions. The radiation that is scattered in the forward directions adds to the intensity of the transmitted radiation. This is called build-up radiation. The intensity of Υ or X-radiation after traversing a distance x through the medium reduces due to attenuation.

$$I(x) \; = \; B(\mu x) \, I(0) \, \exp(-\mu x),$$

where

I(0) is the intensity of radiation at the point of interest without any shielding.

I(x) is the intensity of radiation at the point of interest with shielding of thickness x cm.

μ is the linear attenuation coefficient cm^{-1} and its value depends on the material and the energy of radiation.

$B(\mu x)$ is the build-up factor and is ≥ 1. It depends on the shielding material, thickness, the shape and size of the radiation beam, energy of radiation and distance of the source of radiation from the shield. A narrow radiation beam would lead to minimum build up, in which case, B would be 1. This is called narrow beam geometry. If the beam is extended in size and shape, that is of broad beam geometry, B will be greater than 1.

In making simple hand calculations, the concept of HVL is used as briefly explained below: The thickness of shielding material required to reduce the intensity of radiation to half its original value is defined as the half-value layer (HVL). It is also referred to as half-value thickness (HVT). It can be easily proved that

$$1\,HVL \; = \; 0.693/\mu$$

{For simplicity, assuming narrow beam geometry, $I(x) = I(0) \exp(-\mu x)$.
If $I(x) / I(0) = \frac{1}{2}$, then x is one HVL. That is, $\exp(-\mu x) = \frac{1}{2}$.
Now, $\ln[\exp(-\mu x)] = -\mu x = \ln(0.5) = 0.693$}.

As the definition shows, one HVL reduces the intensity of radiation by a factor of 1/2 and 'n' HVLs would reduce the intensity of radiation by $(1/2)^n$.

Similarly, one Tenth-Value Layer (TVL) reduces the intensity of radiation by a factor of 1/10 and 'm' TVLs would reduce the intensity of radiation by $(1/10)^m$.

Thus, the total shielding comprises 'n' HVLs and 'm' TVLs, and the effective reduction in the intensity of radiation would be $(1/10)^m \cdot (1/2)^n$.

Lead, steel and depleted uranium are good shielding materials for containers of radioactive material, whereas concrete is an ideal material for structural shielding. The required thickness of the protective barriers is calculated using the above shielding principles.

Neutron radiation is an indirectly ionizing radiation like gamma. Neutron shielding is achieved by slowing down fast neutrons to thermal energies and then absorbing the thermal neutrons. For slowing down neutrons, materials of low mass number, such as water, and hydrogenous compounds, for example, paraffin, are effective. Materials such as boron or cadmium that have a high neutron absorption cross-section are utilized to absorb the thermalized neutrons. Absorption of neutrons can lead to prompt gamma emission and induce radioactivity in the absorber.

A judicious combination of time, distance and shielding together with safe work practices would be effective in controlling the dose due to external exposures.

11.2.1.5 Protection Against Internal Exposure

The presence of radioactive contamination on any surface and/or in the environment leads to intake of radioactivity resulting in internal exposure.

11.2.1.6 Contamination of Work Place

Air monitors used for measuring activity in air employ a pump that draws in a known volume of air through a filter paper. The filter paper that has collected particulate airborne activity is taken to a counting set up to estimate the activity in air. Only persons wearing proper protective clothing and respiratory protection are allowed entry to an area where air contamination is likely to occur. In facilities where plutonium is handled, strategically located plutonium air monitors detect air activity and warn of any release. Spot air samples are taken during maintenance operations and also when release of any air activity is suspected. The quantity of activity intake should be limited so as to limit the committed dose equivalent. If an activity intake q Bq would result in a committed dose equivalent equal to the annual dose limit, then q Bq is the Annual Limit of Intake (ALI). For each radionuclide, the ALI values have been derived. For the purpose of controlling activity intake in a nuclear facility, a practical measurable quantity called derived air concentration (DAC) is defined.

DAC (Bq/m^3) = ALI (Bq/year)/[(2000 working hours per year) $\times$ 1.2 m^3 per hour of breathing rate)]

$$DAC = ALI/2400.$$

Control of internal exposure is achieved with reference to the derived air concentrations (DACs) in the facility [8, 9].

Containment Systems

Radioactive materials are handled in containment systems (glove boxes and fume hoods) so as to prevent spread of contamination into working areas. Fume hood is used when the contamination and radiation hazards are not high. The amount of activity handled is limited on the basis of nature of the radionuclide and type of

Fig. 11.6 Photograph of fume hoods (left) and glove boxes (right)

operation. Fume hoods are provided with an airflow of 30–50 linear meters per min across the front face to prevent spread of activity out of fume hood.

Glove box provides total containment and it completely isolates hazardous material from the operators' environment. Material is handled with the help of gauntlet gloves fixed on the glove ports on the walls of the box. It is always kept at a negative pressure of 25–30 mm of water gage (WG). It has two HEPA filters, one at inlet and the other at the outlet. Glove boxes are leak tight. The leak rate should be less than 0.5% of box volume per hour. While handling pyrophoric materials like Pu carbide, the leak rate should be less than 0.05% and an inert atmosphere is maintained inside the glove boxes. Normally, 3–5 air changes are provided in the glove boxes. Working with fume hoods and glove boxes is illustrated in Fig. 11.6.

Personal Protective Equipment (PPE)

For low levels of surface contamination, an ordinary laboratory coat with overshoes and gloves may be sufficient. When the levels of air or surface contamination are high, it is necessary to have a fully enclosed PVC suit with a filter mask or a mask fitted with an air supply. PPE can cause discomfort and interfere with other safety equipment and work efficiency [10].

Surface contamination can be directly monitored using a contamination monitor having end window probe (ZnS scintillator coupled to a PM tube for alpha contamination and by GM tube for beta contamination). Alternately, swipe samples collected from the suspected contaminated surfaces can be checked in a counting setup to determine the level of contamination. Swipe samples are useful for determining only loose contamination levels, whereas direct method gives an estimate of total contamination, that is, sum of removable and fixed contaminations. Hand monitors, portal monitors, etc. are required for monitoring contamination on the body and clothes.

Workers have to wash their hands and face before leaving the facility and refrain from eating, drinking or smoking in the active area. Workers have to shower and change at the end of a working shift to prevent spread of contamination. Work surfaces, walls and floors are made non-porous and smooth so that they can be easily decontaminated.

Decontamination Procedures

Decontamination is removing the contamination to levels within the prescribed limits. Detergents and chemicals such as EDTA are effective in decontaminating laboratory surfaces. For decontamination of hands and other parts of the body, thorough washing with detergents and water is the best general method, regardless of the contaminant. If the contamination is not removed by this, chemicals such as citric acid, EDTA, titanium dioxide, $KMnO_4 + 0.2$ N sulfuric acid solution may be used. If eyes get contaminated, the eyes have to be irrigated with copious quantities of water repeatedly, and then, medical care should be provided at the earliest. In case of contamination of a wound, bleeding should be stimulated, the wound washed with running water and medical help should be sought.

Internal Dosimetry

In the event of intake of radionuclides, assessment of internal exposure can be made on the basis of the measured air activity in working areas and duration of occupancy by the workers. This has to be supplemented by biological monitoring which may be done by direct and indirect measurements. Direct measurement involves determining the activity intake by means of a whole-body counter. Indirect measurement is done by breath monitoring to detect inhaled activity and collecting biological samples such as nasal swabs or nose blows, saliva, blood, urine and excreta of the exposed person.

The concentration of radioactivity and the identity of the radionuclide in the sample are determined using sensitive radiation monitors. From the measured values, the quantity of the radionuclide in the body and the dose rate being delivered by the radionuclide at the time of sampling can be calculated. If no measures are taken to eliminate the radionuclides from the body, the dose would continue to be delivered, though the dose rate may reduce as determined by the effective half-life of the radionuclide. The committed dose equivalent has to be calculated over a specified length of time. Administration of certain chemicals can accelerate the biological elimination of specific radionuclides. Some examples of antidotes for internal contamination by radioisotopes are listed below:

Uranium: sodium bicarbonate.
Strontium and radium: ammonium chloride, barium sulfate.
Cobalt: cobalt gluconate.
Caesium 137: Prussian blue.
Iodine 131: potassium iodide or Lugol solution.
Plutonium: Ca-DTPA, Zn-DTPA.

Americium: Ca-DTPA, Zn-DTPA.

11.2.2   Specific Considerations for Radiation Protection in Nuclear Facilities

The radiation protection considerations in a nuclear facility depend on the hazard associated with the activities carried out in the facility. The safety considerations applicable to mines and mills, fuel fabrication facilities, nuclear reactors, radiological laboratories, hot cells and waste disposal facilities are discussed here.

11.2.2.1   Mine and Mill

Occupational exposure in mines and mills arises from external gamma radiation, the inhalation of long lived radionuclide dust (LLRD) and radon decay products (RDP) and through ingestion and wound contamination. Typical exposure rates of ore stockpiles due to gamma radiation lie between a few μSv/h for low-grade ore (below ~0.1%) and a few tens of μSv/h for higher-grade ores (~1%) [11].

The critical radionuclides for radiation protection in the uranium series are ^{238}U, ^{234}U, ^{230}Th, ^{226}Ra, ^{210}Pb and ^{210}Po, which represent the longer lived radionuclides. Radon (^{222}Rn) and its progeny (^{218}Po, ^{214}Pb, ^{214}Bi, ^{214}Po) can enter the working atmosphere and contribute to occupational exposure. Most of the lung dose comes from the radon progeny and not from radon gas itself. The progeny in equilibrium with 3,700 Bqm^{-3} radon gas in air would release approximately 1.3×10^5 meV of alpha energy in decay. Therefore, the limit on the dose during work in a radon environment is based on the energy that inhaled progeny would deposit in the lung. The limit is expressed as Working Level (WL). The unit Working Level Month (WLM) is frequently used in the risk assessment of occupational radon exposure.

One WLM is the measured alpha energy concentration in unit volume of air multiplied by the time the individual has worked in the environment in one month (= 170 h).

$$1\,WL = 1.3 \times 10^5\,MeV \text{ per liter of air} = 0.0208\,mJ\ m^{-3}.$$

$$WLM = 1\,WL \text{ exposure for } 170\,h.$$

$$1\,WLM = 0.0208\ mJ\ h\ m^{-3} \times 170 = 3.54\,mJ\ h\ m^{-3}.$$

In uranium mining and milling, PPE in the form of respiratory protection is generally used as a protective measure against RDP and LLRD. A well-designed ventilation system can control radon progeny and also LLRD exposure to some degree [12]. Water is effective at reducing the generation of airborne LLRD, so wet drilling and mining methods are preferred to dry techniques. Minimizing spillages and cleaning and decontaminating equipment are also important to control LLRD exposure. At the time of decommissioning, significant exposures will occur around the contaminated plant and land.

11.2.2.2 Fuel Fabrication Facility

In fuel fabrication facilities, the main hazards are potential criticality and releases of uranium hexafluoride (UF_6) and UO_2. The design of a uranium fuel fabrication facility provides against these hazards. Insoluble compounds of uranium such as the uranium oxides UO_2 and U_3O_8 pose hazard because of their long biological half-lives and their relatively small particle size (typically a few micrometers in diameter) in uranium fuel fabrication facilities. Severe chemical hazards having both potential on-site and off-site consequences can arise from release of UF_6 due to the rupture of a hot cylinder, release of HF due to the rupture of a storage tank and large fire, natural disasters and aircraft crash [13, 14]. Where reprocessed uranium is handled, the consequences of an accidental release are likely to be greater.

For ensuring criticality safety, mass and degree of enrichment of fissile material present in a process and in storage, geometry of the processing equipment, concentration of fissile material in solutions and neutron absorbers and degree of moderation available in the facility should be controlled.

The fabricated fuel elements/assemblies should be free from contamination. For measuring loose contamination on the pins, swipe samples are checked in a ZnS(Ag)-based counting system. For measuring fixed contamination, the pin surface should be scanned with a scintillation detector with ZnS(Ag).

In addition to providing adequate shielding, occupancy is restricted in areas used for storing cylinders, in particular, empty cylinders that have contained reprocessed uranium since some by-products of irradiation will remain in the cylinder. For minimizing the radiation exposure of workers from airborne activity, adequate ventilation is provided. Alarm systems are installed to alert operators about high or low differential pressures.

Low Enriched Uranium (LEU)

Natural uranium has isotopes ^{234}U and ^{235}U besides ^{238}U. As the enrichment of uranium increases, the presence of ^{234}U isotope also increases in the bulk material. Generally, in a conversion facility or an enrichment facility, only natural uranium or LEU that has a ^{235}U concentration of no more than 6% is processed. Fabrication of LEU bearing fuel demands extra care during handling of this material because the specific activity of ^{234}U (2.32×10^8 Bq/gm) is very high compared to ^{235}U (8×10^4 Bq/gm) and ^{238}U (1.24×10^4 Bq/gm). The radiotoxicity of LEU is relatively low, and any potential off-site radiological consequences following an accident would be limited. However, the radiological consequences of an accidental release of reprocessed uranium would be likely to be greater, and this has to be taken into account if the facility handles reprocessed uranium [15].

Handling of PuO$_2$

Issues involved in handling of plutonium oxide powder for fuel fabrication are radiological, thermal and criticality in nature. Plutonium is an α emitter and has a high biological half-life. It poses a significant inhalation hazard. So, its handling needs a leak-tight containment such as a glovebox. ^{241}Pu decays into ^{241}Am which emits α and γ. ^{240}Pu emits neutrons during spontaneous fission. Presence of ^{236}Pu causes high γ radiation from its daughter products Thallium (Tl) and Bismuth (Bi). ^{238}Pu and ^{241}Am generate heat during handling because of high specific power. Mass of the fissile material handled should always be maintained well below the critical mass to avoid criticality.

Multiple levels of containment are important for Pu handling. The building should be capable of withstanding seismic and such events. The operating areas and the glove boxes are maintained at a negative pressure. Shielding is required in front of glove box to take care of γ and neutron dose.

Automated pellet inspection system should be used to reduce the handling time of material. Sharp tools are avoided inside the glove box to avoid breach of gloves. In the case of fuel pins, various equipment are used for quality control checks such as radiography, leak testing and visual inspection [16]. Walls of the exposure room should provide adequate shielding against X-rays. The entrance to the radiography room should be electrically interlocked with the X-ray unit's control panel so that if the door is opened, the exposure would be turned off. Warning red indicator lamp should be installed outside the radiography room to indicate when X-ray unit is switched ON.

Thorium

Natural thorium contains ^{228}Th. Daughter products of ^{228}Th are ^{212}Bi and ^{208}Tl which emit high energy γ rays. Storage of ThO$_2$ powder leads to build up of dose over time. Aged ThO$_2$ requires shielding during fuel fabrication. Natural thorium does not contain any fissile material. It has to be converted to ^{233}U which is a fissile material in a nuclear reactor. Recovered ^{233}U will always be contaminated with ^{232}U whose daughter products ^{212}Bi and ^{208}Tl are hard gamma emitters, and hence, ^{233}U has to be handled in shielded facilities.

Handling of ^{233}U bearing MOX poses challenge to fuel fabricators, as reprocessed ^{233}U is always associated with ^{232}U whose daughter product emits strong gamma radiation. The Coated agglomerate pelletization (CAP) process for the fabrication of thoria-based MOX fuel containing ^{233}UO$_2$ or PuO$_2$ is associated with problems such as ^{233}UO$_2$ powder sticking on equipment and glove boxes during coating. The resulting contamination can pose an additional challenge for safe handling.

11.2.2.3 Nuclear Reactor

Safety of nuclear reactor is achieved through design safety and operational control. In a nuclear reactor, the fuel inventory and the fission product inventory combined with the gaseous, volatile and particulate nature of the various fission products warrant implementation of high safety standards. Reactivity control during the operation of the reactor and controlled release of permitted activity and prevention of occurrence of situations that could result in accidental release of radioactivity are measures that are implemented for assuring the high level of safety standards that characterize the nuclear industry [17, 18].

11.2.2.4 Chemical Safety in a Radiological Laboratory

In addition to radiation protection, all necessary precautions should be taken while handling chemicals in a radiological laboratory and the safety procedures should be adhered to. Breach of chemical safety can have radiological consequences.

Flammable Chemicals

Many solids, liquids and gases are known to be flammable substances, for example, acetone, alcohol, ether and chloroform. Only the required quantity should be introduced inside the glove box/ fume hood. No heating device like lamp or oven should be operated during the use of these chemicals. Spillage should be avoided.

Explosive Chemicals

Many peroxides are sensitive to shock, impact or heat and may release sudden energy in the form of heat or explosion. Formation of explosive mixture inside glove box/fume hood should be prevented.

Water-Sensitive Chemicals

Some chemicals such as potassium and sodium metal and metal hydrides react with water to evolve heat and flammable or explosive gases. Hydrogen is produced with sufficient heat to ignite accompanied by explosion. Water source should not be present nearby while handling these chemicals. If water is present inside the glove box, such chemical should not come in contact with water.

Gases Under Pressure

Compressed gas cylinders should be used very carefully so that they are not dropped or they do not strike other objects. The cylinders have to be a kept away from flammable materials. Gas for use inside the glove box/fume hood should be brought through gas lines to avoid storing the cylinders inside the laboratory area. Use of liquid nitrogen inside the glove box needs special attention. Handling liquid nitrogen in a glove box is extremely risky as there is a hazard of contamination of personnel and the laboratory in case of accidental liquid nitrogen spillage inside the glove box. The flooring of the glove box is covered by suitable materials to absorb liquid nitrogen spilled on it to prevent instantaneous expansion of liquid nitrogen. The box has provision of interlocks which can take care of any rise in pressure which may lead to cracking in glass panel or release of radioactivity.

11.2.2.5 Hot Cell

Hot cell-related examination is the most prominent and versatile method of Post-irradiation Examination (PIE) of highly radioactive materials, such as spent fuel elements, structural material and control rods. Hot cell-related facilities make use of different categories of enclosures like concrete/lead shielded cells, glove boxes, fume hoods, etc., depending on the level of activity and the toxicity of the radioactive material required to be handled. Hot cells are shielded enclosures with leak-tight design features to prevent release of radioactive material. Highly radioactive material is handled in isolation using remote handling devices.

Hot cells are provided with lead glass windows, periscopes, CCD cameras, etc., for viewing the interior of the cell. Illumination is provided using sodium/mercury vapor/metal halide lamps. Robotic and automated devices find increasing applications in radioactive material handling owing to their high reliability and productivity, resulting in reduced risk of radiation exposure to operators. Equipment design is modular in construction for easy remote maintenance using robotic devices.

11.2.3 Emergency Management

Emergency is a non-routine situation or event that necessitates prompt action, primarily to mitigate a hazard or adverse consequences for human life, health, property and the environment. Emergency management is an extensive subject in itself. Despite the establishment of design safety and safety protocol to be practiced, an emergency situation may arise under unforeseeable circumstances. Emergencies are classified as Facility emergency, Site area emergency and General emergency.

Facility emergency is an emergency that warrants taking protective actions and other response actions at the facility and on the site but does not warrant taking protective actions off-site. When a facility emergency is declared, actions are promptly taken by the site staff to mitigate the consequences of the emergency and to protect people.

Site area emergency is an emergency that warrants taking protective actions and other response actions on the site and in the vicinity of the site. When a site area emergency is declared, actions are promptly taken: (i) to mitigate the consequences of the emergency on the site and to protect people on the site; (ii) to increase the readiness to take protective actions and other response actions off site if this becomes necessary on the basis of observable conditions, reliable assessments and/or results of monitoring and (iii) to conduct off-site monitoring, sampling and analysis.

General emergency is an emergency that warrants taking precautionary urgent protective actions, urgent protective actions and early protective actions and other response actions on the site and off the site. When a general emergency is declared, appropriate actions are promptly taken, on the basis of the available information relating to the emergency, to mitigate the consequences of the emergency on the site and to protect people on the site and off the site. A general emergency is unlikely to occur in mines and mills. The chances of a general emergency occurring in fuel fabrication facilities are not high. However, nuclear reactors and facilities where fissile materials are handled in significant quantities have to be prepared against general emergencies.

The nuclear facility authorities have to develop comprehensive emergency preparedness and response plans. For managing general emergencies, the concerned State authorities have to develop plans for emergency preparedness and response. A general emergency (e.g., Chernobyl accident that occurred in the erstwhile USSR and Fukushima disaster that occurred in Japan) may warrant shifting residents living in the vicinity of the nuclear power plant to safer locations and administering prophylactics to protect their thyroids and providing medical and general care until they are rehabilitated upon termination of the emergency.

In any country, a national emergency management authority should be established with adequate financial, human and equipment resources with the statutory mandate to develop a national emergency response system. Such a system would incorporate the response actions to be implemented in any emergency situation. A nuclear emergency response system can be fitted into the existing response system for conventional emergencies because of the many common elements. The plant/nuclear facility is required to have its own trained manpower with adequate equipment to handle a nuclear emergency. This expertise would be utilized by the national emergency management authority through the concerned State and Local emergency response centers where trained manpower would have to be available with adequate equipment to manage a nuclear emergency at a facility or during transport of radioactive material. Communication facilities with an established protocol and an approved emergency response manual specifying Standard Operating Procedure are vital for an emergency response system.

11.2.4 Safety Culture

In any nuclear facility, safety culture must be promoted. The elements of safety culture include (a) individual awareness of the importance of safety, knowledge and competence, conferred by training and instruction of personnel and by self-education; (b) commitment, requiring demonstration at senior management level of the high priority of safety, and adoption by individuals of the common goal of safety; (c) motivation, through leadership, the setting of objectives, systems of rewards and sanctions and through individuals' self-generated attitudes; (d) supervision, including audit and review practices, with readiness to respond to individuals' questioning attitudes and (e) responsibility, through formal assignment and description of duties and their understanding by individuals. Where safety culture is inculcated in a facility, the safety goals can be achieved.

Questions

1. What is the dose rate at 2 m from a ^{60}Co source of activity 185 GBq? [The gamma ray constant of cobalt 60 is 0.31 mGyh^{-1} GBq^{-1}.]
2. The dose deposited in different tissues of a worker are as follows: 10 mGy of gamma dose to the skin; 20 mGy of beta dose to the thyroid and 5 mGy of alpha dose to the lungs. Calculate the effective dose received by the worker.
3. Briefly state and explain the following principles of radiation protection:

    (a) Justification of practice
    (b) Optimization of protection and safety
    (c) Dose limitation

4. Distinguish between:

    (a) Stochastic effect and deterministic effect
    (b) Acute exposure and chronic exposure.

5. If the dose rate at a location is 800 mSv/h, what is the shielding required in terms of HVLs and TVLs to reduce the dose rate to 2 mSv/h?
6. What are the annual limits on the effective dose for occupational exposure and public exposure? What is the reason for the difference?
7. What are the typical countermeasures that may be warranted by an general emergency in a nuclear reactor?
8. What are the important elements of design and layout of the plant for minimizing external exposure and contamination?
9. What are the important elements of safety culture in a nuclear facility?
10. Select the correct answer from the options given below:

    (i) The unit of activity is
       (a) Becquerel
       (b) Sievert

(c) mSv/h
(d) Gray

(ii) A person worked at a location for 12 minutes and his pocket dosimeter recorded a total dose of 0.8 mSv. The reading of a radiation survey meter at that place would be
(a) 9.6 mSv/h
(b) 40 mSv/h
(c) 4 mSv
(d) 4 mSv/h

(iii) The energy deposited by radiation in 40 g of tissue is 100 mJ. The absorbed dose received by the tissue is
(a) 0.4 mGy
(b) 2.5 mGy
(c) 2.5 Gy/y
(d) 2.5 Gy

(iv) A person was exposed to radiation and it was noted that he was not wearing a personal monitoring badge. The dose received by him can be estimated
(a) by exposing to radiation again
(b) after the immediate effects manifest
(c) through a CA test
(d) placing a number of TLD badges at the locations where he was working.

11. Classify the biological effects given in the table below appropriately, as per the example:

Biological effect	Nature of effect					
	Stochastic	Deterministic	Somatic	Hereditary	Immediate	Delayed
Cataract	–	x	x	–	–	x
Cancer						
Sterility						
NVD						
Loss of hair						
Blood picture						
Leukemia						
Mongolism						
Erythema						

References

1. International Atomic Energy Agency, Radiation protection and safety of radiation sources: international basic safety standards, general safety requirements, jointly sponsored by: European Commission, Food and Agriculture Organization of the United Nations, International Atomic Energy Agency, International Labour Organization, Oecd Nuclear Energy Agency, Pan American Health Organization, United Nations Environment Programme, World Health Organization, GSR Part 3 (2014)
2. International Commission On Radiological Protection, *The 2007 recommendations of the international commission on radiological protection, Publication 103* (PUBLISHED FOR The International Commission on Radiological Protection, Elsevier, 2007)
3. International Atomic Energy Agency. IAEA Saf. Gloss. (2018)
4. H.Cember, T.E.Johnson, *Introduction to Health Physics*, 4th edn. (McGraw Hill, 2009)
5. International Commission on Radiological Protection, Pregnancy Med. Radiat. Publ. **84** (Pergamon Press, 2000)
6. International Commission on Radiological Protection, The 2007 recommendations of the international commission on radiological protection. ICRP Publ. **103** (Elsevier, 2007)
7. International Atomic Energy Agency, Preparedness and Response for a Nuclear or Radiological Emergency, General Safety Requirements, Jointly Sponsored by: European Commission, Food and Agriculture Organization of the United Nations, International Atomic Energy Agency, International Labour Organization, Oecd Nuclear Energy Agency, Pan American Health Organization, United Nations Environment Programme, World Health Organization, GSR Part 7 (2015)
8. K. Eckerman, J. Harrison, H. Menzel, C. Clement, ICRP publication 119: compendium of dose coefficients based on ICRP publication 60. Annals ICRP **41** (2012)
9. International Commission On Radiological Protection, Annual Limits on Intake of Radionuclides by Workers Based on the 1990 Recommendations. ICRP Publ. **61** (Pergamon Press, Oxford, 1990)
10. G.D. Clayton, F.E. Clayton, *Patty's industrial hygiene and toxicology*, vol. 2A. *Toxicology* (John Wiley & Sons, Inc., Baffins Lane, Chichester, Sussex PO19 1DU, 1981)
11. International Atomic Energy Agency, Occupational radiation protection in the Uranium mining and processing industry. Safety Reports Series No. 100 (IAEA, Vienna, 2020)
12. Atomic Energy Regulatory Board, Radiological safety in Uranium mining and milling. AERB Safety Guidelines No. AERB/FE-FCF/SG-2 (Mumbai, 2007)
13. International Atomic Energy Agency, Safety of Uranium fuel fabrication facilities. Specific Safety Guide Series No. SSG-6, IAEA (Vienna, 2010)
14. Atomic Energy Regulatory Board, Uranium oxide fuel fabrication facilities. AERB Safety Guidelines No. AERB/FE-FCF/SG-3 (Mumbai, 2009)
15. International Atomic Energy Agency, Safety of conversion facilities and Uranium enrichment facilities. Safety Standards Series No. SSG-5 (2010)
16. International Atomic Energy Agency, Safety of Uranium and Plutonium mixed oxide fuel fabrication facilities. Safety Standards Series No. SSG-7 (2010)
17. International Atomic Energy Agency, Safety of nuclear power plants: design, specific safety requirements. IAEA Safety Standards Series No. SSR-2/1 (Rev. 1) (2016)
18. Atomic Energy Regulatory Board, Atmospheric dispersion and modelling. AERB Safety Guide No. AERB/NF/SG/S-1 (Mumbai, 2008)

Further Reading

19. J.E. Turner, *Atoms, Radiation, and Radiation Protection* (John Wiley & Sons, 1995)

20. A. Brodsky (ed.), *Handbook of Radiation Protection and Measurement*, vol. II (CRC Press, Boca Raton, FL, Biological and Mathematical Information, 1982)
21. J. Shapiro, *Radiation Protection for Scientists, Regulators and Physicians*, 4th edn. (Cambridge, Harvard University Press, London, 2002)
22. F.A. Attix, *Introduction to Radiological Physics and Radiation Dosimetry* (John Wiley & Sons, New York, 1986)
23. K.L. Mossman, W.A. Mills (eds.), *The Biological Basis of Radiation Protection Practice* (Williams & Wilkins, Baltimore, MD, 1992)
24. C.B. Braestrup, H.O. Wyckoff, *Radiation Protection* (Charles C Thomas, Springfield, IL, 1958)
25. E.P. Blizzard, L.S. Abbott, *Reactor Handbook*, vol. III B (Shielding. Interscience, New York, 1962)
26. R.G. Jaeger (ed.), *Engineering Compendium on Radiation Shielding* (Springer-Verlag, Berlin, 1968)
27. Pushparaja, *Radiological Protection and Safety a Practitioner's Guide* (Notion Press.com, India, 2019)
28. International Atomic Energy Agency, Optimization of radiation protection in the control of occupational exposure. Safety Reports Series No. 21 (2002)
29. International Atomic Energy Agency, Monitoring and surveillance of residues from the mining and milling of Uranium and Thorium. Safety Reports Series No. 27 (2002)
30. International Atomic Energy Agency, Surveillance and monitoring of near surface disposal facilities for radioactive waste. Safety Reports Series No. 35 (2004)
31. Atomic Energy Regulatory Board, Regulatory control of radioactive discharges to the environment and disposal of solid waste. AERB Safety Guide No. AERB/NRF/SG/RW-10 (Mumbai, 2021)
32. Atomic Energy Regulatory Board, Monitoring and assessment of occupational exposure due to intake of radionuclides. AERB Safety Guide No. AERB/NRF/SG/RP-1 (AERB, Mumbai, 2019)
33. Atomic Energy Regulatory Board, Uranium oxide fuel fabrication facilities. AERB Safety Guidelines No. AERB/FE-FCF/SG-3 (AERB, Mumbai, 2019)
34. United Nations Scientific Committee on The Effects of Atomic Radiation, *Sources, Effects and the Risks of Ionizing Radiation* (UNSCEAR, 2017)
35. National Disaster Management Authority, *National Disaster Management Plan* (New Delhi, 2019)

Index